Ecological Studies, Vol. 133

Analysis and Synthesis

Ecological Studies

Volumes published since 1992 are listed at the end of this book

Springer
Berlin
Heidelberg
New York
Barcelona
Budapest
Hong Kong
London
Milan
Paris
Singapore
Tokyo

D. O. Hessen L. J. Tranvik (Eds.)

Aquatic Humic Substances

Ecology and Biogeochemistry

With 79 Figures and 14 Tables

Springer

Prof. Dr. Dag O. Hessen
University of Oslo
Department of Biology
Div. Limnology
P.O. Box 1027, Blindern
0315 Oslo
Norway

Dr. Lars J. Tranvik
Linköping University
Department of Water and Environmental Studies
58183 Linköping
Sweden

ISSN 0070-8356
ISBN 3-540-63910 Springer-Verlag Berlin Heidelberg New York

Library of Congress Cataloging-in-Publication Data

Hessen, D. O. (Dag Olav), 1956–, Tranvik, L. J. (Lars J.), 1959–
Aquatic humic substances: ecology and biogeochemistry / D.O. Hessen, L.J. Tranvik, (eds.) p. cm. – (Ecological studies / Analysis and synthesis, ISSN 0070-8356; vol. 133) Includes bibliographical references and index. ISBN 3-540-63910-1 (hardcover: alk. paper) 1. Freshwater ecology. 2. Humus – Environmental aspects. H541.5.F7 A6 1998 577.6 dc21 97-048264

Printed in Germany

Cover design: design & production GmbH, Heidelberg
Typesetting: Camera ready by UKT, Reichartshausen
SPIN 10532782 31/3137 5 4 3 2 1 0 – Printed on acid-free paper

Foreword

Understanding of the structure of aquatic ecosystems moved forward swiftly during the past century. The basic premises of food-web structures and feeding relationships among trophic levels have been studied in remarkable detail, and consistent patterns of population and community relationships are emerging for aquatic ecosystems of different physical and chemical environmental properties.

As the metabolism of community components was analyzed with increasing accuracy, however, flux pathways and rates of transfer of organic carbon demonstrated a number of complexitics and inconsistencies that could not be explained within the conventional food-web paradigms. Six pervading alterations of prevailing wisdom have arisen in the past two decades. Firstly, the observed biotic productivity of most lakes and rivers cannot possibly be supported by the autochthonous organic carbon generated by pelagic primary productivity within these waters. The aquatic biotic productivity must be supplemented by external quantities of allochthonous organic matter imported from terrestrial and land-water interface regions. Only recently have a number of diverse and thorough studies demonstrated that these external sources are often dominating, which indicates that most inland aquatic ecosystems are largely heterotrophic ecosystems.

Secondly, a large portion (usually > 90%) of the organic matter imported to these aquatic ecosystems is predominantly in dissolved or colloidal form. Although a portion of the dissolved organic compounds may aggregate and shift to a particulate and hence gravitoidal form that may sediment out of the water, most of the imported dissolved organic matter is dispersed within the water and moved about with the hydrodynamics of the water body.

Thirdly, much of the dissolved organic matter derives from lignin and related structural precursor compounds of higher plants. These substances are abundant, chemically complex, and relatively recalcitrant to rapid biological degradation. These compounds are variously modified by microbial activities during transport and partial decomposition in soils, wetlands, and littoral areas before movement to the receiving lake or river. Because of limited accessibility of large portions of these molecules to enzymatic hydrolysis, the low degradation rates of many of these macromolecules, as well as their common acidic properties, slow turnover times result in long residence times. Dissolved macromolecules are often of considerable age (years) but are mixed with variable and rapidly changing inputs of younger humic substances. In addition, recent studies are demonstrating that humic substances, particularly fulvic acids, are generated by algae and contribute to the multitude of diverse compounds that make up the dissolved humic substances. As a result of the recalcitrance of all of these compounds, the dissolved organic matter can reside *within* lakes and rivers for long periods of time (months, years).

Despite the apparent chemical recalcitrance of humic substances and the belief that these compounds were poorly used by microbiota, a fourth major paradigm is emerging that these compounds can form major energy sources in aquatic ecosystems. Physical processes, such as partial photochemical alteration, can result in changes in availability of portions of the macromolecules by cleavage of simple compounds that are readily degraded by microbes. These photochemical changes can also alter enzymatic accessibility to the residual macromolecules. Despite continued massive loading of dissolved humic substances to aquatic ecosystems, accumulation does not occur and large releases of excessive CO_2 to the atmosphere indicate extensive biological degradation of these compounds. The mechanisms of degradation of these dissolved humic substances represent a major void in our understanding and a place where interdisciplinary collaboration among chemists and biologists is essential in order to progress effectively.

A fifth revelation from continued intensive study of dissolved humic substances in aquatic ecosystems consists of the many ways in which these diverse compounds can interact with inorganic and other organic compounds, associate with inert and living surfaces, alter chemical properties such as redox and pH, and change physical properties such as selective modifications of light penetration. The result is a profound potential alteration of the aquatic environment for biota and their growth and productivity. Examples are many. The predominance of humic acids results in an organic acidity that can influence and at times exceed inorganically derived acidity from natural or anthropogenically induced sources. Humic substances can complex metals and influence biotic availability and toxicity. Humic compounds can inhibit free and surface-bound enzymes and materially alter nutrient recycling and availabilities. Membrane properties, such as lipid hydrophobicity, can be altered by humic substances. Light is selectively attenuated by dissolved humic substances and can change markedly the availability of spectral portions that influence photosynthesis, hormonal activities, and migratory distribution and reproductive behaviors. Absorption of ultraviolet irradiance by humic substances can both protect organisms from genetic damage as well as modify macromolecules and enhance bioavailability of organic substrates.

Lastly, for many years I have promulgated that the large amounts of dissolved humic organic matter and their slow but collectively dominating utilization provide a thermodynamic stability to metabolism within lake and river ecosystems that is essential for the maintenance of efficient nutrient recycling. As a result, most inland aquatic ecosystems are strongly detritus-dependent, and the trophic dynamics are likely predominantly controlled by the microbial interactions with humic compounds that can markedly influence nutrient recycling rates. Despite great resistance to this idea, the same mechanisms likely prevail in the oceans as well. Evidence presented in this book enhances and strongly supports these viewpoints, at least for lakes, rivers, and coastal marine areas.

Aquatic Humic Substances: Ecology and Biogeochemistry is an excellent synthesis of the state of the art of this complex subject regarding the many interactive ways that humic compounds influence the properties, dynamics, and metabolism of aquatic ecosystems. Every topic discussed above, as well as other important facets, have been treated in appreciable detail. Three major aspects result from such a composite treatment. The coupling of the sources of aquatic humic substances, produced both internally and externally, to regulation of properties and metabolism within the waters provides the important ecosystem perspectives that are essential to our understanding as well as management of aquatic systems. The dominant pelagic influence that has so strongly governed study and dogma in limnology and oceanography is gradually relaxing as we gain understanding of the coupled integration of the land to the waters. Secondly, some of the analyses attempt to couple the effects within drainage basins to larger global alterations of climate on hydrology and nutrient cycling. Dissolved humic substances are a primary vehicle of that coupling. Lastly, the messages of this synthesis emphasize the great complexities of the regulation of biological dynamics in aquatic ecosystems. We are progressing beyond the important but simplistic ideas that phosphorus and nitrogen are dominant regulators; it is rather aquatic humic compounds that impose a multiplicity of dynamic controls of metabolism.

Tuscaloosa, Alabama, March 1998

Robert G. Wetzel
Bishop Professor of Biological Sciences
University of Alabama

Contents

Contributors

GEORGE R. AIKEN, US Geological Survey, Water Resource Division, 3215 Marine St., Boulder, Colorado 80303, USA

RONALD H. BENNER, Marine Science Institute, University of Texas at Austin, P.O. Box 1267, Port Aransas, Texas 78373-1267, USA

P. JEFFERSON CURTIS, Division of Science, Okanagan University College, 333 College Way, Kelowna, British Columbia V1V 1V7, Canada

HENK DE HAAN, Provincie Friesland, Ruimte en Milieu, Afd. Waterhuishouding, P.O. Box 20120, 8900 Leeuwarden, The Netherlands

DAG O. HESSEN, Department of Biology, Division of Limnology, University of Oslo, P.O. Box 1027, Blindern, 0316 Oslo, Norway

MATS JANSSON, Department of Physical Geography, University of Umeå, 901 87 Umeå, Sweden

ROGER I. JONES, Division of Biological Sciences, Institute of Environmental and Biological Sciences, Lancaster University, Lancaster LA1 4YQ, UK

DAVID R. S. LEAN, Department of Biology, University of Ottawa, P.O. Box 450, Station A, Ottawa, Ontario, KIN 6 N5, Canada

ESPEN LYDERSEN, Norwegian Institute for Water Research, Brekkeveien 19, P.O. Box 173, Kjelsås, 0411 Oslo, Norway

DIANE M. MCKNIGHT, Institute for Arctic and Alpine Research, University of Colorado, 1560 30th Street, Campus Box 450, Boulder, Colorado 80309-0450, USA

WILLIAM L. MILLER, Department of Oceanography, Dalhousie University, Halifax, Nova Scotia B3H 4J1, Canada

UWE MÜNSTER, Lammi Biological Station, University of Helsinki, 16900 Lammi, Finland

E. MICHAEL PERDUE, School of Earth and Atmospheric Sciences Georgia Institute of Technology, Atlanta, Georgia 30332, USA

LARS J. TRANVIK, Department of Water and Environmental Studies, Linköping University, 58183 Linköping, Sweden

Humic Substances as Ecosystem Modifiers – Introduction

Dag O. Hessen and Lars J. Tranvik

In the early history of limnology, the classification of lakes was a central field of interest. Lakes with high concentrations of humic substances (HS) were identified as a distinct, specific class of lakes, easily recognized by their brown-coloured water. The term "dystrophic" (from Greek, "misfed"), introduced by Thienemann (1925), is widely used to describe such lakes. This expression implies the view of brown-water lakes, bog-lakes or humic lakes as being low-productive, owing to a number of attributes that would slow down metabolism, such as low pH, low light, low oxygen and unfavourable substratum for benthic invertebrates and fish. The humic matter itself was thought to have strong antimicrobial properties, demonstrated by the fact that organic matter was uniquely preserved in bogs. Thienemann (1925) realized the different nature of humic lakes, and introduced the separation between humic lakes and clearwater lakes. In a two-dimensional diagram, there would thus be a continuum along the productivity axis from (ultra)oligotrophy to (hyper)eutrophy, while the humic lakes would stretch along the humus (or dissolved organic carbon) axis, yet at the very oligotrophic end. Similarly, in their survey of Wisconsin lakes, Birge and Juday (1927) made a division between autotrophic lakes, with predominant internal sources of organic matter, and allotrophic lakes, where allochthonous supply of organic matter dominates. Järnefelt (1925) realized that there in fact is a continuum of all lake categories. He introduced the term "chthoniotrophic" for humic lakes, referring to the colour, and also referred to eutrophic humic lakes as "mixotrophic". While most brown-water lakes have low primary production, there are a number of humic lakes that from their nutrient loading would be classified as eutrophic.

In his extensive *Treatise on Limnology*, Hutchinson (1967) concluded that "there is still no agreement on the biological significance of ... [humic matter] ... except in its action in reducing the penetration of light". Since then, there has been a considerable increase in our understanding of the biogeochemistry and ecology of humic lakes. The main purpose of this book is to reflect this development and link some basic properties of HS to large-scale phenomena such as the carbon cycle in general and carbon cycling within the biota in particular. To achieve an understanding of the multitude of ways

Ecological Studies, Vol. 133
Hessen, Tranvik (eds) Aquatic Humic Substances

in which HS affect the biota, we need some basic understanding of *what* HS actually are, and some knowledge of basic physico-chemical features that interact with the biota. From this, we may proceed to the effects on the various biotic compartments and the entire food web structure and stability. It is of central interest to know whether HS modify these processes to an extent that makes those aquatic ecosystems that are strongly influenced by HS behave differently from those along the clearwater axis. Energy is supplied from external sources to aquatic food webs indirectly via indigenous primary production based on the import of inorganic nutrients, and directly via the import of organic matter, such as HS. Thus, loading of inorganic phosphorus (which is normally limiting the primary production of fresh waters) and organic matter (HS) are analogous in the sense that both are means of energy supply to the plankton (cf. Carpenter and Pace 1997). Still, there are distinct differences in the behaviour of systems where these two types of loading dominate.

Classification Problems – What is a Humus Molecule and a Humic Lake? Isolation of HS from a freshwater source was described as long ago as 1809 (Berzelius). From this early attempt to isolate and characterize aquatic humus and onwards, the highly complex and heterogeneous nature of HS has been a major obstacle to giving a proper chemical definition. The heterogeneity also results in a range of different properties of HS. Thus, one must apply highly general characteristics like those of Aiken et al. (1985), who defined HS as "a general category of naturally occurring, biogenic, heterogeneous organic substances that can generally be characterized as being yellow to black in color, of high molecular weight and refractory". The concept of "the structure of the humic acid molecule" is misleading. Instead, a more constructive approach is to infer average structure and functionality from average properties (Perdue 1984). One reason for this complexity is that HS are formed in various ways. These HS-generating processes, as well as the age of HS, are closely examined in this book.

The definition of "humic lake" as opposed to "clearwater lake" is quite arbitrary, as it relates more to the quantity than to the quality of aquatic humus. Even the most remote, ultraoligotrophic alpine or arctic lake is to some extent influenced by humic matter, yet there is a long way to go from these lakes with secchi depths exceeding 20 m and dissolved organic carbon (DOC) concentrations below 0.5 mg l^{-1}, to the highly coloured bog lakes with secchi depths of a few decimetres, DOC exceeding 30 mg l^{-1} and color of 500 mg Pt l^{-1} or more. While a number of highly productive lakes share such properties as low secchi depth and high DOC levels with the humic lakes, the very strong absorbance of UV radiation is a unique feature of HS. Thus, high concentration of DOC with a high absorptivity in the UV region would be a suitable property for the classification of a lake as "humic", as opposed to"clearwater". Oligotrophic lakes may also be "humic" as judged from the quality of their DOC (e.g. UV absorptivity), although the concentration of

DOC is relatively low, due to minimal autochthonous generation of DOC. An operational boundary between clearwater and humic lakes is not imperative for the contents of this book, however, since most chapters deal with qualitative phenomena that may be scaled according to the HS influence. For those particularely interested in the physico-chemical nature and classification of humic matter, we refer to a number of excellent volumes on these topics (e.g. Gjessing 1976; Aiken et al. 1985; Thurman 1985).

ARE HUMIC SUBSTANCES A NEGATIVE OR A POSITIVE CONTRIBUTOR TO ECOSYSTEM PRODUCTIVITY? As the nature of HS has been a matter of controversy, so has their role as an ecosystem modifier. The low light levels, the acidity, the frequent low levels of oxygen, and the frequent unfavourable bottom substrata, with fluffy *dy* sediments, all support the expectations of low diversity and productivity. Yet Hutchinson (1967) noted that the term dystrophic "... suggests a more pathological condition than perhaps exists". This hunch has been supported by more recent works showing that HS in fact may buffer against various anthropogenic hazards such as (further) acidification. Also contradictory to the anticipated antibiotic properties of humus, HS may in fact be a major source of organic carbon for heterotrophic bacteria (Tranvik 1988; Hessen et al. 1990), fuelling overall ecosystem productivity. As long ago as 1918, Naumann considered a direct link from detritus (humus aggregates) to metazooplankton ("... the Entomostraca in these localities gain most of their nutrients from purely allochthonous detritus, while phytoplankton or detritus derived from plankton hardly contributes at all ..." [Naumann 1918]), and similar ideas were put forward by Nauwerck (1963) and Haney (1973). Recent experiments with tracer techniques give support to this view (Salonen and Hammar 1986; Hessen et al. 1990).

HS may significantly contribute to low pH in surface waters, even in poorly buffered lakes impacted by anthropogenic acidification. On the other hand, organic acids play an important role in preventing further pH depressions of acidified systems through their pronounced buffering capacity. Also, a most striking property of humus is the ability to complex and detoxify toxic cations or organic toxins (Hongve et al. 1980), but the same chelating properties may give low access to limiting minerals or metals (Jackson and Hecky 1980). Again, we see the trade-off between positive and negative attributes of HS.

HS play a major role in light attenuation in aquatic ecosystems, and low levels of photosynthetically available light may restrict photosynthesis. On the other hand, HS are in particular an important screen for the biologically detrimental UV radiation, preventing direct radiation damage to biota. However, HS may have a dual role also in this case, as the trapping of high levels of energy in the upper few centimetres results in the production of a number of biologically harmful photoproducts such as free radicals, strong oxidants, and carbon monoxide (Miller and Zepp 1995). This may also be counteracted by the cleavage of HS into more bioavailable organic compounds and by the photochemical liberation of bound minerals and micronutrients, all processes that would stimulate primary production and bacterial secondary production (Lindell et al. 1996).

There is obviously a multitude of antagonistic processes and contrasting views on the overall role of HS in aquatic productivity, and through the following chapters we will see that there indeed is a dichotomy in the way HS affect the biota.

HOW DO HUMIC SUBSTANCES AFFECT THE AQUATIC FOOD WEB? The citation from the work of Naumann suggested in fact a new link in the energy flow, quite distinct from the conventional food web diagram that became established through Lindemans (1941) pioneering work on the trophic levels. This is in spite of the fact that Lindeman developed the concept based on studies in humic Cedar Bog Lake (Minnesota). Lindeman recognized the dominating pool of dead organic carbon ("ooze") of this system, yet did not recognize how this pool was integrated with the biota. The suggestions of Naumann also contradict the conventional structure of food web pyramids that was put forward a few years later (Elton 1927). In fact, pyramidal structures can be largely inverted in humic lakes.

There is a transition from the strict clearwater systems with phytoplankton as the dominating base of the food web and thus a high degree of autotrophy, to the allochthonously fed systems with detritus as the major base of the food web, and an extensive net release of CO_2 to the atmosphere . This decreased production to respiration ratio is clearly linked to external loads of organic C (Hessen et al. 1990; Del Giorgio and Peters 1993), meaning that these lakes support a far higher heterotrophic production than would be expected from primary production alone.

The conventional view of detritus (including humus) as the terminal product of biological production has been challenged (Odum 1968; Wetzel et al. 1972). Not only easily degradable, particulate detritus from recently dead matter, but also recalcitrant detritus in the form of HS re-enters the food web. The observations of Naumann of zooplankton feeding on particulate humic matter were later verified. Even more important, a link between HS and heterotrophic bacteria has been demonstrated (Hessen 1985; Tranvik 1988). Due to the fact that HS serve as a substrate for bacteria, bacterial secondary production in humic lakes may by far exceed the phytoplankton primary production, even in the upper layers that are exposed to photosynthetically available light (Tranvik 1989). Bacterial utilization of HS has a number of important bearings both for the food web structure and biogeochemistry.

The humic-influenced systems thus challenge a number of textbook conventions of relationships between primary and secondary producers, pyramidal structure, trophic cascades and ecosystem stability. These matters will be a main focus of this book.

HOW HUMIC SUBSTANCES INFLUENCE LARGE-SCALE CARBON CYCLING? From the microscale perspective, the line can be drawn to major biogeochemical events (that are mainly driven by bacteria). Soil HS constitute a considerable store of organic carbon, and the way they enter the aquatic systems and interact with biota and the atmosphere could give feedback effects on the entire carbon cycle (McKnight 1991). Ecosystems rich in HS will be net sources of gases like CO_2 and CH_4 to the atmosphere, while weather conditions as well as more systematic climatic changes will have strong feedbacks on the fate of terrestrial primary production and soil stores of organic C. Warmer and drier climate results in less drainage of HS to aquatic ecosystems, and this in turn will affect not only light attenuation and spectral properties of water bodies, but, for reasons given above, also the entire ecosystem in a multitude of ways (Schindler et al. 1996). This book will address these issues, linking small-scale within-lake processes to environmental and large-scale processes.

HS in surface waters are partly buried in sediments and partly oxidized, entering the atmosphere. These processes may occur directly or after the passage of HS into food webs. A significant portion also drains into marine systems via rivers. While we devote this book almost entirely to HS in freshwaters, the fate and effect of HS and dissolved organic carbon in general in the marine ecosystem are addressed in a separate chapter.

The rationale for a book on the linkage of biogeochemistry and ecology in humic lakes is thus based on the special properties described above. These unique properties of systems dominated by HS clearly relate to the biogeochemical and ecological effects of this huge pool of allochthonous DOC. This will affect basic food web characteristics of humus-influenced localities and the entire ecosystem stability (cf. Carpenter and Pace 1997; Wetzel 1995). Chemistry and biochemical aspects of humic matter have only weakly merged with the biological disciplines and vice versa), yet these areas are intimately linked via a series of feedback mechanisms. It is our intention to merge geochemical aspects with biology and ecology in the following pages. Knowledge of the dynamics of the carbon pools is imperative for the understanding of biological processes in these systems. Many of the examples and conclusions derive from fairly small and strongly coloured water bodies, but the same characteristics may be seen even for large lakes influenced by HS. In most surface waters on earth, HS are by far the dominant pool of DOC. By focusing on HS in systems where they most likely have the highest influence, we may gain insight into the general role of these substances in aquatic ecosystems.

References

Aiken GR, McKnight DM, Wershaw RL, MacCarthy P (eds) (1985) Humic substances in soil, sediment, and water: geochemistry, isolation, and characterization. Wiley, New York

Berzelius JJ (1809) Afhandlingar i Physik, Kemi och Mineralogi. 1:124–145 (in Swedish)

Birge EA, Juday C (1927) The organic content of the water of small lakes. Proc Am Philos Soc 66:357–372

Carpenter SR, Pace ML (1997) Dystrophy and eutrophy in lake ecosystems: implications of fluctuating inputs. Oikos 78:3–14

Del Giorgio PA, Peters RH (1993) Balance between phytoplankton production and plankton respiration in lakes. Can J Fish Aquat Sci 50:282–289

Elton C (1927) Animal ecology. Macmillan, New York.

Gjessing ET (1976) Physical and chemical characteristics of aquatic humus. Ann Arbor Science, Michigan, Ann Arbor

Hessen DO (1985) The relation between bacterial carbon and dissolved humic compounds in oligotrophic lakes. FEMS Microbiol Ecol 31:215–223

Hessen, DO, Andersen, T, Lyche A (1990) Carbon metabolism in a humic lake; pool sizes and cycling through zooplankton. Limnol Oceanogr 35:84–99

Hongve D, Skogheim OK, Hindar A, Abrahamsen H (1980) Effects of heavy metals in combination with NTA, humic acid, and suspended sediment on natural phytoplankton photosynthesis. Bull Environ Contam Tox 25:594-600

Hutchinson GE (1967) A treatise on limnology, vol 2. Wiley, New York

Jackson TA, Hecky RE (1980) Depression of primary productivity by humic matter in lake and reservoir waters of the boreal forest zone. Can J Fish Aquat Sci 37:2300–2317

Järnefelt H (1925) Zur Limnologie einiger Gewässer Inlands. Ann Soc Zool Bot Fenn Vanamo 2: 185–352.

Jones RI, Young JM, Hartley AM, Bailey-Watts AE (1996) Light limitation of phytoplankton development in an oligotrophic lake Loch Ness, Scotland. Freshwater Biol 35:533–543

Lindell MJ, Granéli HW, Tranvik LJ (1996) Effects of sunlight on bacterial growth in lakes of different humic content. Aquat Microb Ecol 11: 135–141

Lindeman RI (1942) The trophic dynamic aspect of ecology. Ecology 23:399–418

McKnight DM (1991) Feedback mechanisms involving humic substances in aquatic ecosystems. In: Schneider SH, Boston PJ (Eds) Scientists on Gaia. MIT Press, Cambridge, pp 330–338

Miller WL, Zepp, RG (1995) Photochemical production of inorganic carbon from terrestrial organic matter: significance to the oceanic organic carbon cycle. Geophys Res Lett 22: 417–420

Naumann E (1918) Über die naturliche Nährung des limnischen Zooplanktons. Lunds Univ Årsskrift, N F Avd 2, 14:1–47

Nauwerck A (1963) Die beziehungen zwischen Zooplankton und Phytoplankton im See Erken. Symb Bot Ups 17(5):1–163

Odum EP (1968) Energy flow in ecosystems; a historical review. Am Zool 8:11–18

Perdue EM (1984) Analytical constraints on the structural features of humic substances. Geochim Cosmochim Acta 48:1435–1442.

Salonen K, Hammar T (1986) On the importance of dissolved organic matter in the nutrition of zooplankton in some lake waters. Oecologia (Berl) 8:246–253

Schindler DW, Curtis JP, Parker BR, Stainton M (1996) Consequences of climate warming and lake acidification for UV-B penetration in North American boreal lakes. Nature 379:705–708

Thienemann A (1925) Die Binnengewässer Mitteleuropas. Binnengewässer 1:1–255

Thurman EM (1985) Organic geochemistry of natural waters. Dr W Junk, Boston

Tranvik LJ (1988) Availability of dissolved organic carbon for planktonic bacteria in oligotrophic lakes of differing humic content. Microb Ecol 16:311–322

Tranvik LJ (1989) Bacterioplankton growth, grazing mortality, and quantitative relationship to primary production in a humic and a clearwater lake. J Plankton Res 11:985–1000

Wetzel RG, Rich PH, Miller MC, Allen HL (1972) Metabolism of dissolved and particulate carbon in a temperate hard-water lake. Mem Ist Ital Idrobiol Suppl 29:185–243

Wetzel RG (1995) Death, detritus and energy flow in aquatic ecosystems. Freshw Biol 33:83–89

I Biogeochemical Aspects

1 Sources and Age of Aquatic Humus

Diane M. McKnight and George R. Aiken

1.1 Introduction

As aquatic scientists have recognized the diversity of processes controlled by or dependent upon aquatic humus, it has become important to learn more about the genesis, chemical properties, and concentration of humic substances in aquatic ecosystems. There are three classes of aquatic humus (fulvic acids, humic acids, and humin), all of which share the characteristics of being heterogeneous biomolecules which are yellow to brown or black in color, high to moderate molecular weight, and biologically recalcitrant. Fulvic acids are organic acids which are soluble at any pH; humic acids are soluble above pH 2; and humin is insoluble under the full range of pH. Aquatic humus occurs in both dissolved and solid phases, with molecular weights ranging from about 500 D for dissolved fulvic acid to greater than 100,000 D for humic acids in sediments. Although the heterogeneity of these humic fractions makes rigorous chemical studies challenging, there are sufficient analytical methods at hand to make progress toward understanding the sources, formation pathways, and fate of aquatic humus.

Aquatic humus is present in the sediments and dissolved in the water in all aquatic ecosystems. Aquatic humus is truly ubiquitous; however, its abundance relative to major ions and mineral phases varies greatly among aquatic ecosystems. In the water column, dissolved fulvic acids are much more abundant than dissolved humic acid. Within suspended particulate material, humic and fulvic acids can be equally abundant. Humin is generally not significant in particulate material in the water column. However, in lacustrine and marine sediments, humin can be a significant fraction, and humic acids may be more abundant than fulvic acid (Ishiwatari 1985; Vandenbroucke et al. 1985). In the following consideration of the sources of aquatic humus, we will focus on the nature of aquatic humus present in the water column, and the fluxes of aquatic humus to the water column from lake or stream sediments and from the surrounding watershed.

Aquatic humus can be derived from both external sources of organic material (allochthonous) and from sources produced within the aquatic eco-

Ecological Studies, Vol. 133
Hessen, Tranvik (eds) Aquatic Humic Substances

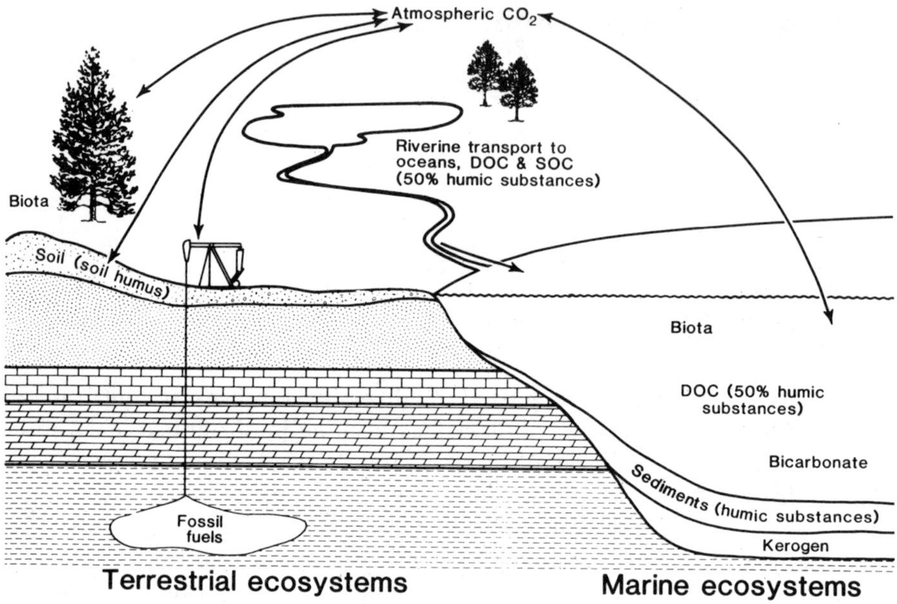

Fig. 1.1. Production and transport of aquatic humus

system (autochthonous). In this context, it is useful to define wetlands and littoral zones as part of the lake or stream ecosystem (Wetzel 1990). Fig. 1.1 illustrates the hydrologic connections between possible sources of aquatic humus, e.g., the transport of dissolved humic substances in shallow groundwater to a stream, which then flows into a lake where additional aquatic humus enters the water column from the sediments. Fig. 1.1 also illustrates that the production, chemical properties, and transformation of aquatic humic substances are intimately linked to the transfer of energy between and within aquatic ecosystems (e.g., Hessen et al. 1990). In aquatic ecosystems, the largest pool of organic material in the water column is dissolved organic material (DOM).

Dissolved fulvic acid accounts for 40–60% of the DOM of many aquatic systems. The importance of dissolved humic substances relative to other fractions of DOM is illustrated in the pie diagram for a water sample collected in the Florida Everglades (Fig. 1.2). For most natural waters, DOM is predominantly organic acids. Oxidative degradation of organic material by microorganisms results in the addition of carboxylic acid functional groups to the aliphatic or aromatic moieties in large organic molecules. These carboxyl groups are the most abundant acidic functional groups of humic substances, an important factor controlling aqueous solubility. Protonated carboxyl groups affect aqueous solubility of organic molecules to roughly the same extent as other functional groups, such as esters, ethers, and amines. Dissociation of carboxylic acid groups, however, can increase aqueous solu-

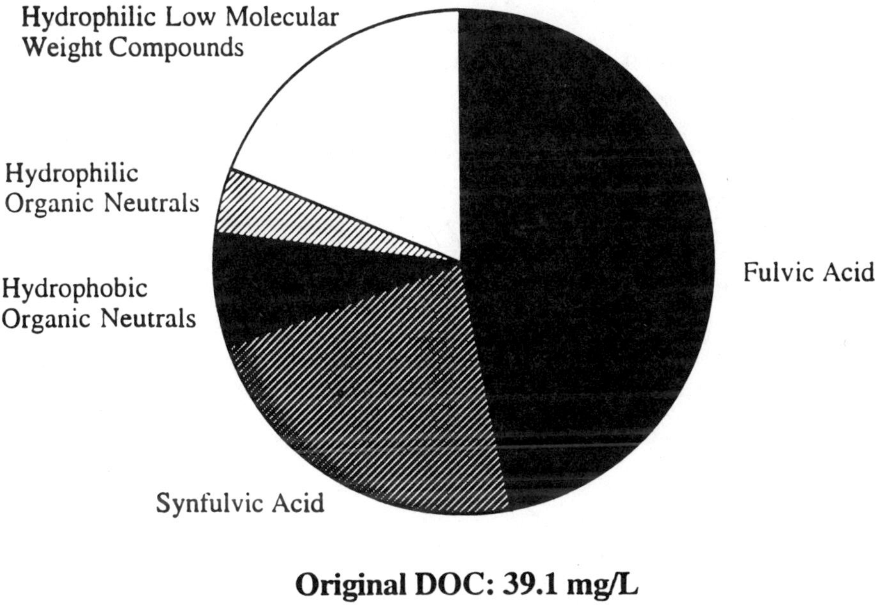

Fig. 1.2. DOM "pie" diagram, showing importance of dissolved fulvic acid

bility by up to four orders of magnitude relative to the protonated form (Thurman 1985).

The microbial degradation of DOM, including dissolved humic substances, is an important process in carbon cycling, directly influencing ecosystem dynamics (Moran and Hodson 1990;Wetzel 1990). In addition, abiotic processes such as photolysis can be important pathways for microbial utilization of humic material by releasing smaller, more labile organic compounds and nutrients (Wetzel et al. 1995). Dissolved humic substances can also influence chemical and physical characteristics of the ecosystem, being important in light attenuation, pH buffering, and complexation of trace metals. These interactions can indirectly influence ecosystem dynamics (Steinberg and Meunster 1985; McKnight 1991). One example is the absorption of UV and visible radiation by fulvic acid which controls the response of freshwater ecosystems to elevated UV-B radiation associated with ozone depletion (Scully and Lean 1994). Another example is the regulation of the bioavailability of metals as nutrients or toxicants by complexation of trace metals by fulvic and humic acids.

The extent of these interactions (microbial uptake, light absorption, etc.) in a lake or stream ecosystem depends upon the chemical properties of the dissolved aquatic humus present in that ecosystem. In order to understand the ecological role of aquatic humus, we need to understand how the sources of aquatic humus influence chemical properties and in turn influence reactivity. The goal is to have a broad understanding such that an aquatic scien-

A

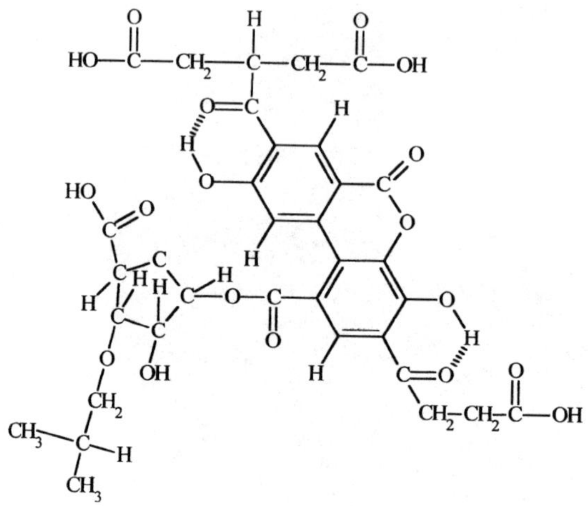

B

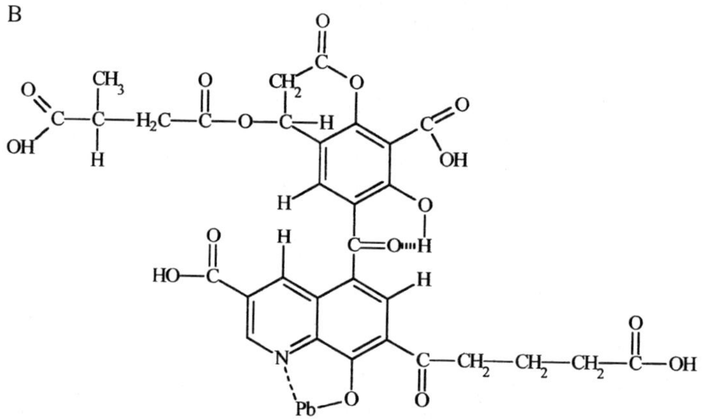

C

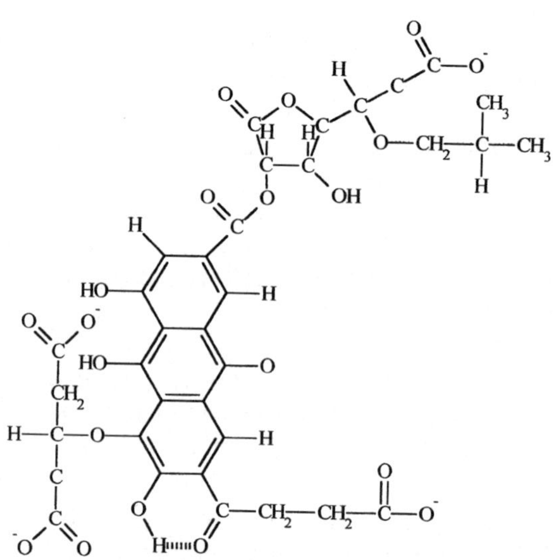

tist or engineer studying a lake, stream, or aquifer would have a sound conceptual framework within which to study the chemical nature and reactivity of the aquatic humus in the system.

A major advance in the study of aquatic humus was the effort of the International Humic Substances Society (IHSS) which made available to the scientific community standard and reference aquatic fulvic and humic acids from the Suwannee River, a blackwater river on the southeastern coastal plain of the USA. The purpose of these samples is to allow for comparison among investigators of methods of isolation and characterization, comparison of samples from different environments, and study of interactions involving humic substances by researchers without capabilities to isolate and purify humic substances from natural water. These reference humic substances were isolated using column chromatography at an "extra"-preparative scale, and were extensively characterized (Averett et al. 1988). Based upon many chemical characterizations of the Suwannee River (Georgia,USA) fulvic acid, Leenheeer et al. (1989) developed structural models that were consistent with the results. They presented an array of structural models, some of which are shown in Fig. 1.3, to emphasize the heterogeneity and differences in reactivity among fulvic acid molecules.

The Suwannee River site was chosen because of the extensive previous study of aquatic humus at the site, and because the high concentration of dissolved fulvic and humic acids made collection of sufficient quantity of material feasible, an important practical consideration. It was recognized then that the Suwannee River humic substances would not be representative of aquatic humic substances in general, although these samples may be representative of blackwater systems. Subsequently, the IHSS has made available reference samples from another aquatic site (the Nordic humic and fulvic acids) which are also from a wetland with colored waters. There are some differences between these samples from the Suwannee River and the Nordic wetland, e.g., in their fluorescence spectra (Mobed et al. 1996). However, these differences do not cover the range of variation that has been reported among aquatic environments (McKnight et al. 1994). It would be advantageous to have additional reference materials from eutrophic and oligotrophic lakes and from coastal and marine environments.

1.2 Isolation and Characterization of Aquatic Humus

Our knowledge of the biogeochemistry of aquatic humus is based upon characterization of samples from diverse aquatic environments. Because of the diversity of natural organic compounds, aquatic humic substances are

← **Fig. 1.3.** Possible structures that are consistent with average properties of the Suwannee River fulvic acid; B containing a nitrogen (anthranilic-acid type) metal-binding site; C containing a semiquinone free radical

operationally defined by methods used for isolation. In comparing the chemical characteristics of aquatic fulvic acids, it is important to be aware of the influences of the isolation procedures employed. One of the first studies to isolate humic substances from water was done by Ossian Aschan in Finland, who precipitated humic substances from colored waters using $FeCl_3$ (Aschan 1932). As with many questions in environmental chemistry, the development of isolation and characterization techniques has been a critical step. There are several different approaches for isolating dissolved fulvic acids from natural waters after filtration (Aiken 1985). One common isolation approach uses chromatographic techniques, e.g., XAD-8 resin, which are based upon hydrophobic character and acidic functional group content (Thurman and Malcolm 1981; Aiken et al. 1992). These methods result in humic fractions comprised of molecules with similar chemical properties. Separation of organic molecules based upon molecular size is also a successful approach employing tangential flow ultrafiltration and reverse osmosis (Serkiz and Perdue 1990; Ranville et al. 1991). Dissolved fulvic acids have a molecular weight range of 500 to 1200 daltons (Aiken and Malcolm 1987), and so membranes with low nominal molecular weight cutoffs of 500 or 1 000 D are required to retain fulvic acids. Molecular weight separation may include organic compounds that are not acidic in a fraction that is chiefly fulvic or humic acids. All these methods can be adjusted to accommodate natural waters with low or high concentrations of dissolved humic substances. In this chapter we will present data for chemical characteristics for fulvic acids and humic acids isolated using XAD-8 resin because we have used this method in many field studies, sometimes in conjunction with ultrafiltration.

Typically, the complicated mixtures of organic compounds obtained by the fractionation of DOM discussed in Section 1.1 are characterized by elemental analysis, molecular weight measurements, acid-base titration, and by ^{13}C-NMR, ^{1}H-NMR, and IR spectroscopy. These measurements provide information about "bulk" properties, i.e., characteristics to which all fulvic acid molecules contribute in determining an average value. Given that humic substances are complex, heterogeneous mixtures the amount of information that can be obtained about the composition of the mixture using these characterization methods is constrained. These techniques do provide valuable structural information that can be used to establish the nature and source of humic substances. A critical overview of the methods used to characterize humic substances can be found in Aiken et al. (1985).

Other measurements quantify specific organic constituents and moieties that are present at low concentrations in a given sample. For example, amino acids, identifiable after hydrolysis, are trace constituents of humic substances, typically accounting for $\approx 5\%$ of the nitrogen content (Hedges et al. 1994). These amino acids are probably structural amino acids bound to the organic matter rather than free amino acids weakly associated with the sample. Similarly, certain aromatic moieties found in aquatic humic substances such as coniferyl alcohol, sinapyl alcohol, and ρ-coumoryl alcohol can be analyzed after oxidizing humic substances. These constituents are lignin derived compounds and have been recognized as chemical characteristics of humic substances indicative of origin from lignin-containing vascular plants (Ertel et al. 1986).

In recent years, there has been a dramatic increase in the use of ^{13}C-NMR spectroscopy, which is an important technique for obtaining structural information on complicated mixtures of organic compounds. The major bands in ^{13}C-NMR spectra for aquatic humic substances are listed in Table 1.1 (Wershaw 1992). When humic substances have been isolated using column chromatography or another chemically based method, the correct interpretation of the ^{13}C-NMR spectrum is that most humic molecules contain most of the functional groups indicated in the spectrum. Although the relative abundance of carbon types may vary, the same bands are present in the ^{13}C-NMR spectra for all samples of humic substances from many environments. For example, carboxylic acids are the major acidic functional groups in fulvic and humic acids, and this is reflected in a prominent C-1 peak. One critical aspect of these analyses is quantitative measurement of carbon atoms in different moieties, which requires optimization of the experimental conditions. The variations in relative intensities of the spectral bands arise from differences in functional group distribution as related to differences in sources of organic material and geochemical transformations. Therefore, ^{13}C-NMR spectra provide a means of discriminating between humic and nonhumic material in aquatic ecosystems and of assessing the origin, fate, and reactivity of aquatic humus.

To illustrate the utility of chemical characterization, especially ^{13}C-NMR characterization, results from characterization of particulates, colloids and fulvic acids in an alpine watershed (Fig. 1.5) are presented in Table 1.2 and Fig. 1.4 (data from McKnight et al. 1997). In the Loch Vale Watershed, Rocky Mountain National Park, Colorado, USA (Fig. 1.5), fulvic acid accounts for 25–65% of the DOC in lake water. Possible processes transporting fulvic acid during snowmelt are (1) flushing through organic-rich lake sediments in the

Table 1.1. Spectral regions for different functional group carbon atoms in ^{13}C-NMR spectra of plant and microorganism tissue and natural organic substances (Wershaw 1992). Chemical shifts are given in parts per million from tetramethylsilane (TMS) resonance at 0 ppm

Designation	Spectral regions (ppm)	Functional group carbons
AL-I	0–62	Aliphatic carbons
AL-II	62–90	Alcoholic carbons in carbohydrates and other aliphatic alcohols and ethers
AL-III	90–110	Anomeric carbons in carbohydrates and some aromatic and olefinic carbons
AR	110–160	Aromatic and olefinic carbons
C-I	160–190	Carboxylate and amide carbons
C-II	190–230	Carbonyl carbons in aldehydes and ketones

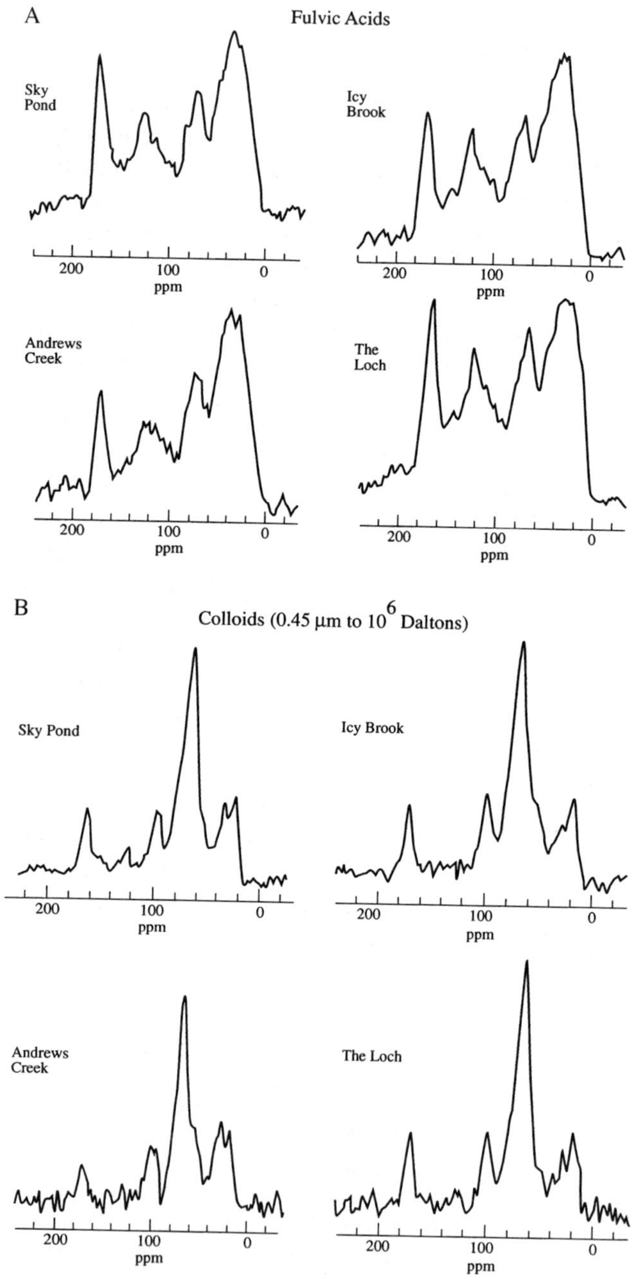

Fig. 1.4. ^{13}C-NMR spectra for fulvic acids and colloidal material from lakes and streams in the Loch Vale Watershed. **A** Dissolved fulvic acids; **B** colloids; **C** small particulates; **D** large particulates

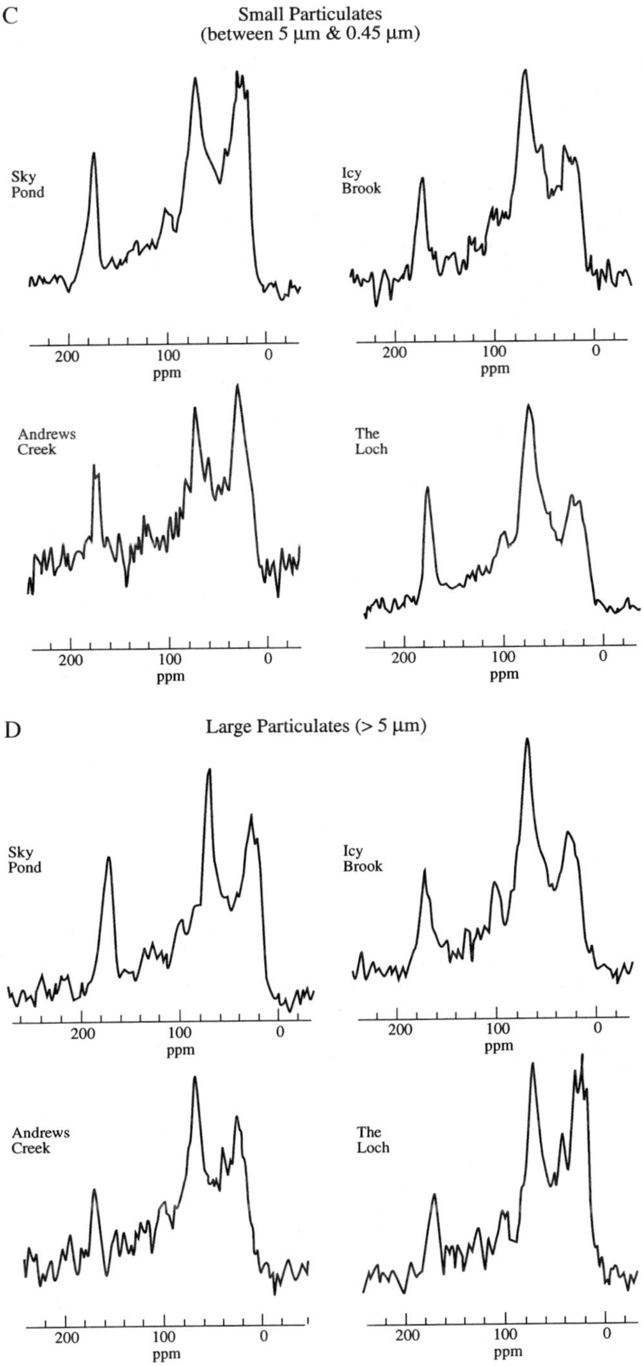
C
Small Particulates
(between 5 μm & 0.45 μm)
Sky
Pond
Icy
Brook
Andrews
Creek
The
Loch
200
100
0
ppm
D
Large Particulates (> 5 μm)
Sky
Pond
Icy
Brook
Andrews
Creek
The
Loch
200
100
0
ppm

Fig. 1.4 C, D

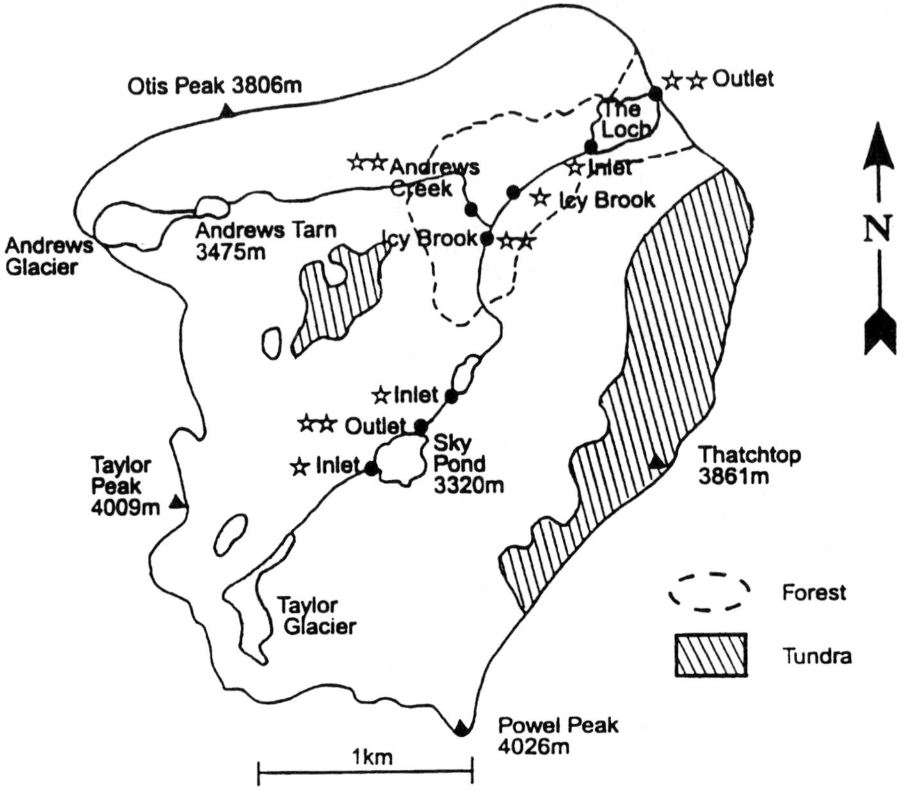

Fig. 1.5. Map of Loch Vale Watershed showing location of sampling sites

Table 1.2. Elemental analysis of dissolved fulvic acid from Loch Vale Watershed. Ash-free values are presented parenthetically. Fulvic acids were isolated following Thurman and Malcolm (1981) from filtrate passing through a 100000-D ultrafilter; a dash indicates insufficient sample for analysis

	Elemental content (and by weight)							C:N	$\delta^{13}C$
	C	H	O	N	S	P	Ash	ratio	(%)
Sky Pond	47.3 (51.0)	4.8 (5.2)	37.3 (40.2)	1.95 (2.1)	–	–	7.16	28.3	–25.3
Icy Brook (50.9)	48.3 (50.9)	4.5 (4.7)	38.7 (40.8)	1.75 (1.8)	0.55 (0.58)	0.033 (0.035)	5.16	32.2	–25.8
Andrews Creek (50.3)	46.3 (50.3)	4.5 (4.9)	38.4 (41.7)	1.46 (1.6)	0.73 (0.79)	0.028 (0.030)	8.0	37.0	–26.5
The Loch	44.3 (51.7)	4.3 (5.0)	33.9 (39.6)	1.5 (1.75)	0.54 (0.63)	0.040 (0.047)	14.35	34.5	–25.8

alpine lake, which is above treeline; and (2) flushing of forest soils and overland and shallow groundwater flow into the subalpine lake and stream (Baron et al. 1991). The ^{13}C-NMR spectra for the fulvic acids show a progressive increase in aromatic carbon content (110–160 ppm) going from the alpine to the subalpine lake. The spectra for the colloids show that they are similar throughout the watershed and primarily composed of carbohydrates (i.e., they are not humic substances). The spectra for the particulates show that they are composed of plant detritus and humic substances sorbed to minerals and do not vary as water draining subalpine forests enters the stream and lake.

1.3 Formation Pathways of Aquatic Humus

We can approach the fundamental, seemingly intractable, question of the *formation of aquatic humus* from two perspectives. One perspective is to consider fundamental chemical processes involving the constituents of plant and microbial material. Another perspective can be gained by comparing the biogeochemistry of humic substances among different aquatic environments. Here, we briefly review chemical and microbial processes which may be involved in aquatic humus formation, and then discuss results from field studies which show that aquatic humic substances are generated from both rapid and long-term degradation processes.

Firstly, we should keep in mind that despite the great diversity of terrestrial and aquatic vegetation, there is a general similarity in the constituent functional groups in detrital organic material. It follows that these functional groups are derived from the partial degradation of structural components of plants and microorganisms (Wershaw 1992). Degradation of vascular plants and microorganisms gives rise to pigments, polysaccharides, degraded lignins, lipids, sterols, and proteins. These constituents vary in their lability with respect to microbial uptake, with carbohydrates probably the most readily assimilated. Because of the absence of lignin in microorganisms, proteins and lipids generally account for a greater proportion of the biomass in microorganisms than in vascular plant tissue.

In the study of humic substances, there has been a long and continuing search for more detailed understanding of sources and formation pathways (Hayes et al. 1989). At the turn of the century, a major breakthrough was recognition of the chemical heterogeneity of humus (reviewed in Stevenson 1985). Since then, there have been two general conceptual models for humus formation. One concept was that humus was formed directly from the lignified tissues of plant material. Another involves polymerization of simple products generated in the degradation of plant material. Cellulose and lignin in plant structural material account for the largest portion of plant biomass on the earth's surface, and as such are likely sources of precursor material for

A

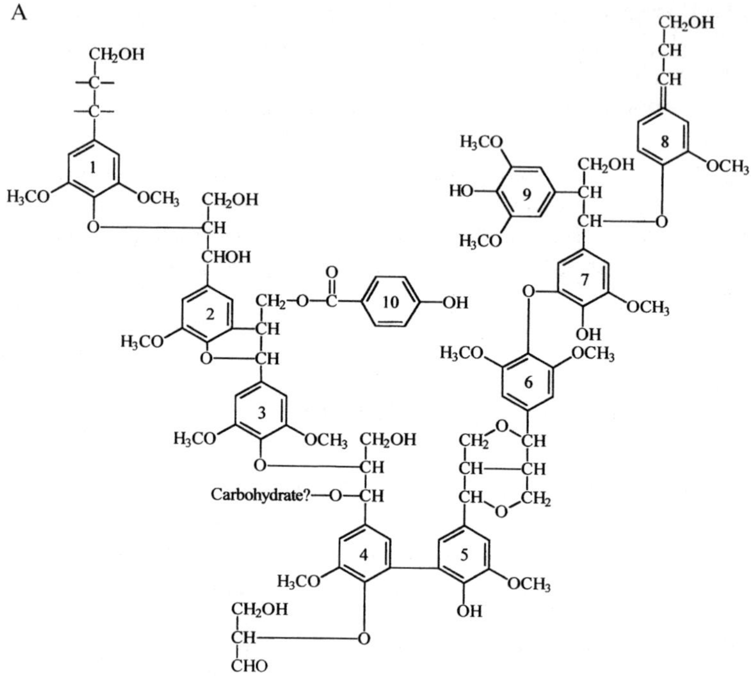

B

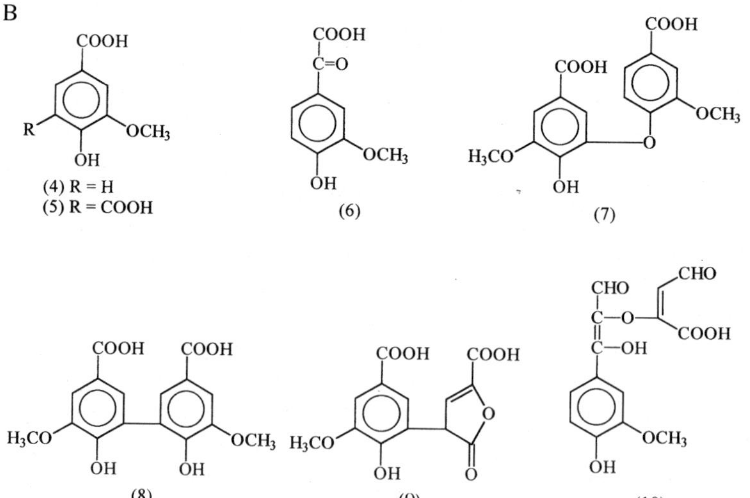

Fig. 1.6. A Schematic formula for a portion of aspen lignin; **B** Carboxylic acids identified in extracts of degraded spruce wood

humus formation. Lignin is unique among biomolecules in several ways (Kirk 1984). The structure of lignin is somewhat irregular because its formation is not highly controlled at the cellular level (Fig. 1.6). Degradation of lignin in plant structural material can be carried out by a limited number of fungal species, and only under aerobic conditions. Lignin is also present in leaf litter, grasses, and stems, although at a lower portion of the total biomass than in structural material. Lignin in litter may be degraded by bacteria as well as by fungi. An important aspect of microbial lignin degradation is that it is a nonspecific process and is not inducible by the presence of substrate. Examples of some products of lignin degradation are shown in Fig. 1.6. It should be noted that these degradation products all contain carboxyl groups resulting from microbial oxidation, which would increase solubility of any larger compound formed by condensation reactions.

Lignin contains abundant aromatic rings and no nitrogen (Fig. 1.6). Dominance of woody debris and other lignin-derived material is a probable explanation for the substantial aromatic peak in the ^{13}C-NMR spectra. To account for the nitrogen in humic acid, Waksman (1936) proposed that modified lignins combined with proteins through the Schiff reaction:

$$C_{52}H_{46}O_{10}(OCH_3)COOH(OH)_4CO + H_2NRCOOH \rightarrow C_{52}H_{46}O_{10}(OCH_3)COOH(OH)_4C{=}NRCOOH + H_2O$$

and this conceptual model was dominant in the first half of this century. The early lignin degradation model can be expanded to a general conceptual model of preferential oxidation and preservation of residual organic materials, which does not require condensation of monomers to produce higher molecular weight organic acids. Examples of other constituents in plant and microbial material are shown in Fig. 1.7 to emphasize that many constituents contain carbon double bonds and nitrogen-containing groups. Through the action of extracellular enzymes and abiotic processes, these constituents can be partially broken down and oxidized, resulting in soluble biomolecules enriched with carboxylic and other oxygen-containing functional groups. These processes would release water-soluble precursor fulvic and humic materials which are mobile in soils and sediment interstitial water. During transport, these materials undergo further modification by chemical and microbial processes and fractionation by sorption and other interactions with inorganic solid phases.

The conceptual model based upon lignin degradation was eventually expanded to include more complex multistage models. The stages in these models include: degradation of plant material producing a variety of simpler monomers, microbial uptake of these monomers producing microbial biomass in soils and sediments, and concurrent polymerization of reactive monomers to form a heterogeneous mixture of larger molecules resistant to further microbial degradation. In the multistage models, polymerization of reactive monomers can occur through many pathways. One class of important reactive monomers are polyphenols (quinones) which are both synthe-

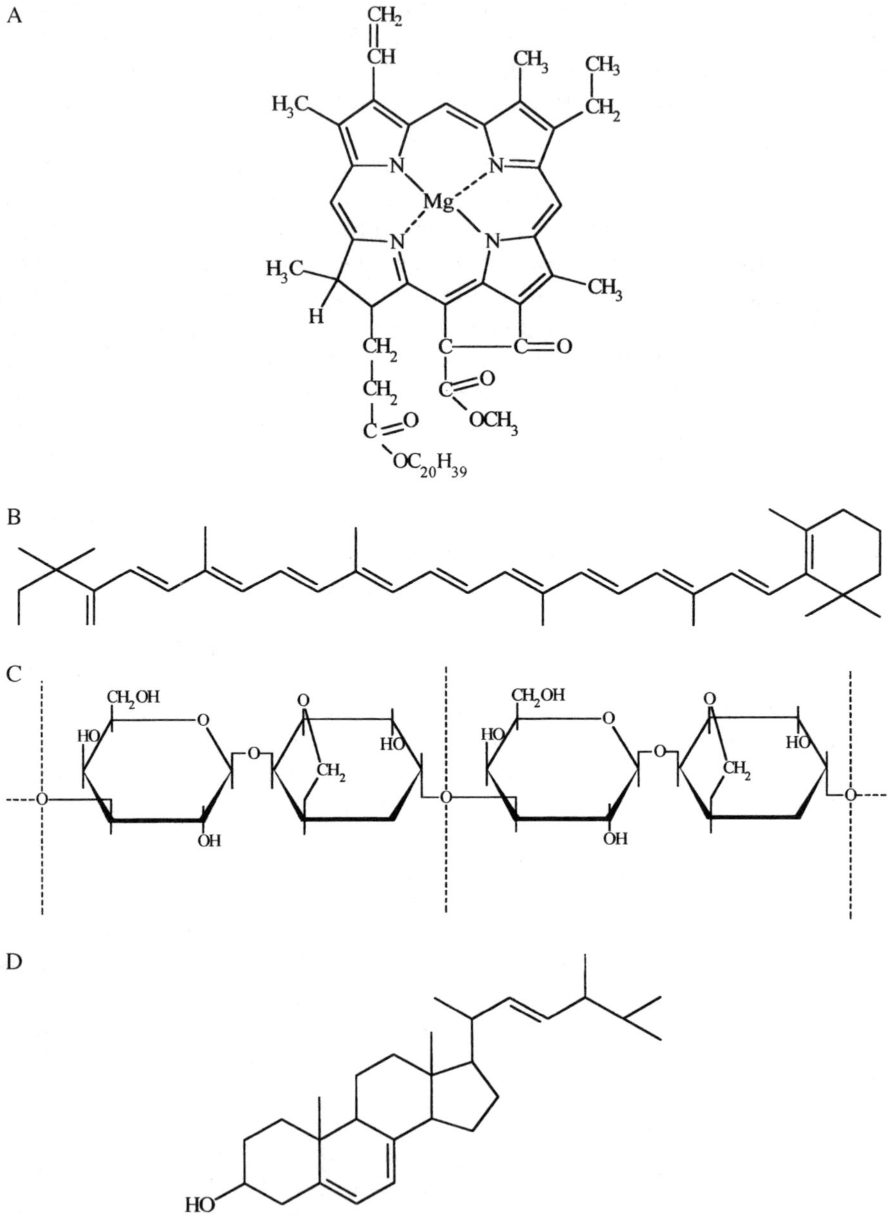

Fig. 1.7. Examples of chemical structures of constituents of plant and microbial material which could serve as precursor organic material in aquatic humus formation. A Chlorophyll a; B carotenoid (B-carotene found in chlorophytes); C polysaccharide (agarose found in marine red algal); D sterol (ergosterol found in some chlorophytes). (Stewart 1974)

sized by microorganisms and released in lignin degradation. Polymerization involving phenols has been shown to be accelerated by the presence of transition metals in solution, by metal oxides, and by the presence of clays (Steinberg and Muenster 1985). In the late 1970s, the possible importance of enzyme mediated oxidations was pointed out (De Haan et al. 1981). An example reaction is shown in Fig. 1.8. Another possible polymerization reaction is the Malliard reaction involving condensation of amino acids and related substances with reducing sugars; however, this reaction is less likely to occur under water-saturated conditions. Clearly, an important consideration in evaluating these multistage models is timeframe. Are these models consistent with the rate at which humic substances are produced from recently senescent plant and microbial material in aquatic environments?

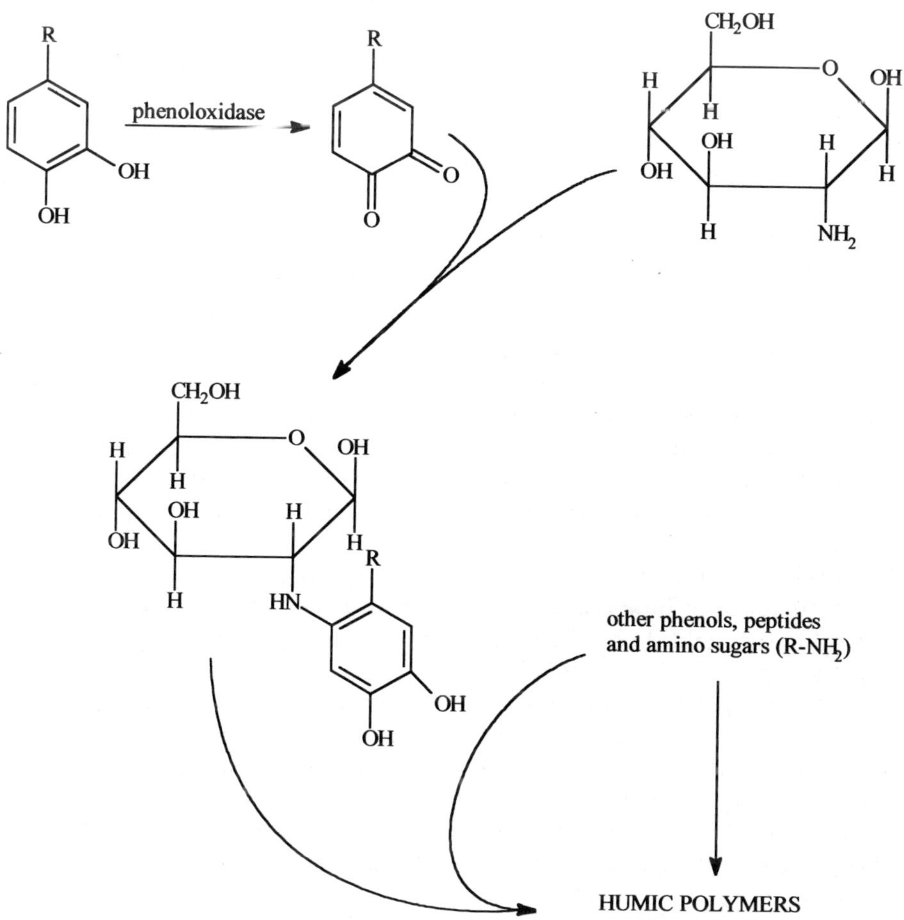

Fig. 1.8. Oxidative polymerization of phenol derivatives involving amino sugar units. (Steinberg and Munter 1985)

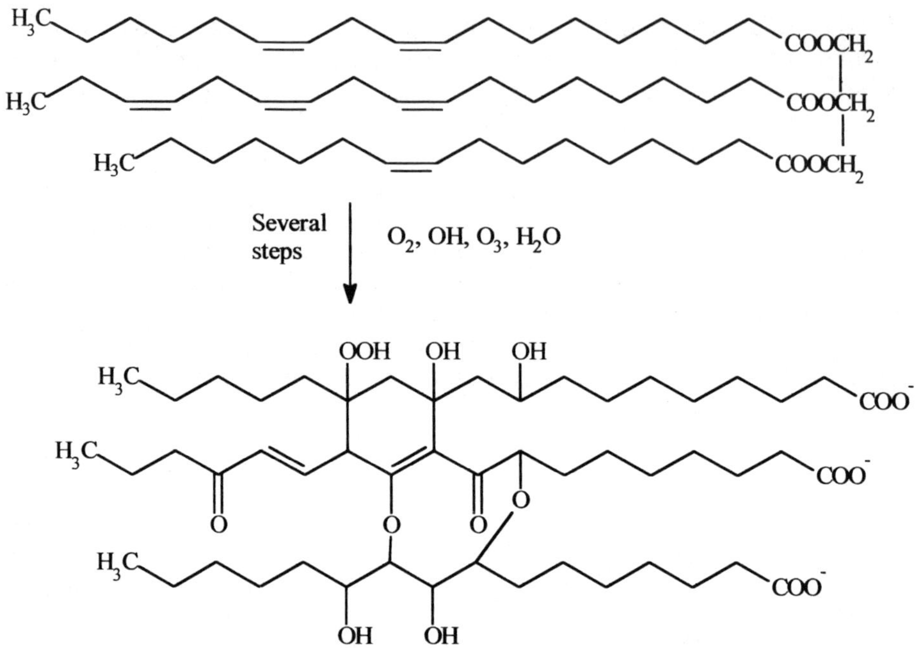

Fig. 1.9. Formation of marine fulvic and humic acids from unsaturated lipids. (Harvey and Boran 1985)

An example of a polymerization model to account for the formation of marine fulvic and humic acids is shown in Fig. 1.9 and was proposed by Harvey and Boran (1985). In this model, humic substances are formed in marine surface waters from the free radical cross-linking of unsaturated lipids released into seawater by algal growth. Similar pathways could be important in large mesotrophic lakes where sources of DOM from the watershed are minimal.

As pointed out by Stevenson (1985), "a completely satisfactory scheme for the occurrence of humic and fulvic acids in diverse geologic environments has yet to be established." The actual importance of processes represented by these various conceptual models would be expected to be influenced by the available organic substrates and environmental conditions. All these models can account for the heterogeneity and major functional groups of humic substances, and all these models can explain a variation in fulvic and humic acid chemistry based upon available organic precursor materials in a particular ecosystem. Given the complexity of the environment, it is difficult to verify the various hypotheses proposed to explain the genesis of humic substances.

In both terrestrial and aquatic ecosystems, humic substances enhance the conditions for growth in many ways, such as binding trace metals and absorbing UV light in aquatic ecosystems, Thus, the formation and persistence of humic substances may further be controlled by an ecological feedback (McKnight 1991). This leads to an important aspect of aquatic humus. Although formed via diverse pathways involving many precursor organic constituents, the resulting fulvic and humic acids share the characteristic of being relatively stable in the aquatic environment, i.e., being recalcitrant to microbial decomposition unless facilitated by oxidative processes such as photolysis (Wetzel et al. 1995). MacCarthy and Rice (1991) present a convincing argument that, from an ecological perspective, the molecular irregularity of humic substances and the lack of biologically induced specificity in their formation are the critical characteristics explaining their persistence and ubiquitous nature. They argue that a complex mixture of compounds in which monomers are not sequenced in a regular order has an inherent resistance to enzymatic decomposition.

1.4 Sources of Aquatic Humus in Aquatic Ecosystems

Much of our knowledge of aquatic humus in lakes, streams, and coastal and marine environments is the result of research conducted in the past 25 years. Dissolved fulvic acids have been studied in a range of surface waters, and consistent differences have been found between fulvic acids from systems dominated by organic inputs from plants and soils and those from algal-rich environments (Malcolm 1990; McKnight et al. 1994). Comprehensive reviews and discussions of these environments are presented in chapters in the book edited by Aiken et al. (1985). In the discussion presented here, we will compare humic substance production among aquatic ecosystems and discuss a few recent studies quantifying the production and transport of aquatic humus in a hydrologic context.

Measurement of $\delta^{13}C$ values can be used as an indicator of fulvic acid source (Schiff et al. 1990), requires only a small sample, and can be used to study the seasonal dynamics of fulvic acid production. As shown in Table 1.2, fulvic acids from all four sites in the Loch Vale Watershed had $\delta^{13}C$ values in the range of –25.3 to –26.5‰ representative of terrestrially-derived organic material. These fulvic acids were lighter, i.e., contained less ^{13}C, than the colloidal organic material, which had $\delta^{13}C$ values ranging from –22.5 to –23.5‰ (Fig. 1.10). The fulvic acids were also depleted in nitrogen relative to the colloids (Fig. 1.10). Baron et al. (1991) found that in the Loch, $\delta^{13}C$ values of fulvic acid remained fairly constant in the spring and summer, whereas in Sky Pond values became heavier following snowmelt, shifting from –33.1‰ in April to –24.5‰ in July. There are several possible interpretations for the changing $\delta^{13}C$ values in Sky Pond: (1) seasonal changes in the $\delta^{13}C$ of the dissolved inorganic carbon (DIC) may influence the $\delta^{13}C$ of microbial sources of fulvic acid; and (2) the sparse vegetation and buried soils around Sky Pond may become a source of fulvic acid during the summer. This example shows

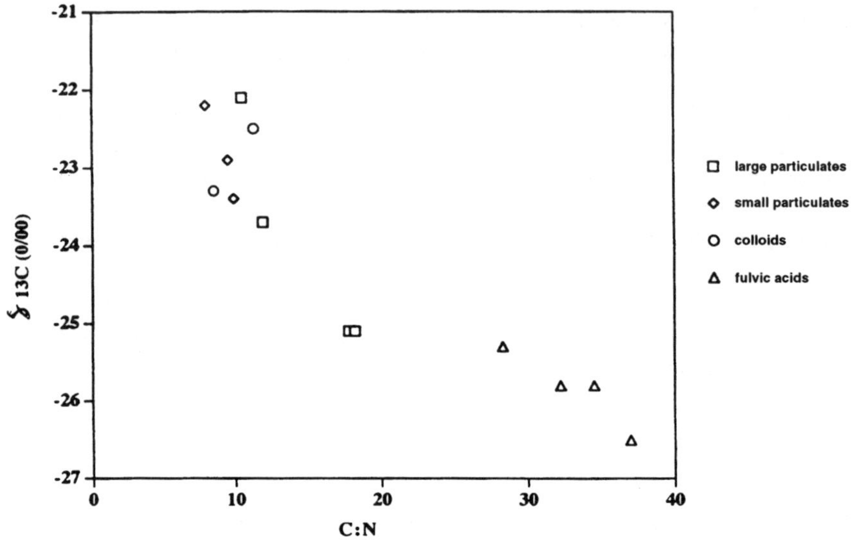

Fig. 1.10. Relationship between C:N ratio and $\delta^{13}C$ for Loch Vale samples

that the carbon isotopic signatures should be interpreted in the context of the biogeochemistry of the aquatic system.

The importance of lignin as precursor material in the formation of aquatic humus can be demonstrated by the presence of recognizable lignin components in dissolved fulvic and humic acids. In a study of dissolved humic substances at 17 stations in the Amazon River system, Ertel et al. (1986) showed that 3–8% of the humic carbon was present in lignin components, with fulvic acids having lower lignin levels than humic acids. The characteristics of the lignin components indicated that the important diagenetic transformations were loss of methoxylated structural units and oxidative degradation of lignin side chains. Greater oxidative degradation of lignin components was indicated in the fulvic acids than in the humic acids from the same sites, which was matched with a greater ratio of carbon to nitrogen in the fulvic acids.

Taken together, other "bulk" properties can be related to sources of aquatic humus. Because lignin contains many aromatic rings and no nitrogen (Fig. 1.6), it can be reasoned that the presence of lignin within the precursor pool of organic material would serve as a source of aromatic moieties and as a diluent with respect to nitrogen. This reasoning is consistent with either polymerization models or residual oxidized product models. The importance of lignin in the precursor pool can thus be tested by comparing samples where lignaceous sources of organic carbon are abundant with samples where lignaceous sources are absent or insignificant relative to other carbon sources.

In Fig. 1.11, fulvic acid samples from different aquatic ecosystems are compared based upon the ratio of aromatic carbon (AR) to aliphatic carbon

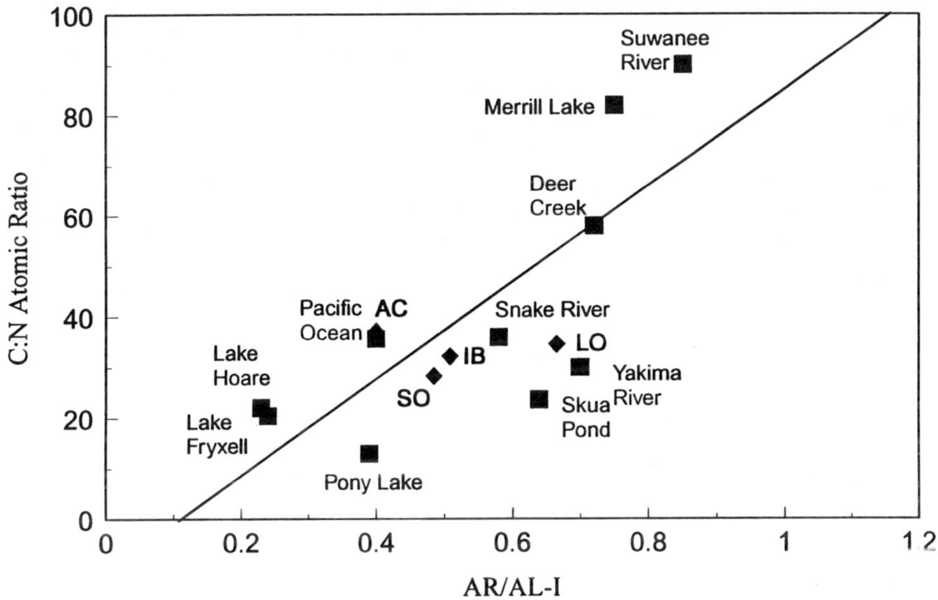

Fig. 1.11. C:N vs. AR:AL-I, showing position of Loch Vale fulvic acids relative to other aquatic fulvic acids. SO Sky Pond Outlet; AC Andrews Creek; IB Icy Brook; LO Loch Outlet. Other fulvic acids are described in McKnight et al. 1994

(AL-I) (as determined by ^{13}C-NMR) and the atomic ratio of carbon to nitrogen (from elemental analysis). This sample set includes samples from three lakes in Antarctica, where the sources of DOM are limited to microbial and algal biomass. One of these lakes, Pony Lake, is extremely eutrophic and the other two are oligotrophic. The samples of fulvic acid from these lakes all plotted in the region of the graph where aromatic content is low and nitrogen content is high. The samples from streams and lakes receiving DOM from a temperate forested watershed were found in a broad region having higher aromatic carbon content and low nitrogen content (i.e., higher carbon to nitrogen ratio). This comparison supports the importance of lignaceous material from terrestrial vegetation in the formation of aquatic humus.

For the microbial end member it is significant that the aromatic carbon content is low but not minuscule. The "aromatic carbon" region of the ^{13}C-NMR spectra includes all sp^2-hybridized carbons bonded to other carbon atoms (Table 1.1). Included in this region are not only carbon atoms associated with aromatic moieties, but also those associated with double bonds that are not aromatic. Further characterization of humic substances by $^{+}$H-NMR confirms the presence of both aromatic and olefinic moieties (McKnight et al. 1991). This result suggests that precursor constituents containing carbon-carbon double bonds are important in the formation of fulvic and humic acids from microbial and algal biomass; possible precursors are certain lipids, amino acids, pigments, and steroids.

In order to adequately partition the sources of carbon in aquatic humic substances, one must know how much variation occurs in the microbial and terrestrial plant/soil endmembers. If samples fall between the two end members in Fig. 1.11, is that a result of microbial and plant sources of organic material being important in formation, a result of mixing of humic substances from different sources during hydrologic transport, or a result of other biogeochemical controls? We can gain some insight related to these questions by considering in more detail the position of different samples in the graph in Fig. 1.11.

1.4.1 The Microbial End Member

Two of the microbial end member samples are clustered close together; these are from two ice-covered closed basin lakes in the McMurdo Dry Valleys, Antarctica. The estimate of the age of these lakes since a prior dry period is between 1000 and 3000 years. These lakes both have very stable water columns and are oligotrophic, but otherwise have many differences. Lake Fryxell has a strong salinity and density gradient and anoxic water from 9 m to the maximum depth of 18 m. Lake Hoare is less saline, is well mixed at the middle depths, and has O_2 concentrations in excess of saturation to a depth of 28 m, below which a small pocket of anoxic water is found (McKnight et al. 1993a). The phytoplankton in Lake Hoare are less abundant than in Lake Fryxell, and are more evenly distributed between chlorophytes, cryptophytes, and cyanophytes than in Lake Fryxell, where cryptophytes account for most of the biovolume (Spaulding et al. 1994). In Lake Fryxell, a much greater area of the sediments is anoxic than in Lake Hoare. In both lakes, the concentration of fulvic acid increases with depth and the chemistry of the fulvic acid is fairly uniform with depth, with an increase in the S content of the fulvic acid in the anoxic deepest water (Aiken et al. 1996). The increasing concentration with depth indicates that the dissolved fulvic acid diffuses upward into the water column from the sediments.

The chemical similarity of the fulvic acids in the two lakes suggests that their chemistry is primarily determined by a microbial/algal source, and possibly the similar lake history. Some differences between the lakes might have been expected to influence the chemistry of the fulvic acid by influencing the precursor materials and polymerization and solubilization processes in the proposed formation pathways. The differences which do not influence fulvic acid chemistry include: (1) oxic vs. anoxic conditions in the source sediments and in the water column (except for sulfur content); (2) phyla of dominant alga and associated differences in pigments, sterols, and other precursor constituents; and (3) salinity, which influences solubility of organic acids.

The positions of the other two microbial end-member samples (Pony Lake and Skua Pond) are shifted toward higher ratios of aromatic to aliphatic carbon (Fig. 1.11). These fulvic acids are from saline ponds which have very abundant phytoplankton or benthic algal mats (McKnight et al. 1994). The fulvic acids are more likely to have been derived from recent algal growth in these ponds than in the dry valley lakes. Examining the ^{13}C-NMR spectra for these samples shows that there is a larger abundance of aromatic and hetero-aliphatic carbon, and a lower abundance of aliphatic carbon compared with the samples from the dry valley lakes. The differences in chemistry indicate that either the precursor pool of organic material contained more material with carbon double bonds and more polysaccharides, or that the short time since formation had not allowed for the loss of moieties with carbon double bonds and carbohydrate moieties from the fulvic acid molecules. Assuming that fulvic acids are not assimilated into bacterial cells, then the loss of carbohydrates would occur through the action of extracellular enzymes. Abiotic photolytic oxidation causes loss of aromatic moieties. The recent age of the fulvic acid does not appear to influence the carbon to nitrogen ratio as much. This implies that the nitrogen incorporated into the fulvic acid molecules, possibly from cell wall proteins, is lost at a similar rate as aromatic and carbohydrate moieties.

Although there is much speculation in the above discussion, which is based upon only four samples of microbial/algal origin, we can conclude that the chemical characteristics of the microbial/algal end member are not tightly constrained and that variation in chemical characteristics may be related to the age of the humic substances in the aquatic environment. If we adopt the terminology of the soil scientists, we can think of the chemistry of the microbially derived fulvic acids changing as "humification" proceeds.

1.4.2 The Plant/Soil End Member

In Fig. 1.11, there are four fulvic acid samples that are most probably derived directly from plant/soil sources in a surrounding catchment; these are samples from the Suwannee River (Georgia, USA), Merrill Lake (Washington, USA), Deer Creek (Colorado, USA), and The Loch (Loch Vale, Rocky Mountain National Park, Colorado, USA). The vegetation source for the Suwannee River is the sphagnum moss and cypress trees in the Okefenokee Swamp, for Merril Lake the source is a Douglas fir forest, and for Deer Creek and The Loch the source is Engelman spruce and subalpine fir forests. Although this sample set represents a range in vegetation and soils, it is by no means comprehensive. Among this sample set, the aromatic to aliphatic ratio varies only from 0.7 to 0.9, whereas, the ratio of carbon to nitrogen varies more widely from 35 to 95. Because the carbon content is always close to 50% of the dry weight, most of this variation is associated with nitrogen content.

If we think about how terrestrial ecosystems differ, the narrow variation in aromaticity and the wide variation in nitrogen content makes sense. As noted in Section 1.3, together cellulose and lignin are the most abundant forms of biomass in all terrestrial ecosystems, comprising most of the plant structural material and a significant amount of leaves, stems, and grasses. Therefore, the presence of decaying plant material implies an abundance of lignaceous precursor material in the production of fulvic acid. On the other hand, the supply of nitrogen in terrestrial ecosystems can vary a great deal and nitrogen is often found to be tightly recycled in soils in well-established forests. It is particularly interesting to note that the Loch fulvic acid has the lowest carbon to nitrogen ratio and that in The Loch Vale Watershed nitrogen is in high concentration because of inputs in deposition associated with the development of the Front Range of the Rocky Mountains (Baron et al. 1994; Williams et al. 1996).

From these considerations, we can expect significant variation among fulvic acids having either a primarily microbial/algal source material or a primarily plant/soil source material. For the microbially derived humic substances, the extent of humification may be a main factor and for plant/soil-derived fulvic acids the availability of nitrogen may be a main factor. Thus, the graph in Fig. 1.11 allows for a meaningful comparison of fulvic acids from different aquatic ecosystems, but is not likely to allow for quantization of the importance of the different sources of fulvic acids from larger rivers, for example. Within a watershed, fulvic acid chemistry may vary in a manner consistent with these trends. For example, in the Loch Vale Watershed the fulvic acid from Sky Pond, the alpine lake which is surrounded by barren talus slopes, did have a lower carbon to nitrogen ratio and a lower aromaticity than the fulvic acid from The Loch, the subalpine lake surrounded by the spruce/fir forest. The Sky Pond fulvic acid had a greater aromaticity than the fulvic acids from the dry valley lakes, possibly reflecting an additional source from buried forest soils from 12000 years past when treeline extended above Sky Pond.

1.5 The Age of Aquatic Humus

It is clear that the relative importance of different chemical and biological processes involved in the formation of aquatic humic substances is strongly influenced by time factors. Production of humic-like material from plant litter can occur very quickly. Fig. 1.12 presents results from an experiment in a mountain catchment in which willow leaves placed under the snowpack were retrieved and leached for only a few hours (McKnight et al. 1993b). While the amount of leachable organic carbon decreased through the winter, there was consistently about 10% of the organic carbon in the leachate that behaved as fulvic acid in the XAD-8 isolation method and was also yellow-colored. During snowmelt, such material could be a significant portion of the pulse of DOC that is commonly observed in mountain streams (McKnight et al. 1993b). Another environment with fresh DOM sources is the littoral zone of lakes, where aquatic macrophytes release large amounts of DOM, some of which has humic characteristics (Wetzel 1990). We also have chemical evidence of young sources of fulvic acids in the eutrophic saline ponds in Antarctica.

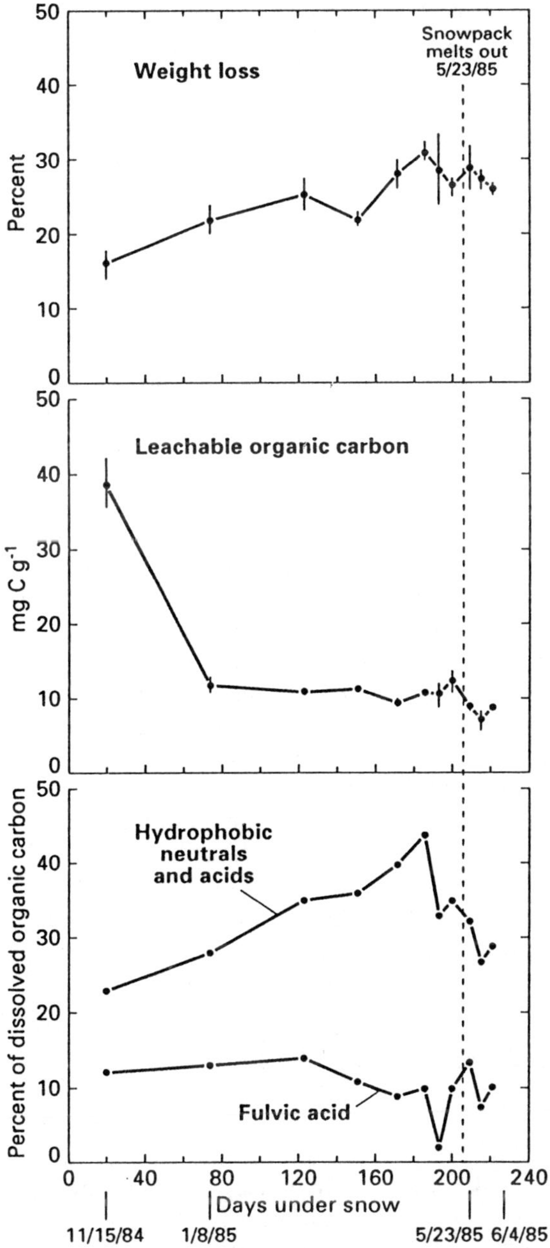

Fig.1.12. Percent weight loss of willow leaf packs under snow pack during winter and spring; concentration of leachable dissolved organic carbon (DOC) per gram dry weight of leafpack; and percent of leachable DOC as hydrophobic neutral acids and as fulvic acid

Table 1.3. Residence time (age) of organic carbon (years)

	Autotrophs	Humic substances
Terrestrial		
Forests	20–200	200–8000$_a$
Prairies	1	200–8000$_a$
Aquatic		
Groundwater		
Biscayne aquifer	–	660$_b$
Fox Hills-Laramie	–	17 000$_b$
Wetland		
Okefenokee swamp	10–200	<20$_b$
Lakes and rivers	<<1	?
Pacific Ocean	<<1	4000$_c$

$_a$ Stevenson (1985) – $_b$ Thurman (1985) – $_c$ Malcolm (1990)

Clearly, an important tool to evaluate the age of aquatic humus is ^{14}C dating. Table 1.3 presents the age range of dissolved humic substances in different aquatic environments as determined by ^{14}C dating. Given that humic substances are a heterogeneous mixture, an age determined by ^{14}C dating represents an average value. At the extremes of postbomb carbon and dead carbon, interpretation is straightforward. However, intermediate ages could reflect a mixture containing very young and very old humic molecules. The results summarized in Table 1.3 are from a limited set of measurements; nonetheless, they do show that fulvic acid is much younger in surface waters than in deep groundwaters and the ocean (Malcolm 1990).

The ^{14}C age is not necessarily definitive by itself. An understanding of the hydrology of the system is required to interpret the data in terms of the length of time the organic carbon was stored as soil humus or as humus in lake sediments before becoming mobile in a shallow groundwater, lake, or stream system. To illustrate the importance of hydrologic context, we present examples from the Deer Creek watershed and from Lake Fryxell.

1.5.1 Fulvic Acid Production and Transport in the Snake River and Deer Creek Watershed

During snowmelt in the Rocky Mountains, DOC concentrations in the Snake River and Deer Creek reach a peak as streamflow rises. A DOC increase during the rising limb of the hydrograph is observed in many forested watersheds (Grieve 1994). The snowmelt flux of DOC represents about 80% of the annual flux of DOC from these watersheds, and one of the largest fluxes of organic carbon from the watershed. Conceptually this can be explained as a flushing of DOC-rich water from the upper soil horizons and the litter layer, caused by the rise in the water table during snowmelt. Hornberger et al. (1994) evaluated this concept by applying a topographic hydrologic model which incorporated a two layer model for DOC production in soils. In the lower layer, the DOC concentration was assumed constant at a low value (1.5 mg C/l) and in the upper layer, soluble organic carbon was accumulated progressively with time during dry periods (when the water table was below the upper layer) and then was flushed from the upper layer using a continuously stirred reactor model during periods when the water table had risen into the upper layer. This modeling approach can account for the timing and the magnitude of the DOC pulse during snowmelt (Fig. 1.13).

The flushing of accumulated soluble organic carbon by the rise in the water table can be justified by observations of decreasing DOC concentrations in soil lysimeters during snowmelt (Fig. 1.14) and by knowledge of hydrologic properties of the watershed, e.g., hydraulic conductivity of the soils

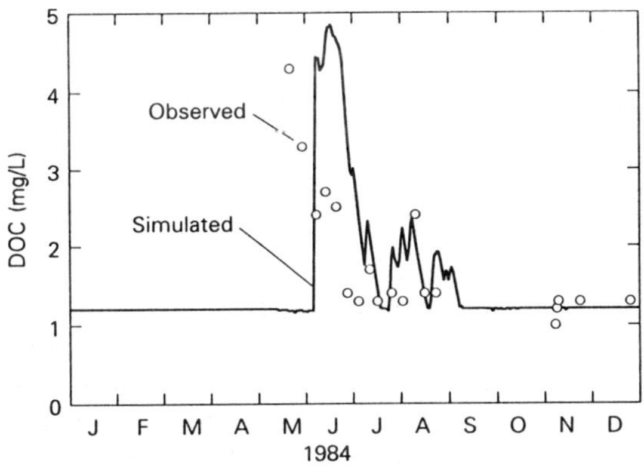

Fig. 1.13. Simulated DOC concentrations for Snake River. Note that lateness of simulated peak is due to lateness of snowmelt as calculated by TOPMODEL

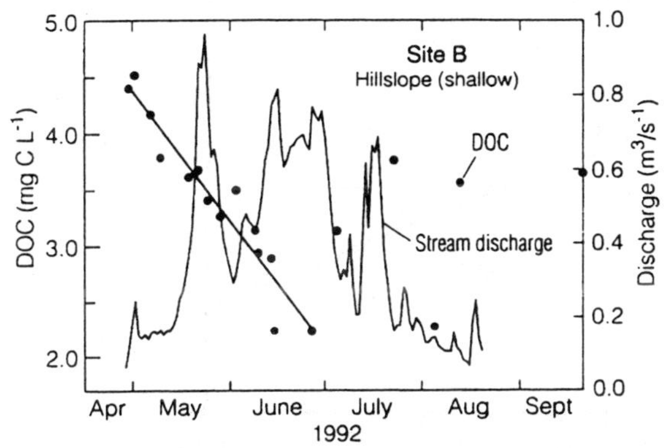

Fig. 1.14. Semilog plot illustrating linear recession of in DOC over time in Deer Creek Watershed; example from site B, shallow hillslope lysimeter. The stream discharge hydrograph is superimposed in the background

(Boyer et al. 1996). However, the progressive production of soluble DOC when the soils are not saturated is more hypothetical because we lack direct observational evidence of production and storage of soluble organic carbon, including dissolved humic substances, in unsaturated plant litter and soils. Some of the dissolved humic substances could come from initial litter leachate with a very short residence time in soil interstitial water. Another process is probably microbial degradation of soil humus generating soluble fulvic acids at a low, steady rate. These two formation pathways represent losses or "leakiness" of organic carbon from the terrestrial ecosystem at two different points in the process of soil development. Humic substances produced from fresh litter and from old soils would be expected to have different chemical properties, and different reactivities en route to the stream. It could be argued that the litter leachate source would be derived more directly from plant pigments and other constituents, whereas the longer term source would be derived more directly from degradation of more resistant plant materials, such as lignin degradation. The soils may be young enough that ^{14}C ages of fulvic acids produced by either pathway would be modern.

1.5.2 Fulvic Acid Production and Transport in Lake Fryxell

In Lake Fryxell the DOC concentrations increase from 3.7 mg C/L under the ice cover to 32 mg C/l at 18 m, the maximum depth. This DOC depth profile was interpreted as resulting from the diffusion of DOC from the sediments into the overlying highly stable water column. This interpretation was supported by the excellent fit of the DOC profile to an equation for diffusion from a plane (Fig. 1.15). In dimictic or monomictic lakes such a stable profile

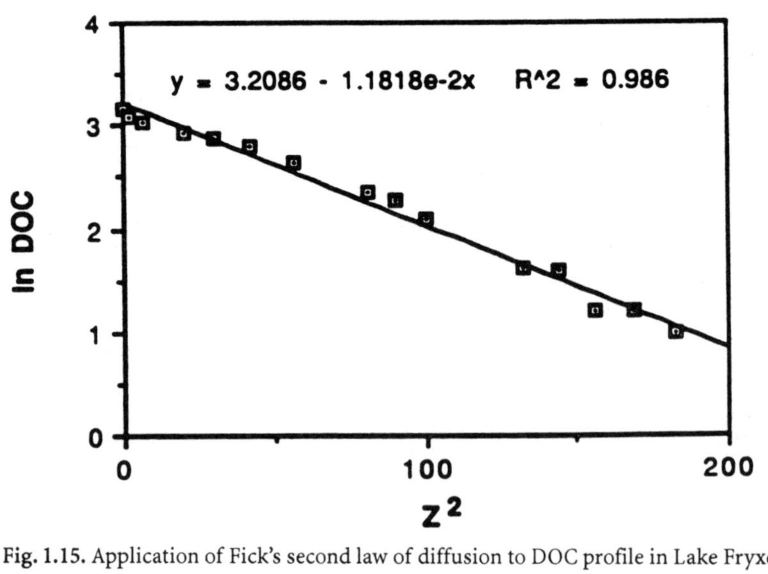

Fig. 1.15. Application of Fick's second law of diffusion to DOC profile in Lake Fryxell

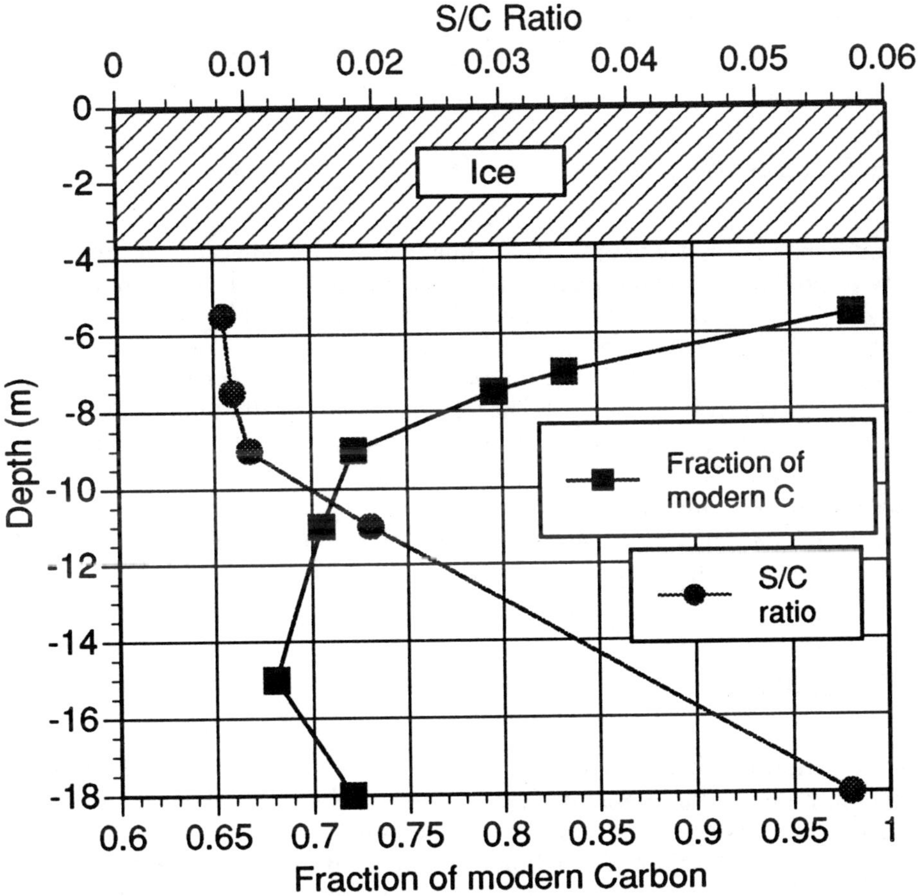

Fig. 1.16. Depth profiles for ^{14}C age and S/C ratios for fulvic acid samples isolated from Lake Fryxell

would not be observed due to surface mixing and turnover of the water column; however, a steady flux of dissolved humic substances from the sediments could be a significant process transporting DOC to the lake.

The measurement of the ^{14}C age of the fulvic acids at different depths revealed a complex set of young and old sources of fulvic acid (Fig. 1.16, Aiken et al. 1996). In the upper water column the fulvic acids have a modern signal and are brought into the lake by the glacial meltwater streams, mixing first in an openwater moat. This young fulvic acid derives from perennial algal mats in the streams, and highest concentrations occur when the mats are first wetted with the beginning of flow. At depth, the fulvic acids are indeed quite old (about 3000 years), perhaps derived from degradation of relict organic material in the sediments. The slight decrease in age between 15 and 18 m can be explained by (1) young fulvic acid being transported to depth as the moat freezes and a denser brine is formed and sinks to the bottom waters; and also (2) the degradation of fresh organic matter from sinking algae.

1.6 Conclusions

Humic substances are ubiquitous in aquatic ecosystems because of the diversity of precursor detrital organic materials and diversity of formation pathways, as well as their poor quality as substrates for microbial growth. Aquatic humus can be derived from plant/soil organic material and from microbial/algal organic material. The importance of these two general sources exerts major control on chemical characteristics of aquatic humus. Dissolved fulvic acid can be produced by processes acting on fresh litter and on senescent algal material and by progressive long-term degradation of organic material in soils and sediments. Thus, in many aquatic ecosystems, dissolved humic substances are probably a mixture of young and old materials, and this mixture changes with the seasonal dynamics of the ecosystem.

1.7 Summary

Aquatic humic substances are a class of heterogeneous, moderate molecular weight, yellow-colored organic acids of biological origin present in all natural waters. This characteristic of being ubiquitous is a result of the diverse sources and pathways of formation of humic substances and their slow degradation by geochemical or microbial processes. Although little is known about the detailed chemical pathways of the formation of aquatic humic substances, we can make strong inferences about dominant sources and the timeframe of formation by studying the biogeochemistry of aquatic humic substances in different aquatic ecosystems.

Lignin is probably a dominant precursor for formation of aquatic humus because of its abundance, complexity, and recalcitrant nature. There are two general conceptual models for the formation of humic substances during degradation of detrital organic material; one is through condensation reactions of low molecular weight compounds and the other is through the solubilization of complex residual organic compounds. Lignin degradation can produce both low molecular weight and complex residual precursors. Degradation of microbial material can also produce precursor compounds. Both conceptual models can involve a range of precursor organic material, and it is probable that the relative importance of formation pathways varies with environment and timescale.

In most temperate regions, the main source of aquatic humic substances is the terrestrial vegetation and soils in the watershed, with the litter layer being an important soil zone for production. Seasonal and spatial variations in humic chemistry and concentration are controlled by the hydrologic linkages between the surface water and the catchment. In aquatic environments where the flux of terrestrial organic material is limited by vegetation in the catchment, or by hydrologic transport, aquatic humic substances can be derived from microbial material and have a distinct chemistry.

References

Aiken GR (1985) Isolation and concentration techniques for aquatic humic substances. In: Aiken GR, McKnight DM, Wershaw RL and MacCarthy P (eds) Humic substances in soil, sediment and water: geochemistry, isolation, and characterization, Wiley, New York, pp 363–385

Aiken GR and RL Malcolm (1987) Moleular weight of aquatic fulvic acids by vapor pressure osmometry. Geochim. Cosmochim. Acta 51: 2177

Aiken GR, McKnight DM, Thorn KA, Thurman EM (1992) Isolation of hydrophilic organic acids from water using nonionic macroporous resins. Organic Geochem 18: 567–573

Aiken GR, McKnight DM, Wershaw RL, Miller L (1991) Evidence for the diffusion of aquatic fulvic acid from the sediments of Lake Fryxell, Antarctica. In: Baker R (ed) Organic substances and sediments, Lewis, Chelsea, Michigan, pp 75–88

Aiken GR, McKnight DM, Harnish R, Wershaw RL (1996) Geochemistry of aquatic humic substances in the Lake Fryxell Basin, Antarctica. Biogeochem 34: 157–188

Aschan O (1932) Water humus and its role in the formation of marine iron ore. Ark. Kemi Miner Geol 10A (15):1–143

Averett RC, Leenheer, JA, McKnight, DM, Thorn, KA (1989) eds. Humic substances in the Suwannee River, Georgia: interactions, properties, and proposed structures, US Geological Survey Open-File Rep 87-557, p 377

Baron J, McKnight DM, Denning AS (1991) Sources of dissolved and particulate organic material in Loch Vale Watershed, Rocky Mountain National Park, Colorado, USA. Biogeochemistry 15:89–110

Baron JS, Ojima DS, Holland EA, Parton WJ (1994) Analysis of nitrogen saturation potential for Rocky Mountain tundra and forest: implications for aquatic systems. Biogeochemistry 27: 61–82

Boyer EW, Hornberger GM, Bencala KE, McKnight DM (1996) Response characteristics of DOC flushing in an alpine catchment. Hydrological Processes (in press)

De Haan H, Halma G, De Boer T, Haverkamp J (1981) Seasonal variations in the compositions of fulvic acid in Tjeukemeer, The Netherlands. Arch. Hydrobiol. 92:11–23

Ertel JR, Hedges JI Perdue EM (1984) Lignin signature of aquatic humic substances. Science 223:485–487

Ertel JR, Hedges JI, Devol AH, Richey JE (1986) Dissolved humic substances of the Amazon River system. Limnol Oceanogr 31(4):739–754

Grieve IC (1994) Dissolved organic carbon dynamics in two streams draining forested catchments at Loch Ard, Scotland. Hydrol Processes 8:457–464

Harvey GR, Boran DA (1985) Geochemistry of humic substances in seawater. In: Aiken GR, McKnight DM, Wershaw RL, MacCarthy P (Eds) Humic substances in soil, sediment and water: geochemistry, isolation, and characterization. Wiley, New York, pp 233–247

Hayes MHB, MacCarthy P, Malcolm RL, Swift RS (eEds) (1989) Humic substances. II. In search of structure. Wiley, New York

Hedges JI, Cowie GL, Richey JE, Quay PD (1994) Origins and processing of organic matter in the Amazon River as indicated by carbohydrates and amino acids. Limnol Oceanogr 39(4): 743–761

Hessen DO, Andersen T, Lyche A (1990) Carbon metabolism in a humic lake: pool sizes and cycling through zooplankton. Limnol Oceanogr 35:84–99

Hornberger, GM, Bencala, KE, McKnight, DM (1994) Hydrological controls on the temporal variation of dissolved organic carbon in the Snake River near Montezuma, Colorado. Biogeochemistry 25:147–165

Huffman EWD Jr., Stuber HA (1985) Analytical methodology for elemental analysis of humic substances. In: Aiken GR, McKnight DM, Wershaw RL, MacCarthy P (eds) Humic substances in soil, sediment and water: geochemistry, isolation, and characterization. Wiley, New York, pp 433–456

Ishiwtari R (1985) Geochemistry of humic substances in lake sediments. In: Aiken GR, McKnight DM, Wershaw RI, MacCarthy P (eds) Humic substances in soil, sediment and water: geochemistry, isolation, and characterization. Wiley, New York, pp 147–180

Kirk TK (1984) Degradation of lignin. In: Gibson DT (ed) Microbial degradation of organic compounds. Microbiology series, vol 13. Marcel Dekker, New York, pp 399–437

Leenheer JA, McKnight DM, Thurman EM, MacCarthy P (1989) Structural components and proposed structural models of fulvic acid from the Suwannee River. In: Humic substances in the Suwannee River, Georgia: interactions, properties, and proposed structures, US Geological Survey Open-File Rep 87–557, pp 335–359

MacCarthy P, Rice JA (1991) An ecological rationale for the heterogeneity of humic substances: a holistic perspective on humus. In: Schneider SH, Boston PJ (eds) Scientists on Gaia (p 339–345). MIT Press, Cambridge, Massachusetts

Malcolm RI (1990) The uniqueness of humic substances in each of soil, stream and marine environments. Anal Chim Acta 232:19–30

McKnight DM (1991) Feedback mechanisms involving humic substances in aquatic ecosystems.. In: Schneider SH, Boston PJ (eds) Scientists on Gaia (p 330–338). MIT Press, Cambridge, Massachusetts

McKnight DM, Aiken GS, Andrews ED, Bowles EC, Harnish RA (1993a) Dissolved organic material in Dry Valley lakes: a comparison of Lake Fryxell, Lake Hoare, and Lake Vanda. In: Green WJ, Friedmann EI (eds) Physical and biogeochemical pocesses in Antarctic lkes, Antarctic research series, vol 59. pp 119–133

McKnight DM, Smith RL, Harnish RA, Miller CL, Bencala KE (1993b) Seasonal relationships between planktonic microorganisms and dissolved organic material in an alpine stream. Biogeochemistry 21:39–59

McKnight DM, Andrews ED, Spaulding SA, Aiken GR (1994) Aquatic fulvic acids in algal-rich Antarctic ponds. Limnol Oceanogra 39(8):1972–1979

McKnight DM, Harnish R, Wershaw RI, Baron JS, Schiff S (1997) Chemical characteristics of particulate, colloidal, and dissolved organic material in Loch Vale Watershed, Rocky Mountain National Park. Biogeochemistry 36:99–124

Mobed JJ, Hemmingsen SL, Autry JL, McGowan LB (1996) Fluorescence characterization of IHSS humic substances: total luminescence spectra with absorbance correction. Environ Sci Technol 30 :3061–3065

Moran MA, Hodson RE (1990) Bacterial production on humic and nonhumic components of dissolved organic carbon. Limnol Oceanogr 35(8):1744–1756

Ranville JF, Harnish RA, McKnight DM (1991) Particulate and colloidal organic material in Pueblo Reservoir, Colorado: influence of autochthonous source on chemical composition. In: Baker RA (ed) Organic substances and sediment in water, vol 1. Lewis, Chelsea, Michigan, pp 47–73

Schiff SL, Aravena R, Trumbore SE, Dillon PJ (1990) Dissolved organic carbon cycling in forested watersheds: a carbon isotope approach. Water Resour Res 26:2949–2957

Scully NM, Lean DRS (1994) The attenuation of ultraviolet light in temperate lakes. Arch Hydrobiol 43:135–144

Serkiz SM, Perdue EM (1990) Isolation of dissolved organic matter from the Suwannee River using reverse osmosis. Water Res 24:911–916

Spaulding SA, McKnight DM, Smith RL, Dufford R (1994) Phytoplankton population dynamics in perennially ice-covered Lake Fryxell, Antarctica. J Plankton Res 16:(5):527–541

Steinberg C, Meunster U (1985) Geochemistry and ecological role of humic substances in lake-water. In: Aiken GR, McKnight DM, Wershaw RL, MacCarthy P (eds) Humic substances in soil, sediment and water: geochemistry, isolation, and characterization. Wiley, New York, pp 105–146

Stevenson FJ (1985) Geochemistry of soil humic substances. In: Aiken GR, McKnight DM, Wershaw RL, MacCarthy P (eds) Humic substances in soil, sediment and water: geochemistry, isolation, and characterization. Wiley, New York, pp 13–52

Stewart WDP (ed) (1974) Algal physiology and biochemistry. University of California Press, Berkeley

Thurman EM (1985) Organic geochemistry of natural waters. Nijhoff/Junk Boston

Thurman EM, Malcolm RI (1981) Preparative isolation of aquatic humic substances. Environ Sci Technol 15:463–466

Vandenbroucke R, Pelet R, Debyser Y (1985) Geochemistry of humic substances in marine sediments. In: Aiken GR, McKnight DM, Wershaw RL, MacCarthy P (eds) Humic substances in soil, sediment and water: geochemistry, isolation, and characterization . Wiley, New York, pp 249–273

Waksman SA (1936) Humus. Williams and Wilkins, Baltimore, Maryland

Wershaw RL (1992) Membrane-micelle model for humus in soils and sediments and its relation to humification. US Geological Survey Open-File Rep 91–513

Wetzel RG (1990) Land-water interfaces: metabolic and limnological regulators. Verh Int Ver Limnol 24:6–24

Wetzel RG, Hatcher PG, Bianchi TS (1995) Natural photolysis by ultraviolet irradiance of recalcitrant dissolved organic matter to simple substrates for rapid bacterial metabolism. Limnol Oceanogr 40(8):1369–1380

Williams MW, Baron JS, Caine N, Sommerfeld R, Sanford R Jr. (1996) Nitrogen saturation in the Rocky Mountains. Environ Sci Technol 30:640–646

2 Chemical Composition, Structure, and Metal Binding Properties

E. Michael Perdue

2.1 Introduction

Humic substances are complex mixtures of organic compounds that occur naturally in soils, sediments, and natural waters, where they account for a large fraction of the nonliving organic carbon. Scientists all over the world have conducted a tremendous number of studies of humic substances in the field and laboratory. There have been several comprehensive critical reviews of the state of the art of the study of humic substances (Christman and Gjessing 1983; Aiken et al. 1985; Hayes et al. 1989; Leenheer 1994; Senesi and Miano 1994). Efforts to characterize humic substances with respect to structure and functionality often rely upon sophisticated methods that are applicable to pure substances, but whose applicability to complex mixtures such as humic substances is less certain. In particular, spectroscopic methods (UV-visible, IR, and NMR), although widely employed, can yield ambiguous results when applied to humic substances (MacCarthy and Rice 1985).

Chromatographic separation of humic substances has been widely attempted in the hope of obtaining pure components whose structures and functional groups can be identified and quantified. Unfortunately, all chromatographic techniques have yielded results which are consistent with the view that humic substances are an exceedingly complex, perhaps unresolvably complex, mixture (Hayes et al. 1989). Scientists can therefore only measure average properties (which are not necessarily representative of any component of the mixture), and they must attempt to infer average structure and functionality from those average properties (Perdue 1984).

In the scientific literature that pertains to humic substances, misleading phrases such as "the structure of the humic acid molecule...," "the molecular weight of fulvic acid ...," "the conformation of the humic acid molecule ...," or "the equilibrium constant for reaction of fulvic acid with ..." are commonly encountered. The ubiquity of such phrases demonstrates that humic substances are often treated conceptually as pure substances, even by those who recognize that they are complex mixtures.

Ecological Studies, Vol. 133
Hessen, Tranvik (eds) Aquatic Humic Substances

Unfortunately, chemists are generally trained to work on pure compounds and to avoid mixtures altogether. The following quote from a general chemistry text by Hammond et al. (1971) clearly states the prevailing view:

> "It is not always easy to determine whether a sample of material is a homogeneous mixture or a single substance. This is an important question to answer, for it is embarrassing for a chemist to find that he has been studying a mixture when he thought it was one substance ... The development of chemistry as a science and the systematic study of chemistry today depends on working with pure substances."

This "pure chemistry" paradigm must be set aside if unresolvably complex mixtures such as humic substances are to be studied in a scientifically acceptable manner. The first step in this direction is the realization that any measurable physical or chemical property of a humic substance is some kind of a weighted-average property, whose value must change as the composition of the mixture changes. The next step is to use experimental methods and mathematical models whose treatment of the weighted-average properties of a humic substance is as rigorous as possible.

This chapter will review the state of current knowledge regarding elemental composition and dominant structural features of humic substances, some recent empirical correlations between composition, structure, and bioavailability of humic substances, and the rapidly converging views of mathematical approaches to descriptions of cation binding by humic substances. The conceptual and mathematical approaches in this chapter are fully consistent with the fact that humic substances are complex mixtures.

2.2 Elemental Composition, Structural Features, and Bioavailability

Among the least ambiguous average properties of a humic substance are its basic compositional properties such as elemental composition, carboxyl content (COOH), and number-average molecular weight (M_n), which are simply weighted averages of the corresponding properties of the individual components of the humic substance, as illustrated by the following equation for weight percent carbon:

$$\%C_{avg} = \Sigma w_i \, (\%C_i) \, / \, \Sigma w_i \,, \qquad (1)$$

where w_i and ($\%C_i$) are the mass and weight percent of carbon in the ith component of the mixture. These basic compositional parameters can, in turn, provide a probabilistic description of the major structural features of a humic substance (Perdue 1984; Wilson et al. 1987).

2.2.1 Analytical Constraints Calculations

The method of analytical constraints calculations and its validation with known chemical structures have been described previously (Perdue 1984; Wilson et al. 1987). Given a set of basic compositional parameters (elemental composition, carboxyl content, and estimated number-average molecular weight), the method yields values of sp^3-hybridized carbon (C_{al}, aliphatic carbon), sp^2-hybridized carbon in aromatic rings (C_{ar}, aromatic carbon), and all remaining sp^2-hybridized carbon (C_{xs}, excess carbon) by simultaneous solution of the following three equations:

$$C_{total} = C_{al} + C_{ar} + C_{xs} + COOH; \quad (2)$$

$$H_{total} = (H/C)_{al}\, C_{al} + (H/C)_{ar}\, C_{ar} + (H/C)_{xs}\, C_{xs} + COOH + N_{total}\,; \quad (3)$$

$$U_{total} = (U/C)_{al}\, C_{al} + (U/C)_{ar}\, C_{ar} + (U/C)_{xs}\, C_{xs} + COOH\,. \quad (4)$$

All concentrations are in units of millimoles/gram. C_{total}, H_{total}, and N_{total} are calculated directly from elemental compositional data. Equation 4 is the conservation equation for unsaturation (rings and/or pi-bonds), and U_{total} can be calculated from the following equation

$$U_{total} = C_{total} + N_{total}/2 - H_{total}/2 + 1000/M_n\,, \quad (5)$$

where M_n is assumed to be approximately 1000 g/mol if a measured value is not available. If the average H/C and U/C ratios for the three forms of organic carbon were known, then Eqs. 2–4 could be solved to obtain C_{al}, C_{ar}, and C_{xs} directly. None of these ratios is known, but a reasonable range for each ratio is readily estimated. For example, $(H/C)_{ar}$ almost certainly lies between 0 and 5/6, and $(U/C)_{ar}$ is probably about 4/6 (the value for benzene rings). Reasonable ranges for all these parameters are given in Table 2.1.

Probabilistic estimates of C_{al}, C_{ar}, and C_{xs} are obtained by randomly generating H/C and U/C ratios within their allowed ranges and solving Eqs. 2–4 for C_{al}, C_{ar}, and C_{xs} until 1000 chemically valid solutions are obtained. From the population of chemically valid solutions, median values, mean values, and standard deviations for all parameters are obtained. The mean values of C_{al}, C_{ar}, and C_{xs} generally correspond fairly well with the actual distribution of organic carbon (as measured by ^{13}C NMR or from known structures of organic compounds).

Table 2.1. Allowed ranges for H/C and U/C ratios in analytical constraints calculations

Molar ratio	Lower limit	Upper limit
$(H/C)_{al}$	1.500	2.500
$(H/C)_{ar}$	0.000	0.833
$(H/C)_{xs}$	0.000	1.500
$(U/C)_{al}$	0.000	0.250
$(U/C)_{ar}$	0.667	0.667
$(U/C)_{xs}$	0.500	1.000

Table 2.2. Composition and structural features of Suwannee River dissolved organic matter[a]

Parameter	Value
Carbon	42.95
Hydrogen	40.66
Oxygen	26.72
Nitrogen	0.75
Sulfur	0.18
Carboxyl groups	5.04
C_{al} (from ^{13}C-NMR)	19.6
C_{ar} (from ^{13}C-NMR)	13.3
C_{xs} (from ^{13}C-NMR)	5.0
C_{al} (calculated)	15.8
C_{ar} (calculated)	15.0
C_{xs} (calculated)	7.1

[a] All parameters are in units of mmol/g.

The approach is illustrated here for a sample of dissolved organic matter which was collected from the Suwannee River (Georgia, USA) by reverse osmosis (Serkiz and Perdue 1990). The dry, ash-free elemental composition, carboxyl content, ^{13}C-NMR estimates of C_{al}, C_{ar}, and C_{xs}, and the values of those parameters that are predicted from analytical constraints calculations are given in Table 2.2. The analytical constraints calculations are in generally good agreement with the NMR results. It can be concluded for this sample that the average structural assignments obtained from ^{13}C-NMR spectrometry are consistent with the elemental composition of the sample.

2.2.2 Compositional Variability and Uncertainty

Because humic substances are complex mixtures, their average properties such as elemental composition, carboxyl content, and average molecular weight vary both spatially and temporally. Two main sources of variability between samples are different types of biomass precursors and different methods of isolation and purification. Even for a single humic substance, sampling errors and analytical errors introduce significant uncertainty into the values of basic compositional properties. Further information regarding elemental analyses can be found in the excellent critique by Huffman and Stuber (1985). A detailed analysis of potentiometric methods for analysis of acidic functional groups is presented by Perdue (1985), and colligative methods for measurement of number-average molecular weights of humic substances are described by Aiken and Gillam (1989).

Rice and MacCarthy (1991) have statistically summarized the elemental compositions of many types of humic substances, including freshwater humic and fulvic acids. They found that elemental compositional data are well

Table 2.3. Statistical description of basic compositional properties of freshwater fulvic acids[a]

Parameter	Mean	SD
Carbon	38.9	3.6
Hydrogen	42.0	7.0
Nitrogen	1.6	1.5
Oxygen	28.7	3.2
Sulfur	0.4	0.3
COOH	5.0	0.5
M_n	1000	100

[a] M_n is in g/mol. All other parameters are in mmol/g.

described by a normal distribution for nearly all elements, element ratios, etc. Statistical summaries are not available for other basic parameters such as COOH and M_n. For these parameters, normal distributions are assumed, with average values that are consistent with current literature, and standard deviations of 10 %. This information is summarized in Table 2.3.

2.2.3 Structural Variability

Given the real and apparent variabilities in basic compositional properties of humic substances, the question is how much corresponding variability is expected in their average structural features? To address this question, analytical constraints calculations have been conducted as described in Section 2.2.1, but with elemental composition, COOH, and M_n randomly varied within prescribed ranges that reflect the natural variability of these parameters in freshwater fulvic acids samples. The goal, in particular, is to obtain a probability distribution of structural features, given a representative range of basic compositional properties.

Given the mean values and standard deviations for the elemental composition, COOH, and M_n for freshwater fulvic acids (Table 2.3), the corresponding distribution of structural features can be obtained from numerous (e.g., 1000) repetitions of the following two steps:

1. Values for %C, %H, %N, and %S are randomly generated, and %O is calculated as 100-%C-%H-%N-%S. This process is repeated until all values are greater than or equal to zero and the corresponding C_{total}, H_{total}, N_{total}, S_{total}, and O_{total} values fall within two standard deviations of their respective mean values. Then values for COOH and M_n are randomly generated, subject to the same constraints.
2. Given a randomly generated set of basic compositional parameters, analytical constraints calculations are conducted as described in Section 2.2.1 to obtain probabilistic estimates of C_{al}, C_{ar}, C_{xs}, and all H/C and U/C ratios.

Table 2.4. Statistical description of structural features of freshwater fulvic acids in analytical constraints calculations[a]

Parameter	Minimum	Maximum	Mean	SD
C_{al}	4.87	26.32	14.69	4.94
C_{ar}	1.10	17.90	10.53	3.67
C_{xs}	1.10	21.61	8.57	2.87
COOH	4.00	5.98	5.01	0.43
$(H/C)_{al}$	1.76	2.04	1.92	0.07
$(H/C)_{ar}$	0.32	0.41	0.35	0.01
$(H/C)_{xs}$	0.22	0.47	0.40	0.03
$(U/C)_{al}$	0.03	0.13	0.11	0.02
$(U/C)_{ar}$	0.67	0.67	0.67	0.00
$(U/C)_{xs}$	0.70	0.88	0.74	0.02

[a] C_{al}, C_{ar}, C_{xs}, and COOH are in mmol/g. All (H/C) and (U/C) values are molar ratios.

Once 1000 different randomly generated chemical compositions have been evaluated by the analytical constraints computations, the resulting distribution of structural features can be constructed. Table 2.4 contains a statistical summary of those results for freshwater fulvic acids. An analysis of (H/C) and (U/C) ratios in Table 2.4 provides some insight into the structural features of freshwater fulvic acids. For example, $(H/C)_{ar}$ values (which, in principle, could vary from 0 to 5/6) are predicted to lie in the narrow range of 0.32–0.41. This corresponds to an average of 1.9–2.5 hydrogens per aromatic ring. In other words, the average aromatic ring is attached to about four substituents other than H atoms or hydroxyl (OH) groups. Values for $(H/C)_{al}$ are predicted to lie within the range of 1.76–2.04, and $(U/C)_{al}$ values are predicted to lie between 0.03 and 0.13. Both of these observations indicate that freshwater fulvic acids contain significant quantities of cyclic aliphatic moieties such as pyranoses, furanoses, and cycloalkyl groups.

On average, C_{al}, C_{ar}, C_{xs}, and COOH are predicted to be about 38, 27, 22, and 13%, respectively, of the carbon in freshwater fulvic acids. It is important to realize, however, that the natural variability of basic compositional properties of freshwater fulvic acids is sufficiently large that substantial variations in their structural features can be anticipated. The overall distribution of the mean values of non-COOH carbon among C_{al}, C_{ar}, and C_{xs} is presented in Fig. 2.1. It is evident from Fig. 2.1 that the compositional and structural variability in freshwater fulvic acids probably arises from variability in the relative amount of C_{al}. In fact, the relative proportions of C_{ar} and C_{xs} are nearly constant over the entire locus of points in Fig. 2.1, except at the lowest C_{al} values. One interpretation is that the components of C_{al} are more biodegradable, and hence more variable, than the components of C_{ar} and C_{xs}.

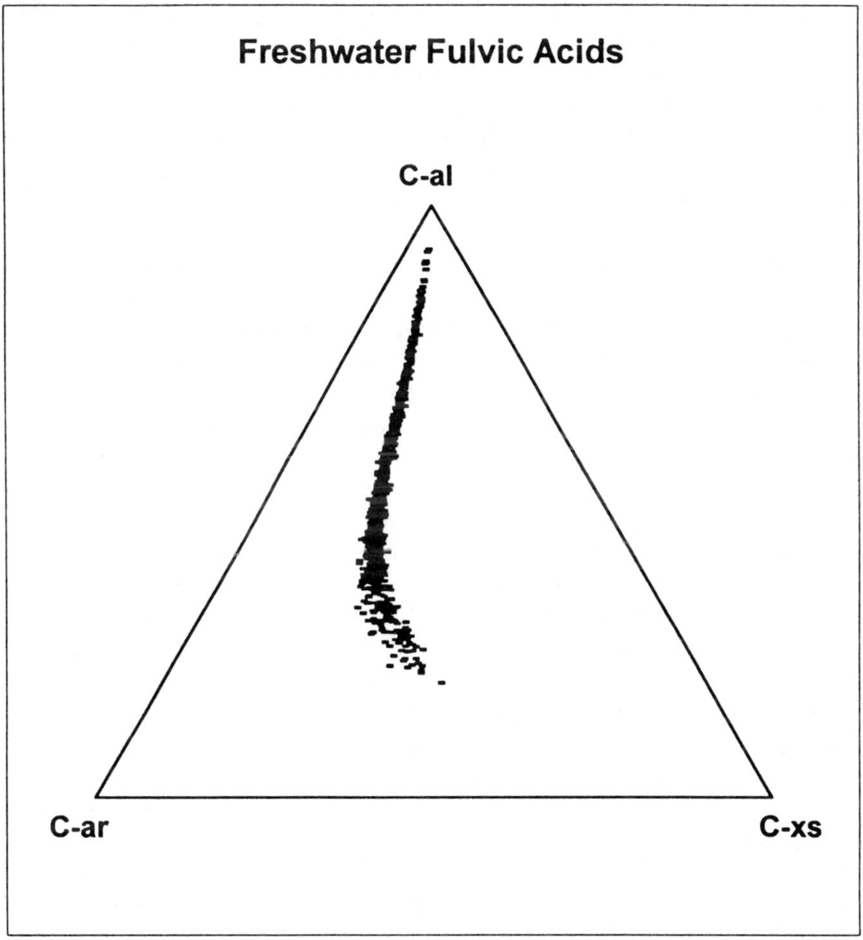

Fig. 2.1. Most probable structural distribution of noncarboxyl organic carbon in freshwater fulvic acids, based on their average compositional data. Apices in this ternary plot correspond to 100% aliphatic carbon (C_{al}), 100% aromatic carbon (C_{ar}), and 100% "excess" carbon (C_{xs}). The opposite edge from an apex represents 0% of that form of carbon

2.2.4 Bioavailability

Sun et al. (1997) have recently studied the relationship between composition, structure, and bioavailability of dissolved organic matter (DOM) from several sites in the drainage basin of the Ogeechee River in Georgia, USA. The headwaters of the Ogeechee River are in the Piedmont Province (sites 1–2), which is an eroded, rolling upland, rich in clay. From its headwaters, the Ogeechee River crosses through the upper Coastal Plain (sites 3–5), which is a rich agricultural area, and the lower Coastal Plain (sites 6–10), which has more swampy wetlands and blackwater tributaries. Allochthonous sources of

DOM include precipitation, leachates from coastal plain soils and riparian plants, buried organic matter, and decomposed particulate organic matter. Macrophytes and benthic algae are autochthonous sources of DOM.

Sampling trips were made in November 1987, March 1988, and April 1989. At a typical sample site, the DOM in 200 liters of water was prefiltered through a 0.4-μm Nuclepore filter and concentrated by reverse osmosis into a volume of 20 l or less at a rate of about 3 l/min. Concentrated samples were filter-sterilized (0.2-μm filter), packed on ice, and transported to the laboratory, where they were stored in a cold room. Representative samples of fresh benthic algal mats, cypress needle litter, pine needle litter, deciduous leaf litter, and macrophyte leaves were also collected from the Ogeechee River and its floodplain. Leaf and needle litters were air-dried, and macrophytes and algae were freeze-dried. A synthetic rainwater solution was used to leach the dried samples in a vacuum extractor in a cold room. All samples, including leachates, were purified further to remove inorganic matter and then freeze-dried. On average, about 80% of the original DOM was obtained from collected samples after all purification processes. The elemental composition of each sample was determined, using a Perkin-Elmer Model 240C elemental analyzer to measure %C, %H, and %N, combustion in a muffle furnace to measure %Ash, and calculation of %O as 100-%C-%H-%N-%Ash.

Bacterial growth was quantified by measuring the amount of bacterial biomass produced in 3-day incubations of Ogeechee River bacteria (collected from site 8) in solutions containing DOM from particulate organic matter (POM) leachates and river water as the sole carbon source (Leff and Meyer 1991). Bacterial growth was determined as the difference between initial and peak (day 3) cell counts minus the mean growth observed in triplicate flasks with no added carbon (Leff and Meyer 1991).

The H/C, N/C, and O/C atomic ratios, COOH contents, and normalized bacterial growth [micrograms of bacterial biomass per milligram of dissolved organic carbon (DOC)] from the 3-day growth experiments are given in Table 2.5. An empirical equation

$$\text{Growth} = a_0 + a_1(\text{H/C}) + a_2(\text{O/C}) + a_3(\text{N/C})$$

provides an excellent prediction of bioavailability of DOM in this study. Multiple linear regression analysis of data in Table 2.2 for the 20 POM leachates and river DOM samples yielded the following regression parameters: $a_0 = 38.4$, $a_1 = 10.6$, $a_2 = -70.9$, $a_3 = 183.2$, $r^2 = 0.933$. The magnitudes of the individual regression parameters are mainly a reflection of the relative magnitudes of the H/C, O/C, and N/C ratios, and, as such, are not very significant. In contrast, the algebraic signs of the regression parameters are quite informative. Bacterial growth is positively correlated with H/C, which itself is indicative of the relative proportions of aliphatic and aromatic carbon in a sample. The negative correlation with O/C may indicate that more highly "weathered" DOM, with its generally higher COOH content, is simply less bioavailable than fresher DOM. The positive correlation between bacterial growth and N/C may indicate that proteinaceous moieties in DOM are preferentially consumed. This particular set of regression parameters is unique to this study, being dependent on the method of isolation and purification of DOM (i.e., which fractions are lost) and especially on the method of measuring bioavailability (e.g., Leff and Meyer 1991).

Analytical constraints calculations were used to estimate the most probable values of C_{al}, C_{ar}, and C_{xs} from the compositional data in Table 2.5. A ternary plot of the most probable distribution of non-COOH carbon is presented in Fig. 2.2. The results are consistent with the more generic analyses that are plotted in Fig. 2.1. More importantly, the highly aliphatic character of the DOM from algal and macrophyte leachates is clearly evident, as is the much lower C_{al} content of the DOM samples from the two blackwater tributaries (sites 6, 7). The remaining results plot along a more or less straight line connecting these "end-members," with DOM from upstream sites (Sites 1–5) and from the other leachates being more aliphatic than DOM from downstream sites (8–10). The relative proportions of C_{ar} and C_{xs} do not change significantly along the line of data points, which suggests that diagenetic processes are mainly causing the removal of C_{al}. There is a good, albeit imperfect, correlation between these calculated structural feaures and the relative bioavailabilities of the samples (compare the bioavailability data in Table 2.5 with locations of points in Fig. 2.2). The observed downstream trends in H/C, C_{al}, and bioavailability are attributable to changing sources of DOM along the river as well as accumulation of refractory C_{al}-depleted DOM from upstream and from riparian swamps and tributaries.

Table 2.5. Atomic ratios, carboxyl contents (mmol/g), and bacterial growth (μg bacterial C/mg DOC) for leachates of particulate organic matter and Ogeechee River samples

Atomic Ratios and Carboxyl Content Bacterial Growth						
Sample	**H/C**	**O/C**	**N/C**	**COOH**	**Mean**	**SD (n=3)**
Algae	1.70	0.61	0.145	4.64	41.1	1.8
Cypress	1.16	0.54	0.017	2.18	17.4	4.0
Hardwood	1.17	0.32	0.050	3.87	40.9	5.8
Macrophytes	1.70	0.19	0.091	5.26	58.9	3.3
Pine	1.32	0.54	0.022	1.70	25.2	3.3
Site 1 (11/87)	1.49	0.76	0.104	1.38	18.4	2.9
Site 4 (11/87)	1.37	0.63	0.053	1.36	12.4	4.6
Site 5 (11/87)	1.17	0.75	0.073	4.22	13.7	3.3
Site 6 (11/87)	0.97	0.72	0.044	3.93	8.4	1.6
Site 7 (11/87)	0.94	0.71	0.058	3.41	5.2	3.8
Site 8 (11/87)	1.10	0.64	0.078	3.24	12.4	3.5
Site 9 (11/87)	1.03	0.66	0.048	3.47	12.0	2.2
Site 10 (11/87)	1.13	0.74	0.042	3.58	5.0	1.3
Site 3 (3/88)	1.25	0.53	0.006	6.31	8.3	3.5
Site 5 (3/88)	1.18	0.70	0.035	3.30	5.1	0.8
Site 1 (4/89)	1.35	0.80	0.070	2.48	12.6	2.6
Site 5 (4/89)	1.10	0.69	0.042	3.77	13.0	0.8
Site 7 (4/89)	0.97	0.71	0.033	4.34	2.4	0.6
Site 8 (4/89)	1.06	0.71	0.037	4.00	5.1	0.7
Site 10 (4/89)	1.02	0.74	0.034	4.31	6.7	0.8

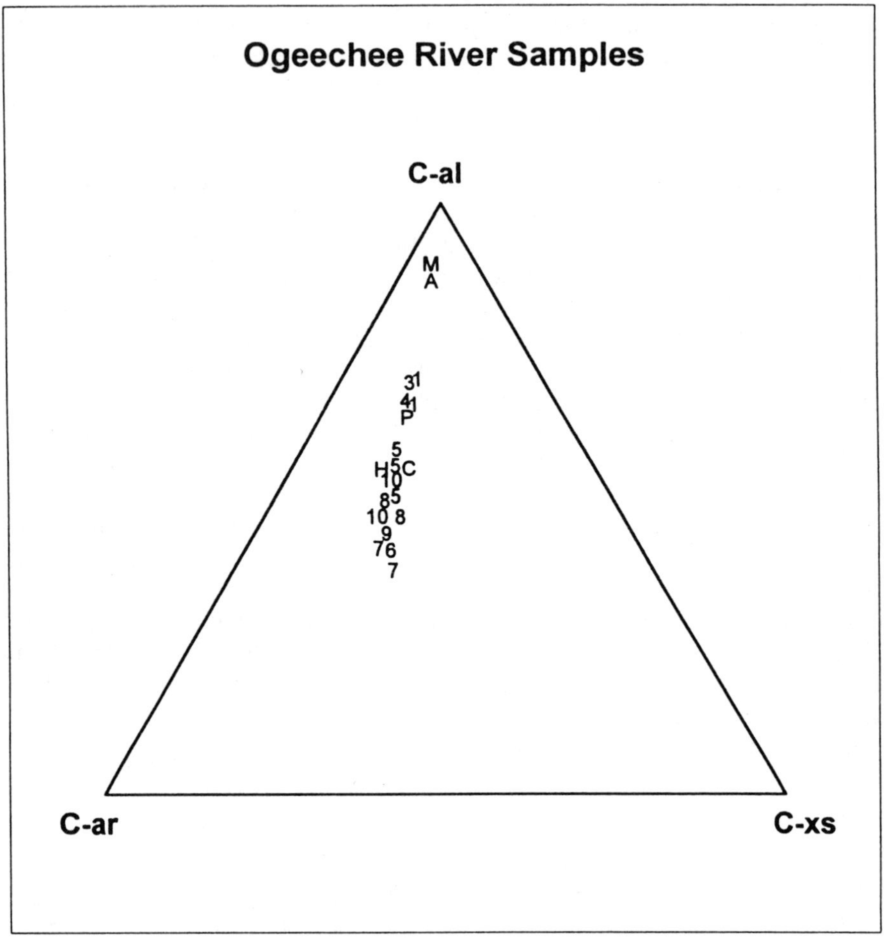

Fig. 2.2. Most probable structural distribution of non-carboxyl organic carbon in biomass leachates (*A* algae; *C* cypress; *H* hardwoods; *M* macrophytes; *P* pine) and in Ogeechee River DOM samples (*sites 1–10*). Apices in this ternary plot correspond to 100% aliphatic carbon (C_{al}), 100% aromatic carbon (C_{ar}), and 100% "excess" carbon (C_{xs}). The opposite edge from an apex represents 0% of that form of carbon

2.3 Mathematical Models of Cation Binding

Metal binding by humic substances is studied in both the field and laboratory. In field studies, experimental conditions are established by the chemical matrix of the natural sample, and metal binding studies generally consist of efforts to distinguish between free and complexed forms of a metal ion. Such a distinction is relevant because of the frequently strong correlation between biological effects of a metal ion and its free concentration (e.g., Sunda and

Guillard 1976; Morel and Hering 1993; Erickson et al. 1996). The major challenge in field studies is analytical methodology, i.e., making measurements of the total concentration of a trace metal ion, distinguishing between free and complexed forms of the metal ion, and conducting all phases of the study using clean techniques to avoid contamination of the sample (Ahlers et al. 1990; Benoit 1994). In laboratory studies, metal binding can be studied over a range of experimental conditions that is hopefully extensive enough with respect to all experimental variables to encompass commonly encountered field conditions.

Both field and laboratory studies often culminate in an effort to summarize the acquired knowledge of metal binding by humic substances through the development of chemical speciation models, the best of which consider competitive effects of H^+, other metal ions, and other ligands (e.g., Cabaniss and Shuman 1988; Dobbs et al. 1989; Tipping and Hurley 1992; Koopal et al. 1994). Those models have reached a substantial level of sophistication, and their predictive capabilities are steadily improving. Even though all models of metal complexation by humic substances are empirical, a condition dictated by the complexity of the system, it is worthwhile to compare and contrast the more successful models. The remainder of this chapter is devoted to a basic overview and comparison of three of the most sophisticated models, focusing on their conceptual similarities and their chemical and mathematical differences.

2.3.1 Modeling Objectives

Any modern model needs to be able to describe proton binding by a humic substance as a function of pH, the concentration of the humic substance, and ionic strength. In addition, the pH dependence of the binding of a metal ion must be well described, as must the competitive binding of several metal ions. Three modern models that have been reasonably successful in meeting these requirements are the competitive Gaussian distribution model, Model V, and the nonideal competitive adsorption (NICA) model (Dobbs et al. 1989; Tipping and Hurley 1992; Koopal et al. 1994). It is generally understood today that humic substances contain many nonidentical binding sites, so the central modeling challenge is to describe quantitatively the relative concentrations and strengths of those binding sites. As it turns out, the perceived distribution of binding sites is directly related to experimental conditions and to the type of conceptual model that is forced on the experimental data (Perdue 1997).

2.3.2 Competitive Gaussian Distribution Model

The competitive Gaussian distribution model (Dobbs et al. 1989; Susetyo et al. 1991) is an extension of the non-competitive Gaussian distribution model (Perdue and Lytle 1983; Perdue et al. 1984). This model was chosen to

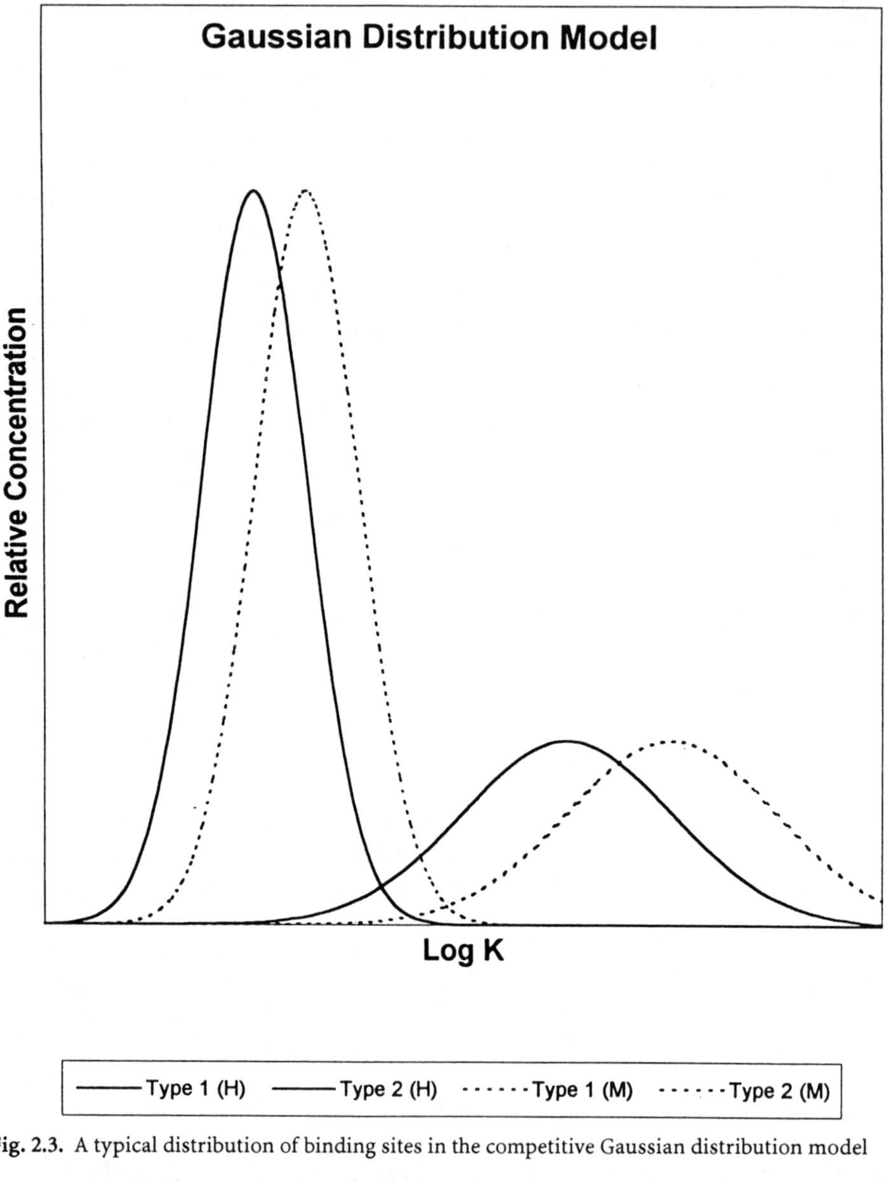

Fig. 2.3. A typical distribution of binding sites in the competitive Gaussian distribution model

describe cation binding by humic substances in MINTEQA2, a well-known computer program for performing chemical equilibrium calculations (e.g., Allison and Perdue 1995). In the Gaussian model, only monodentate (1:1) reactions can occur, i.e., 1 mol of either a proton or a metal ion can react with 1 mol of a binding site. The relative concentration of a binding site is related to its log K value for proton or metal binding by a Gaussian distribution function. Because proton binding studies indicate strongly that humic substances contain two classes of acidic functional groups, i.e., carboxyl and phenolic groups (Perdue 1985), two Gaussian distributions of binding sites are used in the model. For example, the binding site distributions of this model for binding of a proton and one metal ion to two broad classes of binding sites are shown in Fig. 2.3. Each broad class of binding sites is completely defined by the total concentration of binding sites in that class (C), the standard deviation (σ) of the distribution of sites with respect to log K, and the average log K values for proton (μ_H) and metal (μ_M) binding. Both C and σ are assumed to be the same for proton and metal binding. Ionic strength effects are treated empirically by assuming that all empty binding sites have a charge of −2.8 and using the Davies equation to calculate single-ion activity coefficients for all reacting species.

2.3.3 Model V

Tipping and Hurley (1992) described model V, which is the latest in a series of models to be developed by Tipping and coworkers. This model has also been incorporated into WHAM, a computer program for performing chemical equilibrium calculations for natural waters, sediments, and soils (Tipping 1994). In model V, Tipping and Hurley (1992) take a "discrete" approach to the description of the distribution of binding sites in a humic substance. As was the case with the competitive Gaussian distribution model, two broad classes of binding sites are used, and protons react on a 1:1 molar basis with binding sites (monodentate binding). Rather than using a continuous function to describe the relative concentrations and log K values of the binding sites, a discrete approach is used. Each broad class of binding sites consists of four equally abundant binding sites whose log K values for proton binding are symmetrically distributed around an average log K for proton binding by that class of sites. The "width" of the distribution is controlled by a range parameter for that class of sites. The average log K value and range parameter are somewhat analogous to μ_H and σ in the competitive Gaussian distribution model. The total concentration of the second class of sites is defined to be one-half of the total concentration of the first class of sites.

Although only monodentate proton binding is allowed in model V, both monodentate and bidentate (1:2) reactions of a metal with binding sites are allowed. In model V, metal binding equilibria are written for reaction of the metal with a protonated site rather than with an empty site. To facilitate a comparison with the Gaussian and NICA models, all of the reactions in model V have been rewritten here in terms of empty sites. Subsequent discussion is based on this modification of the mass action relationships that were described in model V. To describe monodentate metal binding by a class of sites, four log K values for metal binding are symmetrically distributed around an average log K for metal binding by that class of sites. The average log K is analogous to μ_M in the competitive Gaussian distribution model. Tipping and Hurley (1992) used the same range parameter to control the width of the distribution of binding sites for both proton and monodentate metal binding; however, when their reactions are rewritten in terms of empty sites, the actual range parameter for monodentate metal binding is effectively two times larger than the range parameter for proton binding.

From the total of eight monodentate binding sites in the two classes of sites, twelve arbitrary combinations are used for bidentate binding. The log K values for bidentate binding are a function of the average log K values for proton and metal binding and the range parameters for the two classes of binding sites. Sixty percent of each binding site is used for monodentate binding, and the remaining 40 % of each binding site is used for bidentate binding.

The overall distribution of binding sites in model V is shown in Fig. 2.4. Neglecting the log K values for bidentate metal binding, the distribution of monodentate proton and metal binding sites bears a superficial resemblance to the continuous distribution of binding sites in the Gaussian distribution model. Even though one distribution is discrete and the other is continuous, weaker binding sites are more abundant than stronger sites in both models, and the log K values for metal binding tend to differ systematically from the corresponding log K values for proton binding. One obvious difference is that metal binding constants are spread out over a much larger range of log K than are the corresponding proton binding constants. Another difference is the lower effective concentration of monodentate metal binding sites, which is due to the fact that some of those sites are assumed to form bidentate complexes with the metal. The bidentate reactions have no parallel in the competitive Gaussian distribution model, but they enable model V to more accurately model metal binding at very low concentrations of the metal.

Even though model V uses discrete binding sites with the "intrinsic" log K values that are shown in Fig. 2.4, chemical equilibrium reactions are actually described by electrostatically corrected "effective" log K values. These "effective" log K values are not constants, but are instead functions of the net charge of the average humic substance, which varies continuously as the average charge density of the humic substance is changed through binding of protons and metal ions. This treatment effectively "spreads out" each of the discrete log K values, so that a set of four discrete binding sites behaves like four overlapping, broad, symmetrical distributions (Perdue 1997). The effective log K values are a function of ionic strength, so part of the effect of ionic strength on proton and metal binding reactions is incorporated into the effective log K values.

Ionic strength further impacts reactions in model V, because that model considers "binding" to include both specific binding to sites in the humic substance and nonspecific binding. The latter type of binding is based on the idea that the humic substance (plus a thin shell of solution) is a separate phase whose net charge must be zero. Any charge on the humic substance that is not neutralized by specifically bound protons and metal ions must be balanced by loosely held counterions, including protons and metal ions. Donnan-type expressions are used in model V to account for this accumulation of counterions.

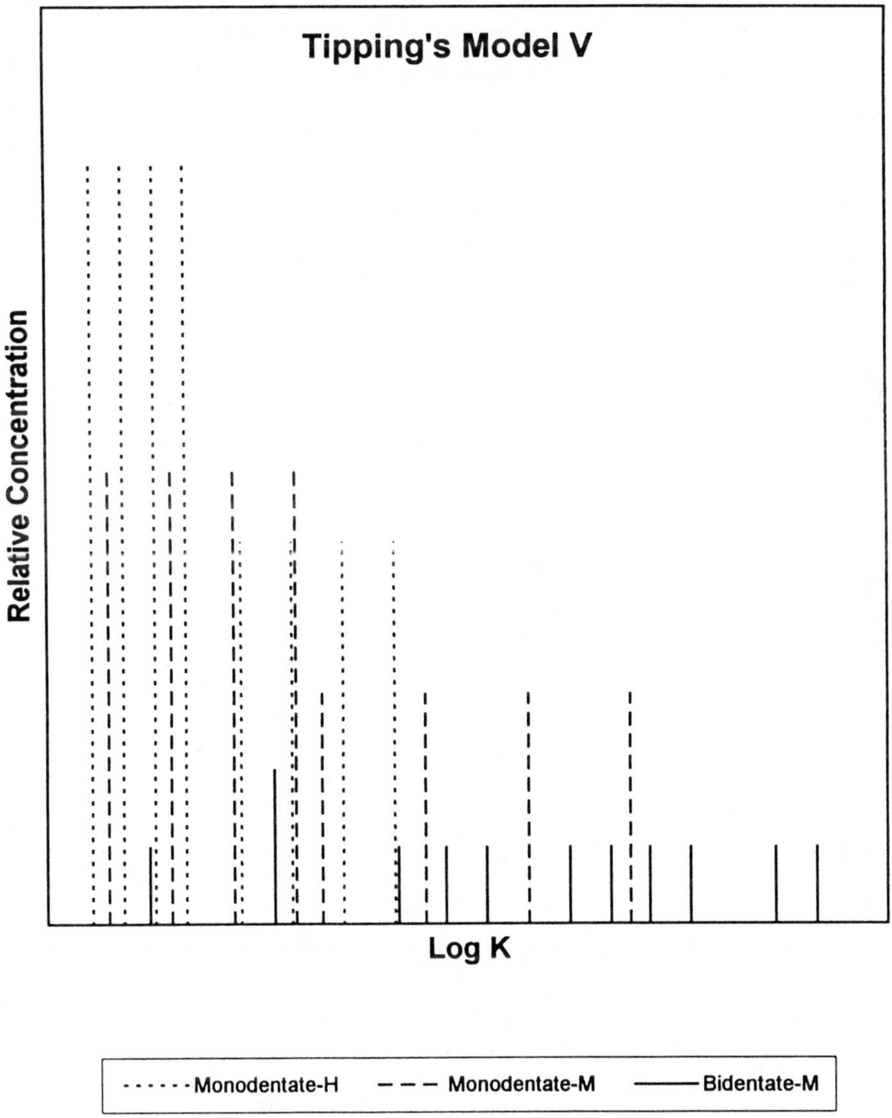

Fig. 2.4. A typical distribution of binding sites in model V

2.3.4 The NICA Model

The NICA (non-ideal competitive adsorption) model resembles the competitive Gaussian distribution model, both in its use of only 1:1 reaction stoichiometries and in the use of continuous symmetrical distributions of binding sites (e.g., Koopal et al. 1994; Benedetti et al. 1995, 1996; Kinniburgh et al. 1996). The model uses two quasi-Gaussian distributions of binding sites for

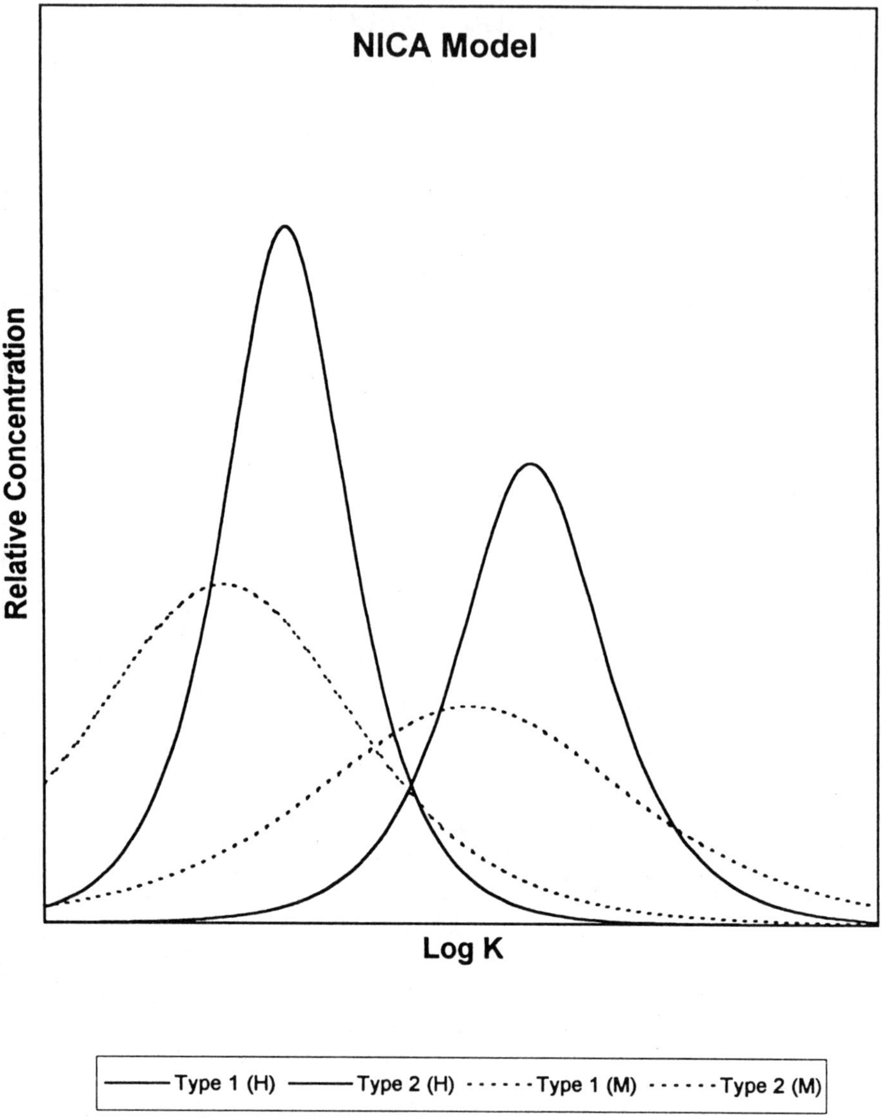

Fig. 2.5. A typical distribution of binding sites in the NICA model

proton and metal binding. Each class of binding sites is described by the total concentration of sites of that type, the median log K value, and a range parameter that controls the width of the distribution. These parameters correlate more or less directly with C, μ, and σ in the Gaussian model. A typical distribution of binding sites is shown in Fig. 2.5. Unlike the Gaussian model, where the width parameter (σ) is the same for proton and metal binding by a class of binding sites, or model V, where the range parameter for metal binding is twice that for proton binding, the range parameters for proton and metal binding of the NICA model are more or less independent of one another.

In the NICA model, the effect of ionic strength on proton and metal binding is attributed entirely to nonspecific binding of counterions. As in model V, the humic substance (plus a thin shell of solution) is treated as a separate phase whose net charge must be zero. Any charge on the humic substance that is not neutralized by specifically bound protons and metal ions must be balanced by loosely held counterions, including protons and metal ions. Donnan-type expressions are used to calculate the concentrations of these counterions in the so-called gel phase. The NICA model then uses the gel-phase concentrations of protons and metal ions to calculate the concentrations of specifically bound protons and metal ions. Both the Gaussian model and model V use the concentrations of protons and metal ions in the bulk aqueous solution for these calculations.

2.3.5 Summary Comparison of Models

To conclude this section on metal binding by humic substances, the major properties of the Gaussian, model V, and NICA models will be reiterated. All three models assume that humic substances contain two broad distributions of binding sites, presumably attributable mainly to carboxyl and phenolic hydroxyl goups. All three models use distributions of proton binding sites that are symmetrically distributed around average log K values, and all three models use some kind of range parameter to control the width of the distribution of log K values for proton binding. All three models thus describe the shape and size of a distribution of binding sites with three parameters (total concentration, average log K, range parameter). These model parameters for proton binding by the two classes of binding sites are independent of one another, except for the assumption in model V that carboxyl groups are twice as abundant as phenolic hydroxyl groups. Both the Gaussian and NICA models assume that each class of proton binding sites contains a continuous distribution of sites, and neither model "cares" whether the distribution results from statistical, electrostatic, or delocalization effects. Model V assumes that each class of sites contains four discrete binding sites; however, that model then "spreads out" the four discrete sites by making their log K values an explicit function of the average charge of the humic substance.

All three models consider monodentate binding of metal ions to binding sites. In the Gaussian and NICA models, all sites can form monodentate bonds to either protons or metal ions; however, in model V, only 60 % of proton binding sites can form monodentate bonds to metal ions. The remaining 40 % of proton binding sites in model V are used to form bidentate metal complexes, a reaction not included in the other two models. All three models have symmetrical distributions of monodentate metal binding sites.

In the Gaussian model, the distribution of log K values for metal binding by a class of sites is the same as it is for proton binding. In model V, the distribution is twice as wide for metal binding, and in the NICA model, the widths of the distributions are more or less independent of one another.

In the Gaussian model, only specifically bound protons and metal ions are considered to be "bound", and the concentrations of specifically bound protons and metal ions are calculated from their respective concentrations in the bulk aqueous solution. The effect of ionic strength on proton and metal binding is modeled entirely using individual ion activity coefficients that are calculated with the Davies equation, assuming that all empty binding sites have a charge of –2.8. In model V, "bound" protons and metal ions include specifically bound ions and nonspecifically bound ions. The concentrations of specifically bound protons and metal ions are calculated from their respective concentrations in the bulk aqueous solution, and the concentrations of nonspecifically bound protons and metal ions are obtained from a Donnan potential-based model. Thus, in model V, ionic strength affects both the spreading of the discrete log K values and the nonspecific binding of protons and metal ions. In the NICA model, both specifically and nonspecifically bound protons and metal ions are considered to be "bound." In contrast to the other two models, the concentrations of specifically bound protons and metal ions are calculated from their respective concentrations in the so-called gel phase and not from their bulk aqueous phase concentrations. The use of gel-phase concentrations is, effectively, the manner in which ionic strength effects are incorporated into the NICA model.

2.4 Summary

This chapter reviews the state of current knowledge regarding elemental composition and dominant structural features of humic substances, some recent empirical correlations between composition, structure, and bioavailability of humic substances, and the rapidly converging views of mathematical approaches to descriptions of cation binding by humic substances.

First, the relationship between the average properties of a humic substance and their quantitative relationship to the corresponding properties of individual compounds in the humic substance is discussed. The concept of unsaturation is then introduced and is shown to be one of the properties that place direct constraints on the average structural features of a humic substance. In fact, given a set of basic compositional parameters (elemental composition, carboxyl content, and estimated number-average molecular weight), it is possible to predict the most probable proportions of sp^3-hybridized carbon (C_{al}, aliphatic carbon), sp^2-hybridized carbon in aromatic rings (C_{ar}, aromatic carbon), and all remaining sp^2-hybridized carbon (C_{xs}, excess carbon) in a humic substance. The known range of elemental compositions of freshwater fulvic acids is found to be consistent with mixtures that have highly variable C_{al}, and nearly constant relative proportions of C_{ar} and C_{xs}. One interpretation is that the components of C_{al} are more biodegradable, and hence more variable, than the components of C_{ar} and C_{xs}.

The relationships between bioavailability, composition, and probable structural features of dissolved organic matter (DOM) from a Georgia river and several biomass leachates were also examined. Using bacterial growth as a measure of bioavailability, it was found that an empirical equation of the form [Growth = $a_0 + a_1(H/C) + a_2(O/C) + a_3(N/C)$] provides an excellent prediction of bioavailability of DOM. When analytical constraints calculations were used to estimate the most probable values of C_{al}, C_{ar}, and C_{xs} for the samples used in this study, the results were consistent with the more generic predictions for freshwater fulvic acids. Biomass leachates and DOM from upstream sites had relatively higher percentages of C_{al} than did the DOM from downstream sites and blackwater tributaries. Downstream changes in elemental composition and the corresponding predicted average structural features of DOM were consistent with the hypothesis that only C_{al} is utilized by bacteria, leaving the relative proportions of C_{ar} and C_{xs} nearly unchanged. Generally, samples with higher bioavailabilities had higher C_{al} values.

Finally, the principal features of three powerful models that have been used to describe competitive cation binding by humic substances were compared. The Gaussian, Model V, and NICA models all assume that humic substances contain two broad distributions of binding sites, presumably attributable mainly to carboxyl and phenolic hydroxyl groups. All three models use distributions of proton binding sites that are symmetrically distributed around average log K values, and all three models use some kind of range parameter to control the width of the distribution of log K values for proton binding. All three models thus describe the shape and size of a distribution of binding sites with three parameters (total concentration, average log K, range parameter). The models use somewhat different methods to generate the basic distributions of proton and metal binding sites, and they also differ significantly in the treatment of the effect of ionic strength on cation binding.

References

Ahlers WW, Reid RM., Kim PJ, Hunter AK (1990) Contamination-free sample collection and handling protocols for trace elements in natural fresh waters. Aust J Mar Freshw Res 41:713–720

Aiken GR, Gillam AH (1989) Determination of molecular weights of humic substances by colligative property measurements. In: Hayes MHB, MacCarthy P, Malcolm RL, Swift RS (eds) Humic substances II. In search of structure. Wiley, Chichester, pp 515–544

Aiken GR, McKnight DM, Wershaw RL, MacCarthy P (eds) (1985) Humic substances in soil, sediment, and water: geochemistry, isolation, and characterization. Wiley, New York

Allison JA, Perdue EM (1995) Modeling metal-humic interactions with MINTEQA2. In: Senesi N, Miano TM (eds) Humic substances in the global environment and implications on human health. Elsevier, Amsterdam, pp 927–942

Benedetti MF, Milne CJ, Kinniburgh DG, Van Riemsdijk WH, Koopal LK (1995) Metal ion binding to humic substances: application of the non-ideal competitive adsorption model. Environ Sci Technol 29:446–457

Benedetti MF, Van Riemsdijk WH, Koopal LK (1996) Humic substances considered as a heterogeneous Donnan gel phase. Environ Sci Technol 30:1805–1813

Benoit G (1994) Clean techniques measurement of Pb, Ag and Cd in freshwater: a redefinition of metal pollution. Environ Sci Technol 28:1987–1991

Cabaniss SE and Shuman MS (1988) Copper binding by dissolved organic matter. I. Suwannee River fulvic acid equilibria. Geochim Cosmochim Acta 52:185–193

Christman RF, Gjessing ET (1983) Aquatic and terrestrial humic materials. Ann Arbor Science, Ann Arbor MI

Dobbs JC, Susetyo W, Carreira LA, Azarraga LV (1989) Competitive binding of protons and metal ions in humic substances by lanthanide ion probe spectroscopy. Anal Chem 61:1519–1524

Erickson RJ, Benoit DA, Mattson VR, Nelson HP Jr, Leonard EN (1996) The effects of water chemistry on the toxicity of copper to fathead minnows. Environ Toxicol Chem 15:181–193

Hammond GS, Osteryoung J, Crawford TH, Gray HB (1971) Models in chemical science - an introduction to general chemistry. Benjamin, New York

Hayes MHB, MacCarthy P, Malcolm RL, Swift RS (eds) (1989) Humic substances II. In search of structure. Wiley, Chichester

Huffman EWD Jr, Stuber HA (1985) Analytical methodology for elemental analysis of humic substances. In: Aiken GR, McKnight DM, Wershaw RL, MacCarthy P (eds) Humic substances in soil, sediment, and water: geochemistry, isolation, and characterization. Wiley, New York, pp 433–455

Kinniburgh DG, Milne CJ, Benedetti MF, Pinheiro JP, Filius J, Koopal LK, Van Riemsdijk WH (1996) Metal ion binding by humic acid: application of the NICA-Donnan model. Environ Sci Technol 30:1687–1698

Koopal LK, Van Riemsdijk WH, de Wit JCM, Benedetti MF (1994) Analytical isotherm equations for multicomponent adsorption to heterogeneous surfaces. J Colloid Interface Sci 166:51–60

Leenheer JA (1994) Chemistry of dissolved organic matter in rivers, lakes, and reservoirs. ACS Adv Chem Ser 237:195–221.

Leff LG, Meyer JL (1991) Biological availability of dissolved organic carbon along the Ogeechee River. Limnol Oceanogr 36:315–323

MacCarthy P, Rice JA (1985) Spectroscopic methods (other than NMR) for determining functionality in humic substances. In: Aiken GR, McKnight DM, Wershaw RL, MacCarthy P (eds) Humic substances in soil, sediment, and water: geochemistry, isolation, and characterization. Wiley, New York, pp 527–559

Morel FMM, Hering JG (1993) Principles and applications of aquatic chemistry. Wiley-Interscience, New York

Perdue EM (1984) Analytical constraints on the structural features of humic substances. Geochim Cosmochim Acta 48:1435–1442

Perdue EM (1985) Acidic functional groups of humic substances. In: Aiken GR, McKnight DM, Wershaw RL, MacCarthy P (eds) Humic substances in soil, sediment, and water: geochemistry, isolation, and characterization. Wiley, New York, pp 493–532

Perdue EM, Lytle CR (1983) Distribution model for the binding of protons and metal ions by humic substances. Environ Sci Technol 17:654–660

Perdue EM, Reuter JH, Parrish RS (1984) A statistical model of proton binding by humus. Geochim Cosmochim Acta 48:1257–1263

Perdue EM (1997) Metal binding by humic substances in surface waters - experimental and modeling constraints. In Allen HE, Garrison W, Luther GW III (eds) Metals in surface waters. Ann Arbor Publishers, Chelsea, Michigan (in press)

Rice JA, MacCarthy P (1991) Statistical evaluation of the elemental composition of humic substances. Org Geochem 17:635–648

Senesi N, Miano TM (eds) (1994) Humic substances in the global environment and implications on human health. Elsevier, New York

Serkiz SM, Perdue EM (1990) Isolation of dissolved organic matter from the Suwannee River using reverse osmosis. Water Res 24:911–916

Sun L, Perdue EM, Meyer JL, Weis J (1997) Use of elemental composition to predict bioavailability of dissolved organic matter in a Georgia river. Limnol Oceanogr 42:714–721

Sunda W, Guilliard RRL (1976) The relationship between cupric ion activity and the toxicity of copper to phytoplankton. J Mar Res 34:511–529

Susetyo W, Carreira LA, Azaragga LV, Grimm DM (1991) Fluorescence techniques for metal-humic interactions. Fresenius J Anal Chem 339:624–635

Tipping E (1994) WHAM – a chemical equilibrium model and computer code for waters, sediments, and soils incorporating a discrete site/electrostatic model of ion-binding by humic substances. Comput Geosci 20:973–1023

Tipping E, Hurley MA (1992) A unified model of cation binding by humic substances. Geochim Cosmochim Acta 56:3627–3641

Wilson MA, Vassallo AM, Perdue EM, Reuter JH (1987) Compositional and solid-state nuclear magnetic resonance study of humic and fulvic acid fractions of soil organic matter. Anal Chem 59:551–558

3 Humus and Acidification

Espen Lydersen

3.1 Introduction

Soil humus may be defined as the product of transformation of plant and animal remains which bear no morphological resemblances to the materials from which they were derived (Kononova 1975; Hayes and Swift 1978). The predominant part of organic material in surface waters is normally allochthonously derived. The material is produced within the catchment and brought to the lakes and rivers by runoff in countless numbers of physico-chemical forms (e.g. Felbeck 1971; Heyes 1991). Types of degradation products as well as the time needed to fully decompose organic matter are extremely variable, depending on the chemical composition of the original material as well on several site-specific physical, chemical and biological factors such as the redox potential (i.e. the oxygen conditions and gas exchange), the biological degradation activity, the hydrological conditions, pH and temperature. Accordingly, it is not surprising to find a river carrying relatively fresh lignin and cellulose from newly fallen debris, as well as series of degradation products derived from these substances where they have been exposed to biochemical decomposition over extended periods of time.

The formation of humic substances is primarily microbiological, where carbohydrates serve as the main microbial source of energy and carbon in the intracellular synthesis of protein and hemicellulose. Carbohydrates and proteins are more available for microorganisms and have a higher rate of chemical decomposition. Therefore lignin is considered to be the most important source of humic compounds (Gjessing 1976). Lignin is modified during degradation, forming humic-like substances of high molecular weights. Further degradation of these compounds results in generation of humic and fulvic acids and various decomposition products (fats, amino compounds and the gases CO_2, H_2, CH_4, N_2, NH_3 and H_2S). Also, autolysis of the microorganisms themselves (including fungi) represents a major source of humic-like substances. A significant part of dissolved humic material is organic acids, and much study has been conducted on these acids, particu-

Ecological Studies, Vol. 133
Hessen, Tranvik (eds) Aquatic Humic Substances

larly regarding their role in water pH, their pH-buffering capacity and metal-complexing properties.

3.2 Water pH and Acid Neutralising Capacity

Acids produced within a catchment and acids deposited from the atmosphere (acid precipitation) are to various degrees buffered and/or neutralised by biogeochemical processes, both within the catchment and within the lakes and streams. The degree depends on the acid neutralising capacity (ANC). ANC is defined relative to a H^+ reference level for each acid-base system in solution (Stumm and Morgan 1981; Sullivan et al. 1989). The H^+ reference level is defined as those species which do not change ANC when added to the solution (although changes in pH will occur). For example, the H^+ reference level of an inorganic C solution includes the ANC components of water (ANC_W) and aqueous CO_2, i.e. the inorganic carbon system (ANC_{IOC}). In dilute natural waters, there are also other proton acceptors, largely represented by dissolved organic carbon (DOC) with functional groups that bind H^+ (ANC_{DOC}), and certain inorganic hydroxy-Al (ANC_{Ali}) and organo-Al complexes (ANC_{Alo}). Thus, total ANC (ANC_T) in low ionic strength natural waters primarily is a sum of the following H^+ donors and acceptors:

$$ANC_T = ANC_W + ANC_{IOC} + ANC_{DOC} + ANC_{Ali} + ANC_{Alo}. \quad (1)$$

Another common expression of ANC often used in the literature (in µeq/L) is:

$$ANC = ([Ca^{2+}]+[Mg^{2+}]+[Na^+]+[K^+]) - ([SO_4^{2-}]+[NO_3^-]+[Cl^-]). \quad (2)$$

This ANC expression is often called the alkalinity concept (Reuss and Johnson, 1986).

To obtain electric neutrality (charge balance) in most freshwaters, ANC primarily expresses:

$$ANC = ([HCO_3^-]+[A^-]) - ([H^+] - [\Sigma Al^{n+}]). \quad (3)$$

$[A^-]$ is an expression of the amount of organic anions. Assuming that all the major ions incorporated in the charge balance equation are measured, the amount of $[A^-]$ can be estimated, so that charge balance is obtained. The concentration of Al is in the expression the sum of positively charged Al ions (ΣAl^{n+}), which can be estimated according to thermodynamic constants (e.g. Schecher and Driscoll 1987, 1988).

One of the most useful aspects of the alkalinity concept is that the base cations and strong acid anions on the right side of Eq. (2) are independent of the CO_2 partial pressure (PCO_2). The concentrations of ions on the right side of Eq. (3) all vary with PCO_2, but must vary in such a way as to maintain the equality of the sum. This is the main reason why this alkalinity concept has been widely used, and often also named ANC. In this chapter, however, the

ANC presented in Eq. (1) is used. This means that the $[H^+]$ concentration is subtracted from Eq. (3). Thus, $[HCO_3^-]$, $[A^-]$ and $[Al^{n+}]$ in Eq. (3) are equal to $ANC_{IOC} + ANC_{DOC} + ANC_{Ali}$ in Eq. (1). These three ANC factors are normally the most predominant in freshwater.

3.3 Quality and Quantity of Organic Acids of DOC

The pH of waters is affected by natural weak inorganic and organic acids as well as strong mineral acids, primarily from atmospheric inputs. Some of the strong acids are buffered and/or neutralised by weak inorganic acids (primarily the HCO_3 system and dissolution of Al) and/or by weak organic acids. Accordingly, the role of organic acids is of considerable interest, especially with respect to their possible role in surface water acidification. A major task is to assess their acid strengths which are a combination of their dissociation constants (pK values) and the numbers of acid functional groups. The latter is normally defined as charge density (CD) expressed as mole sites per mole of carbon.

Numerous methods have been applied to assess pK and CD values. Various base titration techniques (e.g. Johansson 1970; Schnitzer and Khan 1972; Lee and Brosset 1978) can be applied to different operationally defined organic fractions, after removal of interfering compounds as dissolved Al and inorganic carbon species. The fractionation is often an ion exchange or chromatographic technique in combination with an acid or base adjustment (e.g. Leenheer and Huffman 1979; Aiken 1988; Easthouse et al. 1992). Common to most titration procedures is a pre-acidification step in order to fully protonate the organic functional groups. The subsequent base titration is often a modification of the Gran titration (Gran 1952) up to pH 5–7 under N_2 bubbling to exclude CO_2. If Al is not removed from the solution, the buffering by Al can be calculated by use of thermodynamic data. Because natural dissolved organic matter (DOC) is an organic acid cocktail with numerous pK and CD values (Cronan and Aiken 1985; Kramer and Davies 1988; Ephraim et al. 1991), the titration curve will never show very distinct pK values. Many different methods have been used to characterise a few average pK values (often 1–3), and a fixed number of acid functional groups fitted to each actual data set (Table 3.1).

Organic acids have often been overlooked in discussions of the acidity of surface waters highly affected by strong acid inputs (e.g. Galloway et al. 1983; Hedin et al. 1990). An exception was the hypotheses by Rosenqvist (1978) and Krug and Frink (1983), but they do not distinguish between the effects of organic acids in high and low inorganic carbon alkalinity waters. Recent studies, however, clearly demonstrate the importance of organic acids in modifying both the acidity of surface waters and the response to changes in strong acid inputs in waters with low or no bicarbonate alkalinity (e.g.

Table 3.1. Overview of some pK values and the total concentration of acid functional groups, i.e. charge density organics (CD^a mole sites/ mole C; CD^b μeq/mg C) presented in the literature

	pK_1	pK_2	pK_3	CD^a	CD^b	Reference
Monoprotic	4.45			0.135	11.3	Driscoll et al. (1994)
Monoprotic	3.85			0.072	6.0	Cronan and Aiken (1985)
Diprotic	3.37	6.22		0.054	9.0	Schecher and Driscoll (1987,1988)[1]
Diprotic	4.02	6.04		0.082	13.7	Driscoll et al. (1994)
Triprotic	2.10	5.94	6.86	0.043	10.8	Schecher and Driscoll (1987,1988)[1]
Triprotic	2.64	5.66	5.94	0.055	13.8	Driscoll et al. (1994)
Triprotic	4.00	6.10	9.25			Eary et al. (1989)[2]
Fulvic acid	3.5 ± 0.2	6.3 ± 0.2				Leuenberger and Schindler (1986)
Carboxyls	3.66			0.061	5.1	Perdue et al. (1984)
Phenols	12.5			0.064	5.3	Perdue et al. (1984)

[1] These values are used in the Alchemi Version 4.0 speciation program made by the authors.
[2] These values are used in the Integrated Lake-Watershed Acidification Study, i.e. the ILWAS model

Kramer and Davies 1988; Wright 1989; Wilkinson et al. 1992; Driscoll et al. 1994). Even though natural organic acids contains a continuum of acid functional groups, many of which display weak acid characteristics, it has been well documented that an important fraction of the organic acids have pK values < 3.0 (Table 3.1; Munson and Gherini 1993; Ephraim et al. 1991; Leenheer et al. 1995). However, they should not be defined as strong acids per se, because a strong acid is defined as an acid which is totally dissociated when dissolved in water, which means no pK values. However, under most natural pH regimes of freshwaters (pH > 4), these strong organic acids are almost totally deprotonated. Accordingly, they might be evaluated equal to a strong mineral acid.

Based on large surface water data sets, Schecher and Driscoll (1987, 1988 in their Alchemi Version 4.0 speciation program) and Driscoll et al. (1994) assessed best fitted pK values for diprotic and triprotic acid models. The pK values for these two surveys are relatively similar (Table 3.1), with pK_1 describing the strongest organic acids, while the pK_2 and pK_3 lie within a relatively narrow area, i.e. from 5.66 to 6.86. The very high pK_3 value (9.25) presented by Eary et al. (1989) and pK values of phenols (12.3–12.5) by Perdue and Lytle (1983) and Perdue et al. (1984) are irrelevant for most natural waters, since the water pH rarely exceeds 9. Accordingly, the triprotic acid in the Integrated Lake-Watershed Acidification Study (ILWAS) model (Eary et al. 1989) is effectively a diprotic model, and the two first dissociation constants in this model coincide well with other diprotic models (Table 3.1), particularly with Driscoll et al. (1994). In addition, when assessing pK values of organic acids, it is also important to determine the concentration of organic acids (CD) dissolved. Based on the literature, CD largely varies from 5 to 22 μeq/mg C (e.g. Beck et al. 1974; Henriksen and Seip 1980). One important methodical factor for this variation is that many scientists estimates the concentrations of organic anions ($[A^-]$) so that electric neutrality is obtained [See comments to Eq. (3)]. By dividing the concentration of organic charges $[A^-]$ by the concentration of dissolved organic carbon (DOC) measured in streamwater, they often defined this as the CD of organics. However, this is not the CD of organics, but the net charge (NC) per milligram of carbon in the actual water analysed. Since $[A^-]$ is estimated on the basis of charge balance, it gives no information about the total number of negative sites present per milligram of carbon. Thus, when charge balance is used for calculating $[A^-]$, the ratio between $[A^-]$ and DOC should be the NC of organics. Accordingly, when the total number of negative sites are known, the ratio should be called the CD of organics.

Based on operational defined carboxyl contents of fulvic acids and humic acids from various locations in the USA and Canada, Oliver et al. (1983) estimated an average CD of carboxylic acids of 10 μeq/mg humic organic carbon. Further, they made an empirical equation for an average mass action quotient ($\overline{K}$) for fulvic and humic acids:

$$p\overline{K} = 0.96 + 0.90\text{pH} - 0.039\,(\text{pH})^2. \tag{4}$$

This equation offers the possibility of estimating the organic anion concentration $[A^-]$ from pH and the concentration of DOC by the following expression:

$$[A^-] = \frac{\overline{K}\,[C_T]}{\overline{K} + [H^+]}, \tag{5}$$

where $[A^-]$ and $[C_T]$ are the concentrations of ionised and total carboxyl groups, respectively. The ratio between $[A^-]$ and $[C_T]$ is the degree of dissociation of the acids of DOC. This model has commonly been used to determine organic acid concentrations in surface waters. Calibration of the Oliver model by Driscoll et al. (1994) to the Adirondack Lakes Survey Corporation (ALSC) gave the following expression:

$$p\overline{K} = 0.15 + 1.41\text{pH} - 0.078(\text{pH})^2, \tag{6}$$

with a CD of 15.1 μeq/mg C. This is 50% greater than the value obtained by Oliver et al. (1983). The model of Oliver et al. (1983) relies on a far more gradual increase of $[A^-]$ compared with the other model estimates (Fig. 3.1). Accordingly, the Oliver model overestimates $[A^-]$ in the pH interval 4–6 and underestimates $[A^-]$ at pH > 6, compared with the other models. The other three models assume a marked increase in $[A^-]$ in the pH interval 5–7. The discrepancy between the triprotic model by Driscoll et al. (1984) and the Alchemi model (Schecher and Driscoll 1987, 1988) comes primarily from different estimates of CD of DOC and, to a lesser degree, different pK values (Fig. 3.1 and Table 3.1). Both the triprotic models have a very low pK_1, i.e. 2.10

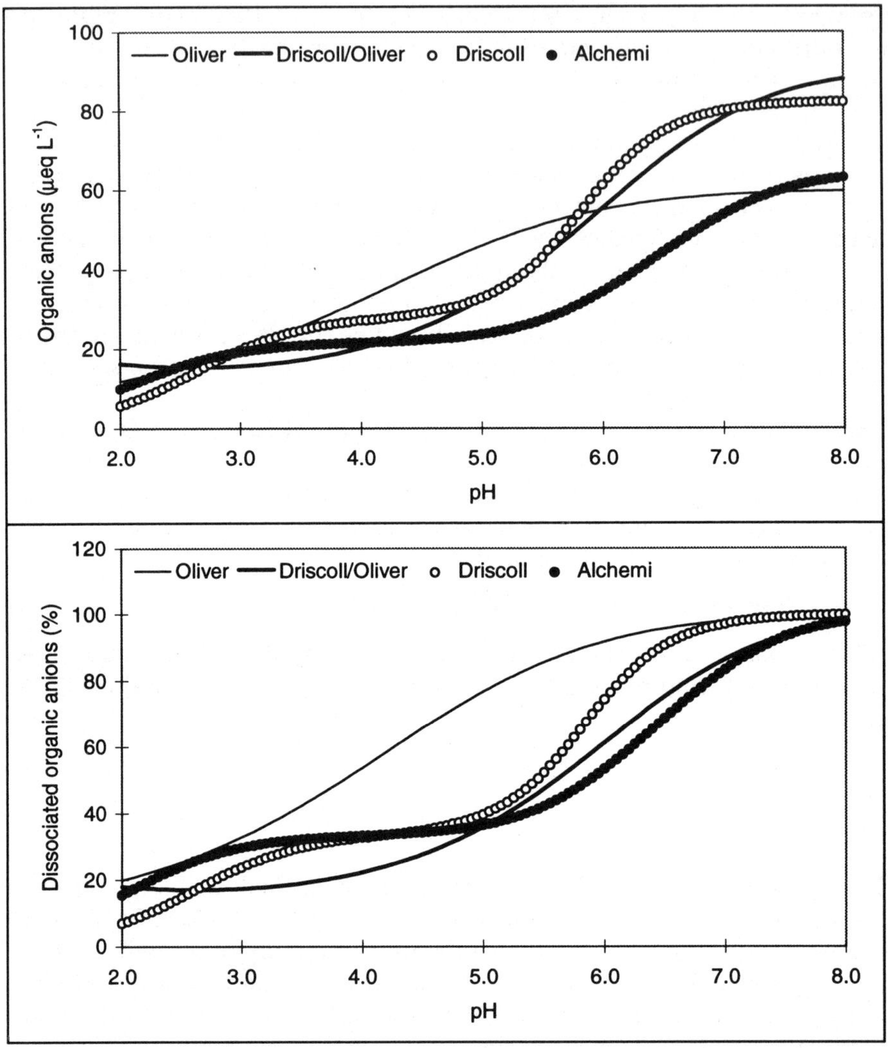

Fig. 3.1. Comparison of predicted values of organic anion concentrations (A^{n-}) for a solution containing 500 µmol C/L (DOC 6 mg C/L) and percentage dissociation of organic acids as a function of pH based on original parameters from Oliver et al. (1983) (—); calibrated values of Oliver et al. (1983) (**—**); based on the ALSC data (Driscoll et al., 1994); triprotic model parameters of Driscoll et al. (1994) (o); triprotic model in the Alchemi Version 4.0 program (Schecher and Driscoll 1987, 1988)(•)

(Schecher and Driscoll 1987, 1988) and 2.64 (Driscoll et al. 1994). Since both models assume an equal concentration of functional groups for all three pK values, the concentration of strong organic acids depends on the estimates of CD. Accordingly, 3.6 and 4.6 µeq/mg C will always act as a strong mineral acid in surface waters with pH > 4 in the Alchemi model and Driscoll model, respectively.

In the Reversing Acidification project In Norway (RAIN project), acid rain was excluded and substituted by clean precipitation. After 4 years of treat-

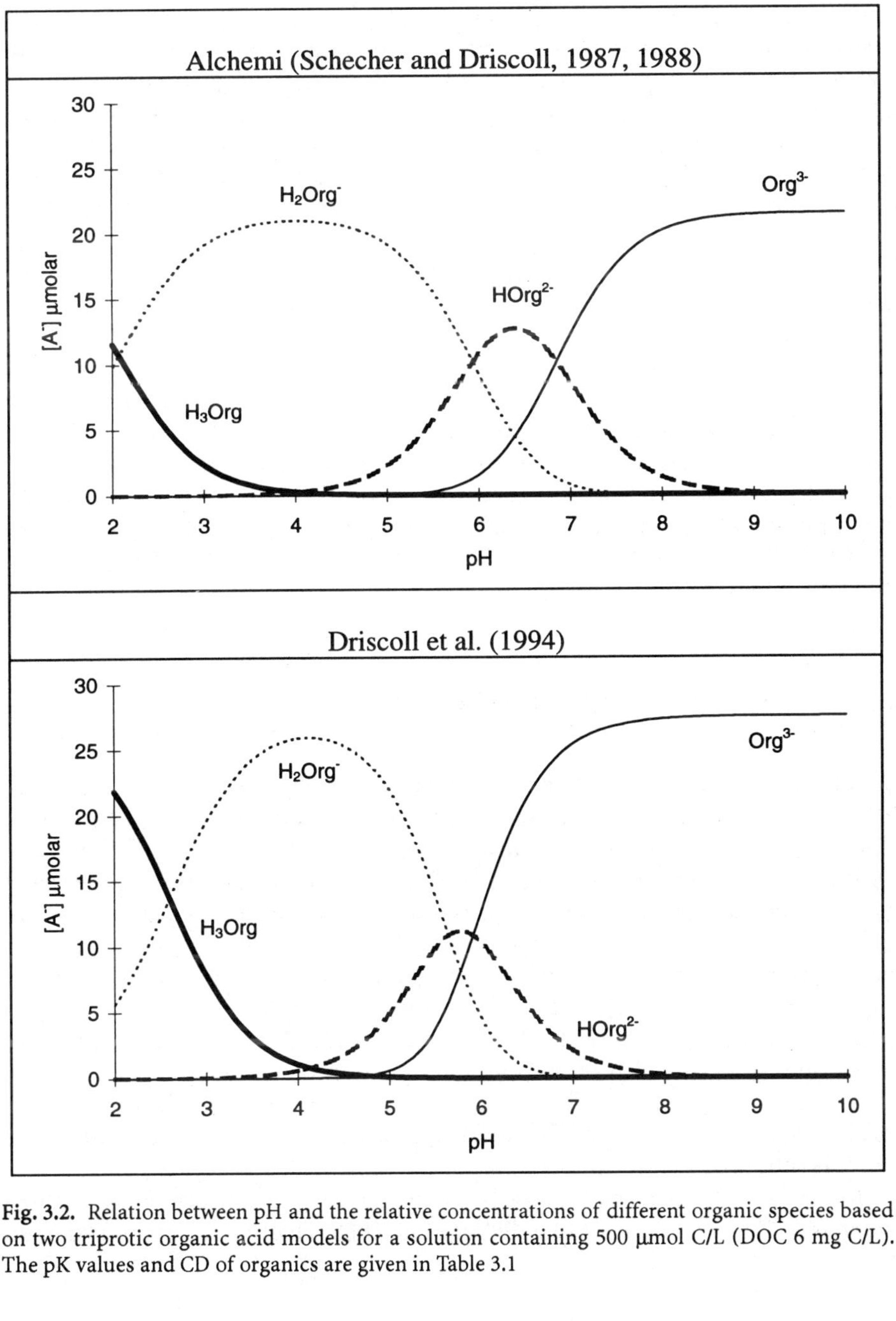

Fig. 3.2. Relation between pH and the relative concentrations of different organic species based on two triprotic organic acid models for a solution containing 500 μmol C/L (DOC 6 mg C/L). The pK values and CD of organics are given in Table 3.1

ment, pH had only increased from 4.0 to 4.1 after a reduction of $\Sigma[SO_4]$, $[NO_3]$ in runoff from 145 to 60 μeq/L (Wright 1989). Organic acids were concluded to be increasingly important to the pH of runoff, not because of a change in DOC concentrations, but due to increased dissociation of organic acids. The maximum CD of organic acids was found to be 4.5 μeq/mg C, with a pK value of about 4. The triprotic models based on empirical data from lakes in the Adirondack area of USA (Fig. 3.2 and Table 3.1) are not able to explain the buffering by organic acids in the RAIN project, because these models have no pK values near pH 4. The buffering effect of a weak acid is ≈ ± 1 pH unit from the pK value. However, pK values from soil leachates of 3.8–4.0 and the CD values of 4.5–6.0 μeq/mg C presented by Dempsey and O'Melia (1983) and Cronan and Aiken (1985) are close to the data reported from the RAIN project.

Since the pK values for both the referred triprotic models are fitted on the basis of large lake surveys data, the fitted pK values obviously reflect the pH interval of the surveyed lakes. Exception is of course the predefined very low pK_1 value (Table 3.1), taken care of the existence of strong organic acids (Munson and Gherini 1993; Ephraim et al. 1991; Leenheer et al. 1995). This indicates a clear relationships between water pH and the pK values of a major fraction of dissolved organic acids. Accordingly, many discrepancies in pK and CD values in the literature depend on the water types studied and not only on analytical factors such as various isolation techniques and/or acid/base titration procedures and differences according to whether CD or NC is calculated.

3.4 Dissolution of Organic Acids with pH

Acidification may cause an increase in transparency of water (Almer et al. 1974; Schindler and Turner 1982), a decrease in water colour (Dickson 1978) and decrease in total organic carbon (TOC) or DOC in lakes (Davis et al. 1985). A corresponding increase in DOC by a pH-increase has been reported (Christophersen et al. 1982). These physico-chemical effects can be, or are often, explained by two major processes: (1) the pH-buffering role of the organic acids of DOC; (2) the pH-dependent dissolution of essential complexing cations such as Al^{n+}, Fe^{n+} and Ca^{2+}, or ionic strength.

In process 1, a decrease in pH leads to decreased dissociation of organic acids due to increased protonation. Acidification of a catchment may therefore cause decreased solubility of the weakest acids because they are totally protonated as a result of the H^+ increase. Thus, their ability to be present as DOC is reduced.

In process 2, acidification leads to increased concentrations of cations such as Al^{n+}, Ca^{2+} and Fe^{n+} (e.g. Fuller et al. 1988; Hedin et al. 1990; Lydersen et al. 1996). Accordingly, more organic acids will precipitate as metal humates.

Despite these fundamental chemical properties, several works are not able to document any effects on the concentration of DOC after a pH change (e.g. Fuller et al. 1988; Wright 1989; Hedin et al. 1990; Lydersen et al. 1996), while others have even reported increases in DOC by acidification (e.g. Krug and Isaacson 1984; McColl and Pohlman 1986). There are many explanations for these discrepancies. One is to evaluate the two major processes (1 and 2) together, by considering that H^+ competes with metal ions for anionic binding sites of organics, while OH^-competes with organic anions for the cationic metal ions. As the pH decreases, the organic acids become less available for complexation because the organic acids are less dissociated, while metal cations become more available due to increased dissolution and/or increased cationic charge. An intermediate pH favours complexation between organic acids and metal ions (Fig. 3.3) Thus, the pH and the quality and quantity of inorganic and organic ions are decisive for whether the acidification will lead to a change in DOC in freshwater systems or not.

Regarding the quality and quantity of DOC, it is well documented that large vertical differences exist depending on which soil horizon DOC derives from (e.g. Krug and Isaacson 1984; Easthouse et al. 1992). Acidification may cause complex changes in both the organic and inorganic composition of leachate (Ritchie and Posner 1982; Stevensen, 1982) in addition to the two major processes already mentioned. For example, acidification has been

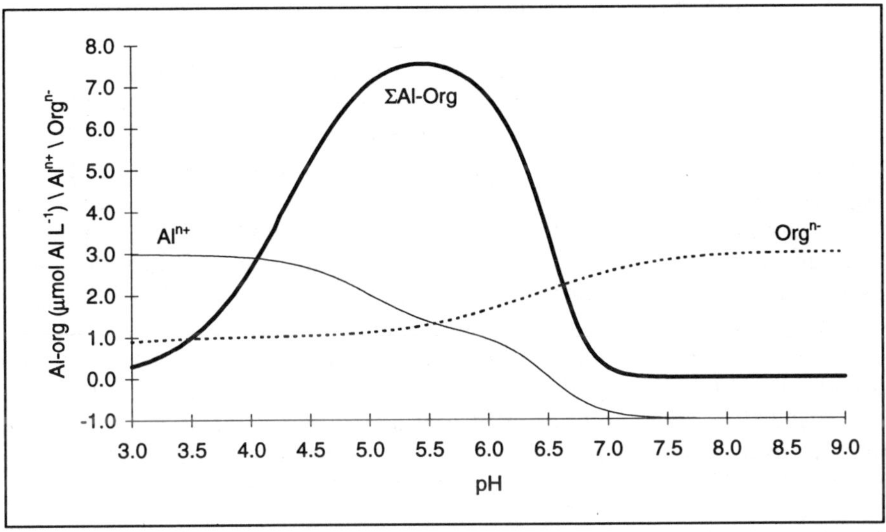

Fig. 3.3. Concentration of organic Al complexes, average charge of Al and a triprotic organic acid as a function of pH. Organic Al complexing constants, Al hydrolysis constants and pK values and CD values are those given in the Alchemi Version 4.0 program (Schecher and Driscoll 1987, 1988). Amounts of Al-Org are based on solution containing 500 µmol C/L and 10 µmol Al/L

shown to increase the solubility of organic matter from plant residues in *organic* soil horizons by hydrolysis of carbohydrates and amino acids (e.g. Stevenson 1982). On the other hand, acidification may cause reduced aqueous solubility of organic matter due to a combination of increased H^+ content of humic macromolecule and lower pH. This may increase the role of inter- and intramolecular hydrogen bonding in the macromolecules, which in turn may enhance aggregation of organic compounds (e.g. Hayes and Swift 1978; Ritchie and Posner 1982). Acidification may also increase the solubility of organic matter from *mineral* soil horizons by cleavage of metal ion-organic bonds or from disaggregation of metal-humic complexes (e.g. Ritchie and Posner 1982). This illustrates both the enormous complexity of DOC as well as the complexity of physico-chemical processes involved.

Evidence for significant changes in DOC primarily comes from studies with unrealistically high treatment levels of mineral acids. In more realistic experiments, only significant increases in base cations and Al^{n+} have been observed, with no accompanying changes in the concentration of DOC, neither in lake and streamwaters (e.g. Fuller et al. 1988; Hedin et al. 1990; Lydersen et al. 1996) nor in soil solutions (e.g. Fuller et al. 1988). The only general change in DOC due to acidification, has been a certain and very often significant decrease in $[A^-]$, reflecting the acid-neutralising capacity of DOC (Hedin et al. 1990; Lydersen et al. 1996). The same patterns were also observed by Wright (1989) after removal of strong acids in precipitation, where no change in the concentration of DOC was observed, while $[A^-]$ increased and base cations and Al decreased.

The unrealistically high levels of mineral acid often added in many experiments (e.g. Krug and Isaacson 1984) may, however, be an indicator of the tight interactions between inorganic and organic chemistry present, where the quality and quantity of organic and inorganic compounds in the different soil horizons, types and volumes of soil horizons, quantity and quality of precipitation, runoff pattern, temperature, origin of organic compounds and stage of degradation/humification, are all essential factors. The sum of all these factors gives the observed water chemistry, including the water pH. Accordingly, an essential part of the organic acids in solution have pK values closely related to the pH of the actual water, and thereby a significant pH-buffering role for each particular system. As mentioned in Section 3.3, fitted pK values of organic acids on the basis of different surveys are therefore closely related to the pHs of the waters incorporated. The only exception is the assumption of an important strong acid organic fraction ($pK < 3$), a fraction always totally deprotonated in most natural waters, i.e. at $pH > 4$, and thereby with an acid strength equal to a strong mineral acid.

3.5 Effects of DOC on pH of Norwegian Lakes

The effects of DOC on the water pH of Norwegian lakes are based on data collected from nearly 1000 lakes during autumn 1995 (Skjelkvaale et al. 1997). The statistical criteria for the selection are presented in Henriksen et al. (1996). In this survey, total organic carbon (TOC) was analysed. As the particular carbon fraction is generally low (< 10%), TOC ≈ DOC. Alkalinity was measured by acid titration to fixed pH 4.5 ($ALK_{4.5}$). In Norwegian lakes, $ALK_{4.5}$ comprises a sum of many proton accepting processes within the pH area from the actual pH to pH 4.5. The most essential are: ANC_{IOC}, ANC_{DOC}, ANC_{Ali} and ANC_{Alo}. To estimate the concentration of different ANC factors, the Alchemi Version 4.0 program has been used (Schecher and Driscoll 1987, 1988). This means that the pK values and CD values of organics, Al-organic constants and Al-hydrolysis constants applied are those present in this program. Only lakes with ANC < 350 µeq/L were incorporated in this study, i.e. 931 lakes (93.8% of all 993 investigated lakes).

The ANC and TOC in Norwegian lakes are generally low (Table 3.2). About 73% of the lakes have ANC < 100 µeq/L, and nearly 60% have TOC < 4 mg C/L. By the Alchemi program, almost all lakes were calculated to be supersaturated with respect to PCO_2 in air. It was not measured but calculated by the Alchemi program, assuming open CO_2 atmosphere and variable pH. There was no significant relation between TOC and degree of supersaturation of CO_2, in contrast to Hope et al. (1996) who found a strong positive correlation. However, their study relied on summer samples, while the Norwegian lakes were sampled in the autumn. In the Norwegian lake survey, a significant positive correlation was found between supersaturation with CO_2 and pH (Fig. 3.4). However, as expected, at pH < 5.4, where ANC_{DIC} is minimal, no significant trend was found between CO_2 saturation and pH. For all lakes, the median CO_2 saturation was 217%, while for lakes with pH < 5.4, median saturation was 133%. This saturation status is in accordance with earlier lake studies (e.g. Logan et al. 1982; Wright 1983; Norton and Henriksen 1983).

No significant correlation was found between TOC and pH when all lakes were evaluated together (Fig. 3.5). However, based on median values, there are tendencies of somewhat lower pH as the TOC concentration increases (Table 3.2). Within the ANC range 0–50 µeq/L, however, a significant corre-

Table 3.2. Numbers, percentage distribution, median ANC and pH in Norwegian lakes with ANC < 100 µeq/L. Total number of lakes: 993

TOC	No. of lakes	Percentage of total lakes	ANC	pH
0–4 mg C/L	569	57.3	9.8	5.98
4–8 mg C/L	120	12.1	41.8	5.71
> 8 mg C/L	38	3.8	54.3	5.60

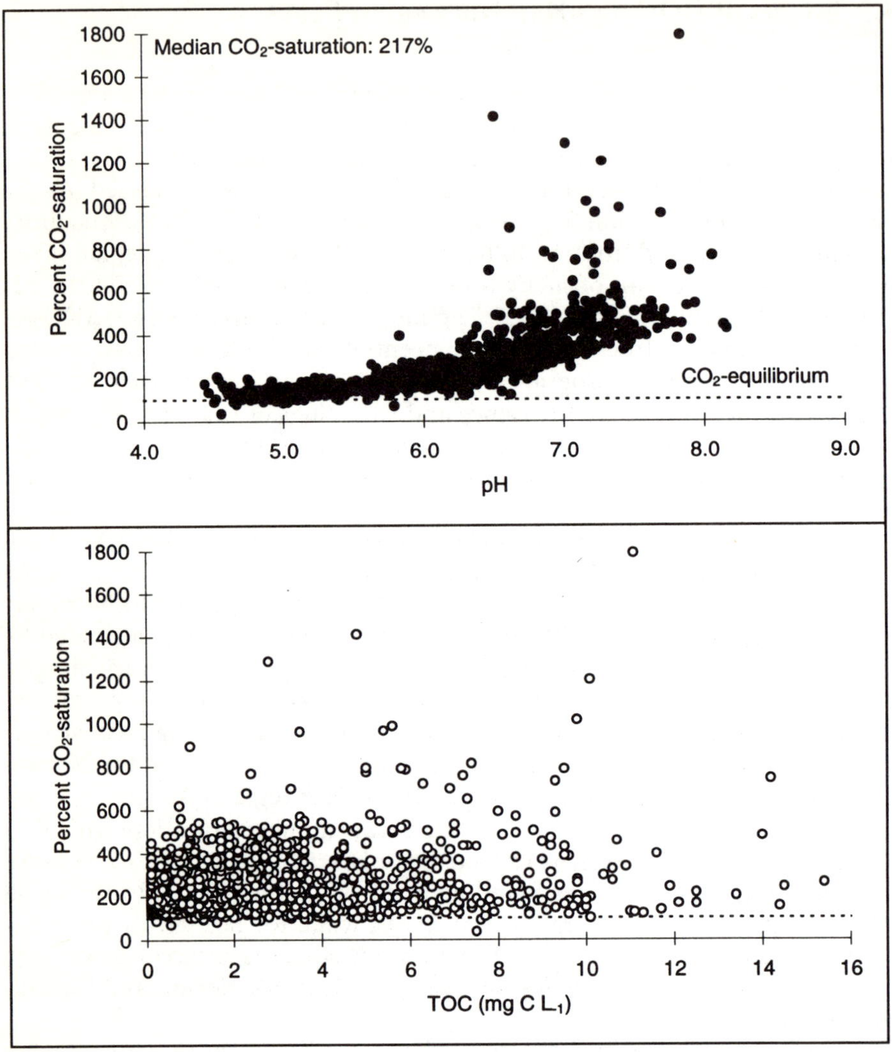

Fig. 3.4. Norwegian lake survey 1995. Relationships between CO_2 saturation and pH and CO_2 saturation and concentration of TOC. 100% saturation is at equilibrium with PCO_2 in air. Also, lakes with ANC > 350 μeq/L are presented. (Data from Skjelkvaale et al. 1997)

lation was found between TOC and pH (Fig. 3.5). Within this ANC range, the influence from ANC-factors other than ANC_{DOC} is relatively small (Fig. 3.8).

The influence of supersaturation of CO_2 ($ANC_{DICdiseq}$), organic acids (ANC_{DOC}), inorganic Al (ANC_{Ali}) and organic Al (ANC_{Alo}) on solution pH is illustrated in Fig. 3.6. As illustrated in Fig. 3.7, supersaturation of CO_2 is an essential factor for depression of water pH. The depression ranges from 0 to more than 2 pH units. This is almost identical to the Adirondack lakes survey

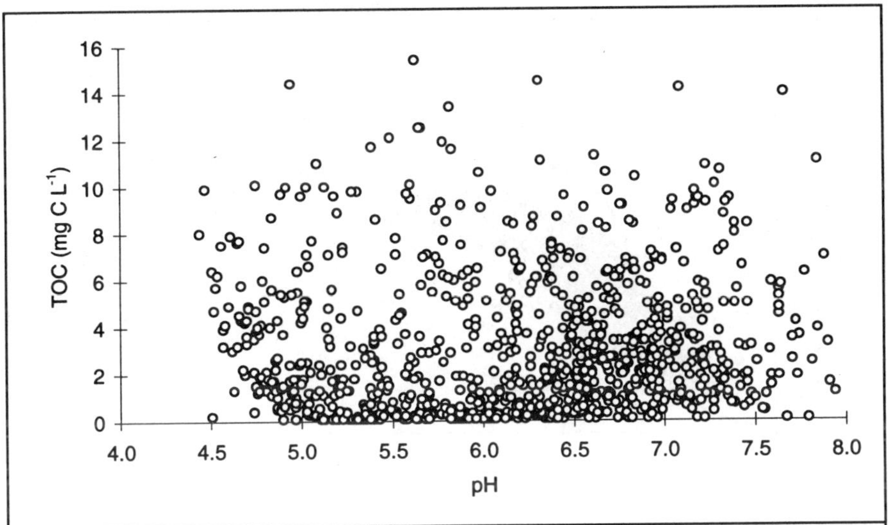

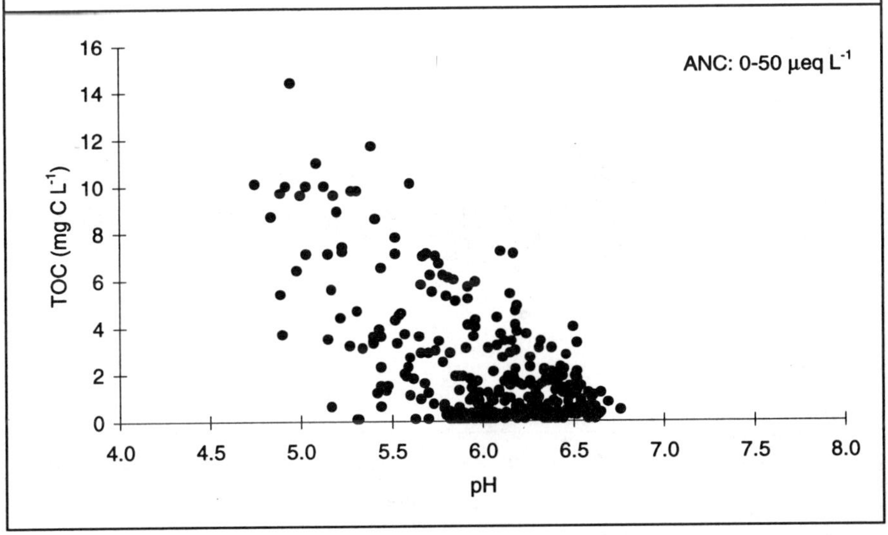

Fig. 3.5. Norwegian lake survey 1995. Relationship between pH and TOC for all Norwegian lakes (above) and all lakes within the ANC-range: 0–50 µeq/L (below). (Data from Skjelkvaale et al. 1997)

(Munson and Gherini 1993). Norton and Henriksen, (1983) found that aeration of samples caused an increase in pH of about 0.5–0.7 after the samples were purged with N_2 for 10 min. Their results are not directly comparable with the work by Munson and Gherini (1993) and the Norwegian lake calculations, because these two works rely on theoretically air equilibrated pH values.

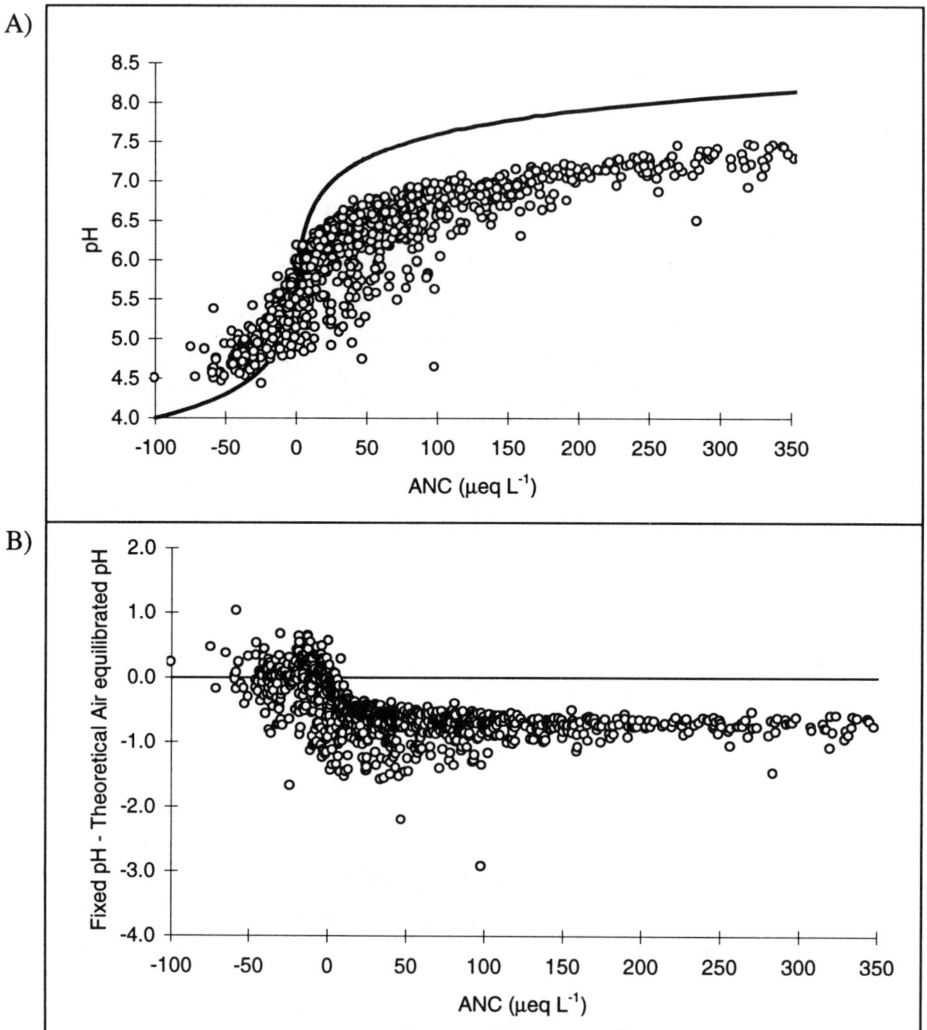

Fig. 3.6. Norwegian lake survey 1995. Influence of organic acids (ANC_{DOC}), CO_2 disequilibrium with PCO_2 in air ($ANC_{DICdiseq}$), inorganic Al hydrolysis (ANC_{Ali}) and Al-complexation with organics (ANC_{Alo}) on solution pH. A Observed field pH versus ANC. Solid curve represents theoretical expected pH of a solution containing just mineral acids and bases and atmospheric CO_2. Vertical deviation from theoretical curve represents influence from other acids and bases. B Direct plot of deviation from the solid curve in A. (Data from Skjelkvaale et al. 1997)

Other ANC factors also play an important role in water pH, i.e. from an increase in pH of almost 1 pH unit to a depression of about 1 pH unit (Fig. 3.7). As illustrated in Fig. 3.8, the two major factors, besides ANC_{DIC} and $ANC_{DICdiseq}$, for solution pH are ANC_{Ali} and ANC_{DOC}. The effect of organic Al complexation (ANC_{Alo}) is minor (Fig. 3.8C), at least in this survey. Increases

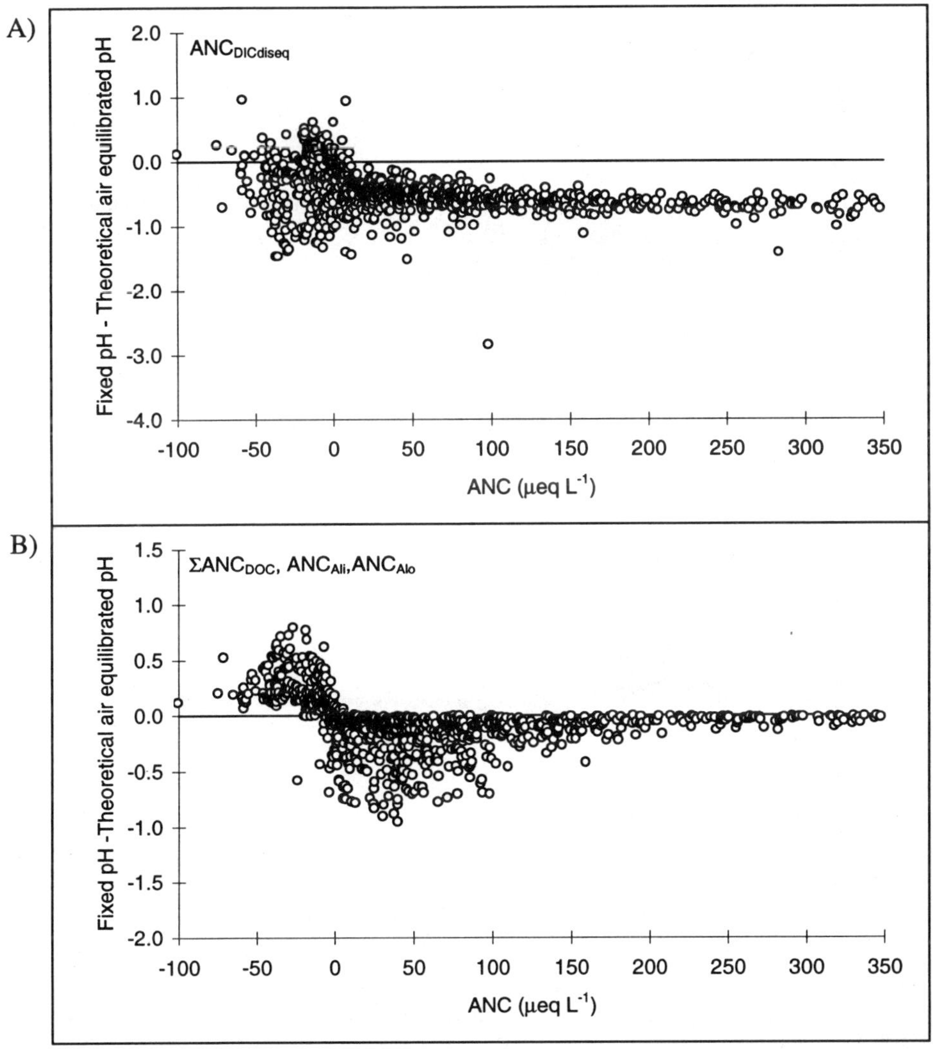

Fig. 3.7. The Norwegian lake survey, 1995. Influence of (A) volatile, mostly CO_2 ($ANC_{DICdiseq}$) and non-volatile acids and bases, primarily organic acids (ANC_{DOC}) and inorganic Al (ANC_{Ali}) on solution pH. Data from Skjelkvaale et al. (1997).

in pH are primarily because of dissolution/hydrolysis of inorganic Al (ANC_{Ali}), which illustrates the pH buffering role of Al. ANC_{Ali} is most important for solution pH in water systems with negative ANC values. This is generally in lakes with both low pH and TOC concentrations(Fig. 3.9B). Low pH is essential for significant Al dissolution. In the Norwegian lakes, organic acids depress water pH up to about 1 pH unit in the ANC range of 0–50 μeq/L. Within the same ANC range, Baker et al. (1990) concluded that organic acids depress the pH of Adirondack lakes by 0.5–2.5 pH units. They

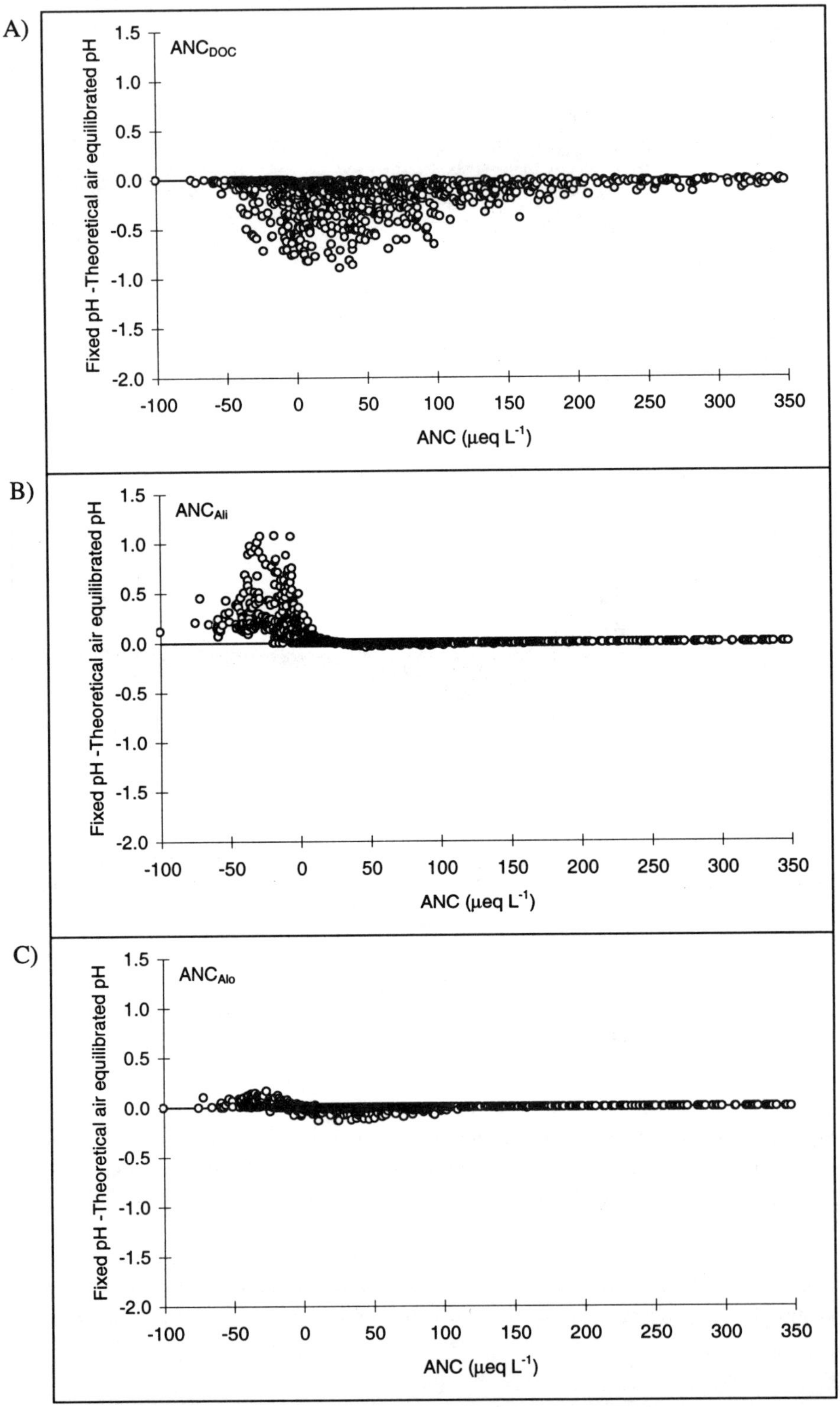
A)
ANC$_{DOC}$
Fixed pH - Theoretical air equilibrated pH
1.5
1.0
0.5
0.0
-0.5
-1.0
-1.5
-2.0
-100
-50
0
50
100
150
200
250
300
350
ANC (μeq L^{-1})
B)
ANC$_{Ali}$
Fixed pH - Theoretical air equilibrated pH
ANC (μeq L^{-1})
C)
ANC$_{Alo}$
Fixed pH - Theoretical air equilibrated pH
ANC (μeq L^{-1})

← **Fig. 3.8.** Norwegian lake survey 1995. A Influence of organic acids (ANC_{DOC}), B inorganic Al hydrolysis (ANC_{Ali}) and C Al complexation with organics (ANC_{Alo}) on solution pH. The calculation is made according to pK and CD values of organics and hydrolysis and Al complexing constants presented in Alchemi Version 4.0 (Schecher and Driscoll 1987, 1988). (Data from Skjelkvaale et al. 1997)

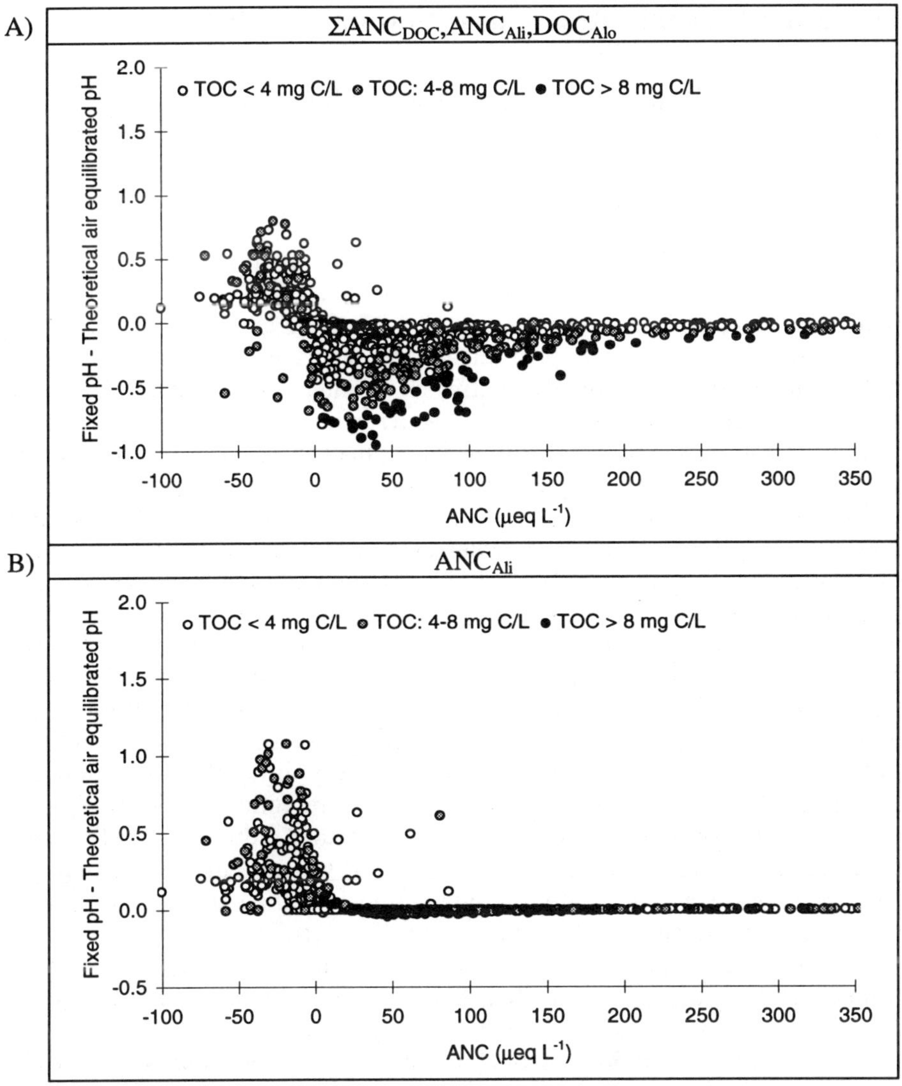

Fig. 3.9. Norwegian lake survey 1995. Influence of ANC factors on solution pH for three concentration categories of TOC. A ΣANC_{DOC}, ANC_{Ali}, ANC_{Alo}; B ANC_{Ali} only. (Data from Skjelkvaale et al. 1997)

further concluded that organic acids could explain over 90% of the H^+ in 41% of the 1469 lakes studied. As also well illustrated in Figs. 3.7. and 3.8, at ANC > 100 μeq/L, the inorganic carbon buffering system (ANC_{DIC}), often termed the bicarbonate system, really starts to predominate. Therefore, at high ANC-values, other ANC factors are of minor importance.

3.6 Cation Exchange Reactions with Humic Matter

Because the organic humic material to a large extent acts as a complex acidic cation exchange resin, cation-exchange (CE) reactions are predominant processes for humic material in soil and DOC in soilwater and surfacewater. However, significant changes in water chemistry by CE do primarily occur as a result of extreme physico-chemical impacts on the catchment. It is therefore difficult to reveal significant effects of CE on solution pH by large lake surveys, where only one sample is taken from each lake more or less at the same time. This will only give a static picture of the chemistry of lakes. An illustration of extreme CE impact on surface water chemistry is sea-salt episodes where high inputs of pH-neutral NaCl enter the catchment. In many areas this process is relevant to surface water acidification, because a certain amount of cations in sea-salts, primarily Na^+, may exchange for other cations in the soil. In acid soils, significant amounts of the exchanged cations will be the acid cations H^+ and ΣAl^{n+} (Wicklander 1975; Skartveit 1980; Reuss and Johnson 1985; Hindar et al. 1994; Lydersen and Henriksen 1995). A chemical description of a seasalt episode might be obtained by evaluating the concentration of Na relative to Cl. When the equivalent ratio of Na/Cl is less than that of seawater, the concentration of non-marine Na is negative, which indicates that CE reactions have occurred. Divalent hardness metals (Ca^{2+} and Mg^{2+}) are concluded being much more successful in the competition compared with common monovalent cations such as Na^+ and K^+. (Pagenkopf, 1983). However, to quantify this difference is very difficult since cation exchangers exhibit low or moderate preference for one cation species compared with another (Bolt 1979), a preference that is further reduced when the temperature is low (Boyd 1970), as in most natural waters. This means that the concentration of a cation often is more important than the type of cation concerning the possibility of taking part in CE reactions.

Several cases of episodic acidification of stream water have been reported from coastal regions of Norway (e.g. Skartveit 1980; Mulder et al. 1990; Hindar et al. 1994; Lydersen and Henriksen 1995) and eastern USA (Kahl et al. 1985; Sullivan et al. 1988). Particularly during autumn rainstorms, when seasalt-enriched precipitation is percolating acidic soils, the acidification of surface water coincided with retention of Na ions. Although episodic acidification of surface water due to NaCl deposition may occur, there is so far little support for the neutral sea-salt effect being an important long-term acidifying process (Seip 1980; Sullivan et al. 1988). Because short-term acidification of streams and lakes often is the main cause of fish kill and other undesirable biological effects (Schofield 1977; Muniz and Leivestad 1980; Harvey and Whelpdale 1986), sea-salt episodes may play an important role concerning fish kill in surface waters (Hindar et al. 1994).

3.7 Aluminium Toxicity to Biota and Effects of DOC

Acidification of soil water systems often causes an increased transfer of Al from edaphic to aquatic environments (e.g. Cronan and Schofield 1979, Dickson 1980, Seip et al. 1989) primarily by dissolution and CE processes. It is well known that Al might be present in toxic amounts in soils (e.g. Hartwell and Pember 1918; Magistad 1925; Hardy 1926) and aquatic environments (Penny and Adams 1863; Weigelt et al. 1885; Thomas 1915; Ebeling 1928; Oshima 1931; Ellis 1937, Sanborn 1945; Pulley 1950; Murdock 1953; Wallen et al. 1957; Jones 1964; Dickinson Burrows 1977). In general, these older studies have been performed under conditions where the Al concentration exceeded the solubility of Al and little attention has been given to the effects of precipitates. Furthermore, the studies have often been performed at pH levels where pH itself could have been the lethal factor. Recent concerns about the effects of soil acidification on the biogeochemistry of Al have broadened the scope of interest in Al toxicity, and the recent toxicity studies have therefore been performed at more relevant concentrations of Al, i.e. at concentrations in agreement with values reported for many acidified surface waters (< 10 µmole Al/L).

Even though Al toxicity to fish was demonstrated almost 70 years ago (Ebeling 1928), Schofield (1977) was the first to report the link between acidic Al-rich freshwater and fish mortality. Similar results were obtained independently by Dickson (1978). From that time, numerous scientists have observed toxic effects of Al on many species of freshwater fishes. Toxicity of Al has also been reported in oversaturated solutions where unstable inorganic Al chemistry is present, as after liming of acidic Al-rich waters (e.g. Dickson 1978). Lydersen (1991) found that generally, for all later corresponding studies, high toxicity in unstable Al-chemistry water only occurs when pH is raised compared with the initial water pH. Thus, he concluded that there has to be a prehistory of lower pH for significant amounts of unstable toxic Al polymers to form. This was later supported by Ougthon et al. (1992) by use of ^{26}Al-tracer. They found that Al on gills of fish exposed to limed water only derived from Al present in the acidic water where Al primarily was present as inorganic low molecular weight species prior to liming. Later studies have supported these observations and further concluded that the most toxic Al conditions exist soon after a pH increase in Al-rich acidic waters (Weatherley et al. 1991; Rosseland et al. 1992; Poléo et al. 1994; Lydersen et al. 1994), and that the toxicity decreased by ageing of the polymers (Lydersen et al. 1994). Today, there is no doubt that the increased concentration of inorganic Al during soil and surface water acidification is the primary cause of fish kill in acidic freshwaters, and that highly toxic conditions may occur soon after a pH increase (e.g. by liming) of acidic Al-rich waters.

Al toxicity to aquatic organisms other than fish has also been reported. Several studies have demonstrated toxicity of inorganic Al to amphibians (e.g. Clark and Hall 1985; Clark and LaZerte 1985; Cummins 1986; Leuven et al. 1986; Freda and McDonald 1990; Freda et al. 1990). Others have shown that Al has lethal and sub-lethal effects on several invertebrates (Correa et al. 1985; Malley and Chang 1985; Burton and Allan 1986; Fjeld et al. 1988). Fjeld et al. (1988), and McCahon et al. (1989) observed increased Al on gills of invertebrates (crayfish and zooplankton). Many have documented Al toxicity to different species of green algae (e.g. Helliwell et al. 1983; Folsom et al. 1986; Törnqvist 1989), while Gensemer (1991) showed Al toxicity to an acidophilic diatom.

It is mainly strong mineral acids (primarily H_2SO_4) through acid rain that most frequently have been the focus as the main cause of soil acidification and subsequently dissolution of inorganic Al. This is understandable, because there is a co-occurrence in time and space between increased acid rain input and severe decline in many fish populations in large acid sensitive areas of the northern hemisphere. Natural acidic water systems with high concentration of inorganic Al also exist, despite minor or no impacts from acid rain (Krug 1991). These lakes might contain both low and high concentrations of DOC. A major difference between these naturally acidic environments and the acid-rain-acidified environments, however, is the biological possibility of physiological adaptation. In natural acidic environments, the biology may have gradually been adapted to the water chemical changes through hundreds or maybe thousands of years. The significant changes in water chemistry during a few decades as a result of strong impacts from acid rain have proceeded too fast in relation to the time many aquatic organisms need to adapt. Accordingly, large biological effects have been documented as a result of acid rain impact.

There are only a few studies concerning Al toxicity when Al is complexed to natural humic substances. Several works have, however, demonstrated that complexation of Al with citrate essentially eliminates toxic effects of Al to fish (e.g. Baker and Schofield 1980, 1982; Driscoll et al. 1980; Fivelstad and Leivestad 1984). Skogheim et al. (1986) found that the toxicity of Al to salmon was significantly reduced when Al was complexed to humic material. The same conclusion was drawn for stream invertebrates when organic substances where added to the water (Burton and Allan 1986). Later, Lydersen et al. (1990) and Witters et al. (1990) both observed no toxic effect of Al (within 10 days) when complexed to humic substances, even at relatively high Al organic concentrations (> 7–8 μmole/L).

Many attempts to relate toxicity to levels of Al in aquatic systems have met with variable degree of success, probably due to insufficient understanding of the importance of Al speciation when assessing its biological availability and toxicity. The toxicity of Al is highly dependent on the Al species present where pH, ionic strength, type of ligands and water temperature are essential factors. Based on present knowledge, it is reasonable to conclude that *acute* Al toxicity primarily is related to inorganic Al chemistry, at least for aquatic gill organisms, while organic Al-complexes per se have at least *no acute* biological effect. Nevertheless, organic Al complexes might represent a potential toxic pool, because the equilibrium between inorganic and organic Al species may change due to other physico-chemical factors. Relevant examples of this are acidic episodes related to seasalt events (Section 3.6), where Na^+ ions are able to cation exchange for H^+ and inorganic Al (Al^{n+}) complexed to organic sites.

There are several plausible reasons why organic Al-complexes exhibit no acute toxic effect to aquatic gill organisms. The gill surface is covered with a layer of protecting mucus, consisting of glycoproteins, mycopolysaccarides, amino acids (Fletcher et al. 1976, Van de Winkel et al. 1986, Lumsden and Ferguson 1994) and 95% water (Wold and Selset 1977). The glycoproteins are rich in sialic acid with a pK value less than 3 (Clamp et al. 1978), which means that at most freshwater pHs sialic acid is primarily negatively charged. This means that both the gill surface and DOC act as complex acid cation exchange resins. Accordingly, they compete for cations present in solution. In other words, more Al is able to bind to gill surfaces in solution with no or low concentrations of DOC. In waters with high concentrations of DOC, however, a major part of Al is already complexed to DOC when it enters into surface waters where the gill organisms normally live. Accordingly, the strength of the cation exchanger present on the gill surface is decisive to which extent Al ions complexed to DOC are able to resist this competition. Lydersen et al. (1990) cation exchanged water from a natural DOC rich lake with high Al concentrations by use of a strong acidic Amberlite IR-120 resin, according to the procedure described by Driscoll (1984). The fish were then exposed to the eluate. Even though this is an operational technique, most likely the predominant part of the Al present in the eluate is Al organic complexes, which have passed through the cation exchanger and still kept their Al complexed to DOC. Because no fish died (during 10 days of exposure) when exposed to the eluate, it is reasonable to conclude that Al organic complexes have no acute toxic effects on fish. The reason why this analytical cation exchange procedure has been shown to be so applicable to distinguish between toxic and non-toxic Al forms has to be related to the cation exchange properties on the surface of fish gills. This confirms that acute Al toxicity primarily is related to cationic and thereby predominantly inorganic Al forms.

There are also other works within other scientific fields that may support the hypothesis of cationic properties of inorganic Al and toxicity. Numerous cationic polymers are known being highly acute toxic to fish (Biesinger and

Stokes 1986; Goodrich et al. 1991; Scott Hall and Mirenda 1991). The physiological and pathological description of fish exposed to many different cationic electrolytes are almost identical to the description of Al toxicity to fish (e.g. compare Biesinger and Stokes 1986 with Rosseland et al. 1992 and/or Poléo et al. 1994). Cationic polyelectrolyte studies conclude that the lowest molecular weight polymers (e.g. Goodrich et al. 1991) and the cationic polymers with highest positive charge densities (Scott Hall and Mirenda 1991) are the most toxic. This agrees well with the chemical and biological relationships in the Al toxicity study reported by Lydersen et al. (1994). There is good evidence that the toxicity of cationic polymers can be eliminated by addition of an anionic polymer (Biesinger and Stokes 1986). Accordingly, as also proposed by them, natural humic acids and clays might be used for detoxifying cationic Al polymers.

There may also be other possible reasons why DOC may eliminate Al toxicity. Karlsson-Norrgren et al. (1986) concluded that even though humic substance additions reduced or totally inhibited gill lesions among brown trout exposed to Al at pH 5.5, there was no significant difference in bioaccumulation compared to control fish exposed to Al without added humic substances. The reduced or eliminated toxicity of Al to fish in the presence of dissolved organic carbon could potential result from the changed coordination chemistry of the Al bound to the gill surface (Wilkinson et al. 1993). On the other hand, binding of organic acids at the gill surface itself can be expected to induce physiological effects due to modifications in the

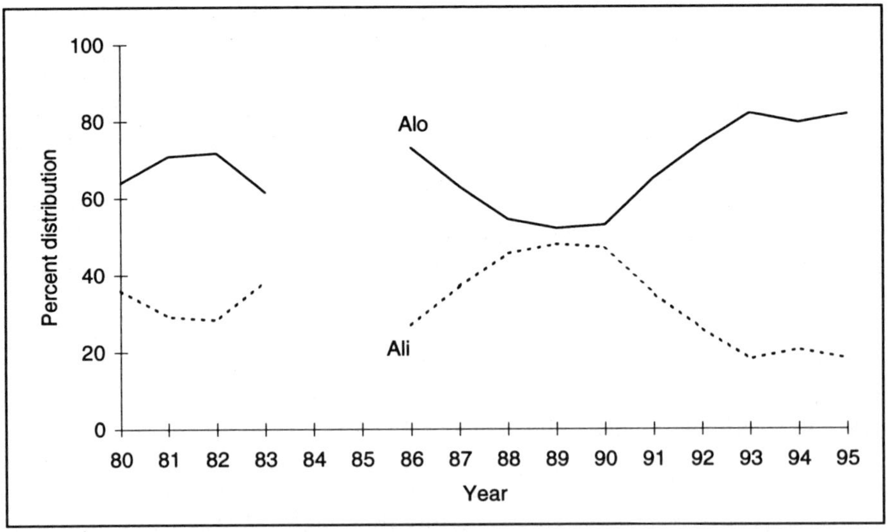

Fig. 3.10. Distribution between inorganic Al (Al_i) and organic Al (Al_o) in Lake Langtjern during 1980–1995. During same period, no significant change in RAl ($Al_i + Al_o$) has occurred. During same period SO_4^{2-}in runoff has decreased by ≈ 40%, which further has caused an increase in ANC and a slight increase in pH. TOC in Lake Langtjern has not changed during same period. Average weighted TOC is 9.1 ± 1.1 mg C/L. (Data are taken from Skjelkvaale 1996)

structure, fluidity, or permeability of the gill membrane which may contribute to increased tolerance towards Al toxicity (Visser 1982). DOC generally plays an important role in reducing or completely detoxifying toxic cations in water. This means that reactions responsible for death of fish exposed to metals often take place outside the body, i.e. primarily at the surface of gills. This was already claimed by Erichsen Jones in 1939. Accordingly, the cation exchange capacity (CEC) properties of DOC are of major importance to reduce the negative impacts of CEC reactions with toxic cations at the gill surface. Since acute Al toxicity primarily is related to inorganic Al species, it is of major interest that it seems to occur a decrease in inorganic Al (Al_i) and accordingly an increase in organic Al-forms (Al_o) in many surface waters during the last years. This is a result of declining SO_2-emission and thereby reduced inputs of SO_4^{2-}(Fig. 3.10).

3.8 Summary

A significant part of dissolved humic material (DOC) is organic acids. Accordingly, many studies have been conducted on these acids, particularly regarding their role in water pH, their pH-buffering capacity and metal-complexing properties. Due to the extreme complexity of ligands in humic substances, a large part of DOC is an organic acid cocktail with numerous pK and CD values. Thus, many different methods have been used in order to find one to three average pK values and a fixed number of acid functional groups (CD), often fitted to each actual data set. In later years, it has been relatively well documented that an important fraction of the organic acids have pK values < 3, and accordingly, several models operate with a first dissociation constant mirroring this fact. The other dissociation constants (often 1 or 2) are normally fitted to actual data sets, and thereby the pH of the waters incorporated (normally within pH 4–7). Accordingly, fitted pK values normally vary within the pK interval of 4–7. Based on the literature, the concentration of organic acids (CD) varies from 5 to 22 μeq/mg C, often with an average CD of carboxylic acids of ≈ 10 μeq/mg C.

Until recently, organic acids were often considered unimportant for the acidity of surface waters strongly affected by strong acid inputs. More recent studies, however, clearly demonstrate the importance of organic acids in modifying both the acidity of surface waters and the response to changes in strong acid inputs in waters with low or no bicarbonate alkalinity. Large lake surveys show that organic acids are able to depress water from 0.5 to 2.5 pH units in the ANC range of 0–50 μeq/L, which means that organic acids might explain a significant part of the pH of natural waters. However, it should not be forgotten that organic acids at the same time are highly important to prevent the system from further pH depressions, e.g. by strong acids in precipitation, through their large buffering capacity.

Water pH depends on numerous interactions between inorganic and organic compounds in different soil horizons, where types and volumes of the soil horizons, quantity and quality of precipitation, temperature, runoff pattern, origin of organic compounds and stage of degradation/humification are essential factors. The interplay between such factors is decisive for the water pH. Thus, it seems reasonable that an essential part of the organic acids in solution have pK values close to the water pH. Because significant buffering capacity only occurs within $\pm$ 1 pH unit of each pK value, it means that DOC plays a significant pH-buffering role for each unique system where DOC is present. Within realistic acidification levels, acidification normally leads to small changes in the transparency of water, water colour and concentration of DOC in lakes. The only significant change in DOC generally observed due to acidification is a decrease in $[A^-]$, reflecting the acid-neutralising capacity of DOC.

Since organic humic material is a complex acid cation exchange resin, CE reactions often predominate in soil and surface waters. Results of CE processes have been well documented during seasalt episodes in coast-near areas where significant amounts of Na^+ (from NaCl) have been cation exchanged for H^+ and inorganic Al (ΣAl^{n+}). Undesirable biological effects, such as fish kill, have been well documented during such episodes. However, it is reasonable to assume that such episodes are unimportant for the long-term acidifying process in catchments.

DOC generally plays an important role in reducing or completely detoxifying toxic cations in water. This is well documented for Al, where acute toxicity only is reported for different inorganic forms. However, organic metal complexes, like Al organic complexes, might represent a potential toxic pool, because the equilibrium between inorganic and organic species may change according to water physico-chemical changes, as during sea-salt episodes. Within the DOC-concentration of most natural surface waters, it is unlikely that humic acids, fulvic acids or other species of natural organic matter themselves are toxic.

References

Aiken GR (1988) A critical evaluation of the use of macroporous resins for the isolation of aquatic humic substances. In: Frimmel FH Christman RF (eds) Humic substances and their role in the environment. Wiley, New York, pp 15–28

Almer B, Dickson W, Ekström C, Hornström E, Miller U (1974) Effects of acidification of Swedish lakes. Ambio 3:30–36

Baker JP, Schofield CL (1980) Aluminium toxicity to fish related to acid precipitation and Adirondack surface water quality. In: Drabloes D Tollan A (eds) Ecological impacts of acid precipitation, SNSF-project. Proc Int Conf Sandefjord, Norway. Grefslie Ltd, Mysen, pp 292–293

Baker JP, Schofield CL (1982) Aluminium toxicity to fish in acidic waters. Water Air Soil Pollut 18:289–309

Baker JP, Gherini SA, Christensen SW, Driscoll CT, Gallagher J, Munson RK, Newton RM, Reckhow KH, Schofield CL (1990) Adirondack lakes survey: interpretive report, Adirondack Lakes Survey, Ray Brook, New York

Beck KC, Reuter JH, Perdue EM (1974) Organic and inorganic geochemistry of some coastal plain rivers of southeastern United States. Geochim Cosmochim Acta 38:341–364

Biesinger KE, Stokes GN (1986). Effects of synthetic polyelectrolytes on selected aquatic organisms. J Water Pollut Contam Fed 58:207–213

Bolt GH (1979) Soil chemistry. B. Physico-chemical models. Developments in soil science, vol 5B. Elsevier, Amsterdam

Boyd GE (1970) Thermal effects in ion-exchange reactions with organic exchangers: Enthalpy and heat capacity changes. In: Hall GR (ed) Ion exchange in the process industries. Conf at Imperial College of Science and Technology, London, 16–18 July 1969. Society of Chemical Industry, London, pp 261–269

Burton TM, Allan JW (1986) Influence of pH, aluminium, and organic matter on stream invertebrates. Can J Fish Aquat Sci 43:1285–1289

Christophersen N, Stuanes AO, Wright RF (1982) Runoff chemistry at a mini-catchment watered with "unpolluted precipitation". Nord Hydrol 13:115–128

Clamp JR, Allen A, Gibbons RA, Roberts GP (1978) Chemical aspects of mucus. Br Med Bull 34: 25–41

Clark KL, Hall RJ (1985) Effects of elevated hydrogen ion and aluminium concentrations on the survival of amphibian embryos and larvae. Can J Zool 63:116–123

Clark KL, LaZerte BD (1985) A laboratory study of the effects of aluminium and pH on amphibian eggs and tadpoles. Can J Fish Aquat Sci 42:1544–1551

Correa M, Coler RA, Yin C-M, Venables BJ (1985) The impact of depressed pH and elevated aluminium concentrations on specific dynamic action in *Somatochlora cingulata* (De Selys). Comp Biochem Physiol [C] 82:199–201

Cronan CS, Aiken GR (1985) Chemistry and transport of soluble humic substances in forested watersheds of the Adirondack Park, New York. Geochim Cosmochim Acta 49:1697–1705

Cronan CS Schofield CL (1979) Aluminium leaching response to acid precipitation: effects on high-elevation watersheds in the northeast. Science 204:304–306

Cummins CP (1986) Effects of aluminium and low pH on growth and development in *Rana temporaria* tadpoles. Oecologia 69:248–252

Davis RB, Anderson DS, Berge F (1985) Palaeolimnological evidence that lake acidification is accompanied by loss of organic matter. Nature 316:436–438

Dempsey BA O'Melia CR (1983) Proton and calcium complexation of four fulvic acid fractions.

In: Christman RF Gjessing ET (eds). Aquatic and terrestrial humic materials. Ann Arbor Sci, Michigan, pp 239–274

Dickinson Burrows W (1977) Aquatic aluminium: chemistry, toxicology, and environmental prevalence. CRC Crit Rev Environm Control, vol 7, pp 167–216

Dickson W (1978) Some effects of the acidification of Swedish lakes. Verh Int Verein Limnol 20:851–856

Dickson, W (1980) Properties of acidified waters. In: Drabloes D Tollan A (eds.), Ecological impacts of acid precipitation, SNSF-project. Proc Int Conf Sandefjord, Norway. Grefslie Ltd, Mysen, pp 75–83

Driscoll CT (1984) A procedure for the fractionation of aqueous aluminium in dilute acidic water. Int J Environ Anal Chem 16:267–283

Driscoll CT, Baker JP Jr, Bisogni JJ, Schofield CL (1980) Effects of aluminium speciation on fish in dilute acidified waters. Nature 284:161–164

Driscoll CT, Lehtinen MD, Sullivan TJ (1994) Modeling the acid-base chemistry of organic solutes in Adirondack, New York, lakes. Water Resour Res 30:297–306

Eary LE, Jenne EA, Vail LW, Girvin, DC (1989) Numerical models for predicting watershed acidification. Arch Environ Contam Toxicol 18:29–53

Easthouse KB, Mulder J, Christophersen N, Seip HM (1992) Dissolved organic carbon fractions in soil and stream water during variable hydrological conditions at Birkenes, southernmost Norway. Water Resour Res 28:1585–1596

Ebeling G (1928) Über die Giftigkeit einiger Schwermetallsalze an Hand eines Falles aus der Praxis. Z Fischerei 26:49–61

Ellis MM (1937) Detection and measurement of stream pollution. Bull 22, US Bureau of Fisheries. Bull Bur Fish Bull 48:365–437

Ephraim JH, Reddy MM, Marinsky JA (1991) Ion binding by humic substances: considerations based on the solution chemistry and heterogenety of humic substances. In: Allard B, Borèn H, Grimvall A (eds) Humic substances in the aquatic and terrestrial environment. Lecture notes in earth science, vol 33, Springer, Berlin Heidelberg New York, pp 263–276

Erichsen Jones JR (1939) The relation between the electrolytic solution pressures of the metals and their toxicity to the stickleback (*Gasterosteus aculeatus* L.). J Exp Biol 16:425–437

Felbeck GT Jr (1971) Structural hypotheses of soil humic acids. Soil Sci 111:42–48

Fivelstad S, Leivestad H (1984) Aluminium toxicity to Atlantic salmon (*Salmo salar* L.) and brown trout (*Salmo trutta* L.): mortality and physiological response. Inst Freshwater Res, Drottningholm, vol 61:69–77

Fjeld E, Hessen DO, Roos N, Taugbøl T (1988) Changes in gill ultrastructure and haemolymph chloride concentrations in the crayfish, *Astacus astacus*, exposed to de-acidified aluminium-rich water. Aquaculture 72:139–150

Fletcher TC, Jones R, Reid L (1976) Identification of glycoproteins in the goblet cells of epidermis and gill of plaice (*Pleuronectes platessa* L.) flounder (*Platichtys flesus* (L.)) and rainbow trout (*Salmo gairdneri* Richardson). Histochem J 8:597–608

Folsom BR, Popescu NA, Wood JM. (1986) Comparative study of aluminium and copper transport and toxicity in an acid-tolerant freshwater green alga. Environ Sci Technol 20:616–620

Freda J, McDonald DG. (1990) Effects of aluminium on the leopard frog, *Rana pipiens*: life stage comparisons and aluminium uptake. Can J Fish Aquat Sci 47:210–216

Freda J, Cavdek V, McDonald DG (1990) Role of organic complexation in the toxicity of aluminium to *Rana pipiens* embryos and *Bufo americanus* tadpoles. Can J Fish Aquat Sci 47: 217–224

Fuller RD, Simone DM, Driscoll CT (1988) Forest clearcutting effects on trace metal concentrations: spatial patterns in soil solutions and streams. Water Air Soil Pollut 40:185–195

Galloway JN, Norton SA, Church MR (1983) Fresh water acidification from atmospheric deposition of sulphuric acid: a conceptual model. Environ Sci Technol 17:541–545

Gensemer RW (1991) The effects of pH and aluminium on the growth of the acidophilic diatom *Asterionella ralfsii* var. *americana*. Limnol Oceanogr 36:123–131

Gjessing ET (1976) Origin, formation and distribution of humus. In: Gjessing ET (ed) Physical and chemical characteristics of aquatic humus, Ann Arbor Sci, Michigan, pp 3–10

Goodrich, MS Dulak LH, Friedman MA, Lech JJ (1991). Acute and long-term toxicity of water soluble cationic polymers to rainbow trout (*Oncorhynchus mykiss*) and the modification of toxicity by humic acid. Environ Toxicol Chem 10:509–515

Gran G (1952) determination of the equivalence point in potentiometric titrations II. Analyst 77:661–671

Hardy F (1926) The role of aluminium in soil infertility and toxicity. J Agric Sci 16:616–631

Hartwell BL, Pember FR (1918) The presence of aluminium as a reason for the difference in the effect of so-called acid soil on barley and rye. Soil Sci 6:259–279

Harvey HH, Whelpdale DM (1986) On the prediction of acid precipitation events and their effects on fishes. Water Air Soil Pollut 30:579–586

Hayes MHB (1991) Influence of the acid/base status on the formation and interactions of acids and bases in soils. In: Ulrich B, Sumner ME (eds) Soil acidity. Springer, Berlin Heidelberg New York, pp 80–96

Hayes MHB, Swift RS (1978) The chemistry of soil organic colloid. In: Greenland DJ, Hayes MHB (eds) The chemistry of soil constituents. Wiley, Chichester, pp 179–320

Hedin LO, Likens GE, Kimberley MP, Driscoll CT (1990) A field experiment to test whether organic acids buffer acid deposition. Nature 345:798–800

Helliwell S, Batley GE, Florence TM, Lumsden BG (1983) Speciation and toxicity of aluminium in a model fresh water. Environ Technol Lett 4:141–144

Henriksen A, Seip HM (1980) Strong and weak acids in surface waters of southern Norway and southwestern Scotland. Water Res 14:809–813

Henriksen A, Skjelkvaale BL, Lien L, Traaen TS, Mannio J, Forsius M, Kämäri J, Mäkinen I, Berntell A, Wiederholm T, Wilander A, Moiseenko T, Lozovik P, Filatov N, Niinioja R, Harriman R, Jensen JP (1996) Regional lake surveys in Finland, Norway, Sweden, Northern Kola, Russian Karelia, Scotland, Wales 1995. Coordination and design. Report 40/1996, Serial No 3420-1996. Norwegian Institute for Water Research, Oslo.

Hindar A, Henriksen A, Toerseth K, Semb A (1994) Acid water and fish death. Nature 372: 327–328

Hope D, Kratz TK, Riera JL (1996) Relationship between PCO_2 and dissolved organic carbon in northern Wisconsin lakes. J Environ Qual 25:1442–1445

Johansson A (1970) Automatic titration by stepwise addition of equal volumes of titrant. Analyst 95:535–540

Jones JRE (1964) Fish and river pollution. Butterworth, Washington, DC

Karlsson-Norrgren L Björklund I Ljungberg O Runn P (1986) Acid water and aluminium exposure: experimentally induced gill lesions in brown trout, *Salmo trutta* L. J Fish Dis 9:11–25

Kahl JS, Anderson JL, Norton SA (1985) Water resource baseline data and assessment of impacts from acidic precipitation. Acadia National Park, Maine. Technical Report 16. National Park Service, North Atlantic Region, Boston

Kononova MM (1975) Humus of virgin and cultivated soils. In: Gieseking JE (ed) Soil components, vol 1. Springer, Berlin Heidelberg New York, pp 475–526

Kramer JR, Davies SS (1988) Estimation of non-carbonato protolytes for selected lakes in the eastern lakes survey. Environ Sci Technol 22:182–185

Krug EC (1991) Review of acid-deposition-catchment interaction and comments on future research needs. J Hydrol 128:1–27

Krug EC, Frink CF (1983) Acid rain on acid soil: a new perspective. Science 221:520–525

Krug EC, Isaacson PJ (1984) Comparison of water and dilute acid treatment on organic and inorganic chemistry of leachate from organic rich horizons of an acid forest soil. Soil Sci 137:370–378

Krug EC, Isaacson PJ, Frink CR (1985) Appraisal of some current hypotheses describing acidification of watersheds. J Air Pollut Control Assoc 35:109–114

Lee YH, Brosset C (1978) The slope of Grans's plot: a useful function in the examination of precipitation, the water-soluble part of airborne particles, and lake water. Water Air Soil Pollut 10:457–469

Leenheer JA, Huffman EWD (1979) Analytical method for dissolved organic carbon fraction. US Geol Surv Water Resour Invest, Report 79–4

Leenheer JA, Wershaw RL, Reddy MM (1995) Strong-acid, carboxyl-group structures in fulvic acid from the Suwannee River, Georgia. I. Minor structures. Environ Sci Technol 29:393–398

Leuenberger B, Schindler PW (1986) Application of integral pK spectrometry to the titration curve of fulvic acids. Anal Chem 58:1471–1474

Leuven RSEW, Den Hartog C, Christiaans MMC, Heijligers WHC (1986) Effects of water acidification on the distribution pattern and the reproductive success of amphibians. Experientia 42:495–503

Logen RM, Derby JC, Duncan LC (1982) Acid precipitation and lake susceptibility in the central Washington cascades. Envrion Sci Technol 16:771–775

Lumsden JS, Ferguson HW (1994) Isolation and partial characterisation of rainbow trout (*Oncorhynchus mykiss*) gill mucin. Fish Phys Biochem 12:387–398

Lydersen E (1991) Aluminium in dilute acidic freshwaters. Chemical, analytical and biological relevance. PhD thesis, University of Oslo

Lydersen E, Henriksen A (1995) Seasalt effects on the acid neutralising capacity of streamwaters in southern Norway. Nord Hydrol 26:369–388

Lydersen E, Polèo ABS, Muniz IP, Salbu B, Bjoernstad HE (1990) The effects of naturally occurring high and low molecular weight inorganic and organic species on the yolk-sack larvae of Atlantic salmon (*Salmo salar* L.) exposed to acidic aluminiumrich lake water. Aquat Toxicol 18:219–230

Lydersen E, Kroglund F, Nandrup Pettersen M, Polèo ABS, Rosseland BO, Riise G, Salbu B (1994) The importance of "in situ" measurements to relate toxicity and chemistry in dynamic aluminium freshwater systems. J Ecol Chem 3:357–265

Lydersen E, Fjeld E, Gjessing ET (1996) The humic lake acidification experiment (HUMEX): main physico-chemical results after five years of artificial acidification. Environ Int 22: 591–604

Magistad OC (1925) The aluminium content of the soil solution and its relation to soil reaction and plant growth. Soil Sci 20:181–225

Malley DF, Chang PSS (1985) Effects of aluminium and acid on calcium uptake by the crayfish, *Orconectes virilis*. Arch Environ Contam Toxicol 14:739–747

McCahon CP, Brown AF, Poulton MJ, Pascoe D. (1989) Effects of acid, aluminium and lime additions on fish and invertebrates in a chronically acidic Welsh stream. Water Air Soil Pollut 45:345–359

McColl JG, Pohlman AA (1986) Soluble organic acids and their chelating influence on Al and other metal dissolution from forest soils. Water Air Soil Pollut 31:917–927

Mulder J, Christophersen N, Haus M, Vogt RD, Andersen S, Andersen DO (1990) Water flow paths and hydrochemical controls in the Birkenes catchment as inferred from a rainstorm high in seasalts. Water Resour Res 26:611–622

Muniz IP, Leivestad H (1980) Acidification –effects on freshwater fish. In: Drabloes D, Tollan A (eds) Ecological impacts of acid precipitation, SNSF-project. Proc Int Conf Sandefjord, Norway. Grefslie Ltd, Mysen, pp 84–92

Munson RK, Gherini SA (1993) Influence of organic acids on the pH and acid-neutralising capacity of Adirondack lakes. Water Resour Res 29:891–899

Murdock HR (1953) Some data on toxicity of metals in wastes to fish life are presented. Ind Eng Chem 45:99A–102A

Norton SA, Henriksen A (1983) The importance of CO_2 in evaluation of effects of acidic deposition. Vatten 39:346–354

Oliver BG, Thurman EM, Malcolm RL (1983) The contribution of humic substances to the acidity of coloured natural waters. Geochim Cosmochim Acta 47:2031–2035

Oshima S (1931) On the toxic action of dissolved salts and their electrolytes upon young eels (*Anguilla japonica*). J Imp Fish Exp Sta 2:191–193

Oughton DH, Salbu B, Bjoernstad HE (1992). Use of aluminium-26 tracer to study the deposition of aluminium species on fish gills following mixing of limed and acidic waters. Analyst 117:619–621

Pagenkopf GK (1983) Gill surface interaction model for trace-metal toxicity to fishes: role of complexation, pH and water hardness. Environ Sci Technol 17:342–346

Penny C, Adams C (1863) Report of experiments made upon fish, and of observations in connection with alleged pollution of the river Leven by discharges from the dyeworks at Levenbank, Levenfield, Dillichip, and Dalmonock. River Pollution Commission: Evidence. Correspondence and Reports, vol 2, part 4, Scotland, pp 377–391.

Perdue EM, Lytle CR (1984) Distribution model for binding of protons and metal ions by humic substances. Environ Sci Technol 17:654–660

Perdue EM, Reuter JH, Parrish RS (1984) A statistical model of proton binding by humus. Geochim Cosmochim Acta 48:1257–1263

Polèo ABS, Lydersen E, Rosseland BO, Kroglund F, Salbu B, Vogt RD, Kvellestad A (1994) Increased mortality of fish due to changing Al-chemistry of mixing zones between limed streams and acidic tributaries. Water Air Soil Pollut 75:339–351

Pulley TE (1950) The effect of aluminium chloride in small concentration on various marine organisms. Texas J Sci 3:405–411

Reuss JO, Johnson DW (1985) Effect of soil processes on the acidification of water by acid deposition. J Environ Qual 14:26–31

Reuss JO, Johnson DW (1986) Acid deposition and the acidification of soils and waters. Ecological studies, vol 59, Springer, New York Berlin Heidelberg Tokyo

Ritchie GSP, Posner AM (1982) The effect of pH and metal binding on the transport properties of humic acids. J Soil Sci 33:233–247

Rosenqvist IT (1978) Alternative sources for acidification of river water in Norway. Sci Tot Environ 10:39–49

Rosseland BO, Blakar IA, Bulger A, Kroglund F, Kvellestad A, Lydersen E, Oughton DH, Salbu B, Staurnes M, Vogt RD (1992) The mixing zone between limed and acidic river waters: complex aluminium chemistry and extreme toxicity for salmonids. Environ Pollut 78:3–8

Sanborn NH (1945) The lethal effect of certain chemicals on fresh water fish. Canning Trade 67 (49):10–12

Schecher WD, Driscoll CT (1987) An evaluation of uncertainty associated with aluminium equilibrium calculations. Water Resour Res 23:525–534

Schecher WD, Driscoll CT (1988) An evaluation of the equilibrium calculations within acidification models: the effects of uncertainty in measured chemical compounds. Water Resour Res 24:533–540

Schindler DW, Turner MA (1982) Biological, chemical and physical responses of lakes to experimental acidification. Water Air Soil Pollut 18:259–271

Schnitzer M, Khan SU (1972) Humic substances in the environment. Dekker, New York

Schofield CL (1977) Acid snow-melt effects on water quality and fish survival in the Adirondack Mountains of New York State, US Research Technical Completion Report A-072-NY. Office of Water Research and Technology, Dept of the Interior, Washington, DC

Scott Hall, W Mirenda RJ (1991). Acute toxicity of wastewater treatment polymers to *Daphnia pulex* and the fathead minnow (*Pimephales promelas*) and the effects of humic acid on polymer toxicity. J Water Pollut Contam Fed 63:895–899

Seip HM (1980) Acidification of freshwater –sources and mechanisms. In: Drabloes D, Tollan A (eds) Ecological impacts of acid precipitation, SNSF-project. Proc Int Conf Sandefjord, Norway. Grefslie Ltd, Mysen, pp 358–366

Seip HM, Andersen DO, Christophersen N, Sullivan TJ, Vogt RD. (1989) Variations in concentrations of aqueous aluminium and other chemical species during hydrological episodes at Birkenes, southernmost Norway. J Hydrol 108:387–405

Skartveit A (1980) Observed relationships between ionic composition of precipitation and runoff. In: Drabloes D, Tollan A (eds) Ecological impacts of acid precipitation, SNSF-project. Proc Int Conf Sandefjord, Norway. Grefslie Ltd, Mysen, pp 242–243.

Skjelkvaale BL (1996) Monitoring of long-range transported polluted air and precipitation (in Norwegian). Norwegian State Pollution Control Authority (SFT), Oslo, report 671/96

Skjelkvaale BL, Henriksen A, Faafeng B, Fjeld E, Traaen T, Lien L, Lydersen E, Buan AK (1997) Regional lake survey 1995. A water chemical survey of 1500 Norwegian lakes. (in Norwegian). Monitoring of long-range transported polluted air and precipitation. Norwegian State Pollution Control Authority (SFT), Oslo, report 677/96

Skogheim OK, Rosseland BO, Hoell E, Kroglund F (1986) Effects of humic acid on acute aluminium toxicity to smolts of Atlantic salmon (*Salmo salar* L.) in acidic soft water. In: Rosseland BO, Skogheim OK (eds) Acidic soft water and neutralisation: effects on fish physiology, fish toxicology and fish populations. Directorate for Nature and Management, Fish Research Division, Trondheim

Stevenson FJ (1982) Humus chemistry. Wiley-Interscience, New York

Stumm W, Morgan JJ (1981) Aquatic chemistry, Wiley, New York

Sullivan TJ, Driscoll CT, Eilers JM, Landers DH. (1988) Evaluation of the role of sea salt inputs in the long-term acidification of coastal New England lakes. Environ Sci Technol 22:185–190

Sullivan TJ, Driscoll CT, Gherini SA, Munson RK, Cook RB, Charles DF, Yatsko CP (1989). Influence of aqueous aluminium and organic acids on measurements of acid neutralising capacity of surface waters. Nature 338:408–410

Thomas A (1915) Effects of certain metallic salts upon fishes. Amer Fish Soc Trans 44:120–124

Törnqvist L (1989) Studies of aluminium toxicity to the green algae *Monoraphidium dybowskii* and *Stichococcus* sp. with emphasis on phosphate metabolism. Acta Univ Ups Abstr, Uppsala Diss Sci 193

Van de Winkel JG, Van Kuppevelt THMSM, Janssen HMJ, Lock RAC (1986) Glycosaminoglycans in the skin mucus of rainbow trout (*Salmo gairdneri*). Comp Biochem Physiol [B] 85: 473–475

Visser SA (1982) Surface active phenomena by humic substances of aquatic origin. Rev Fr Sci Eau 1:285–296

Wallen IE, Greer WC, Lasater R (1957) Toxicity to *Gambusia affinis* of certain pure chemicals in turbid waters. Sewage Ind Wastes 29:695–711

Weatherley NS, Rutt GP, Thomas SP, Ormerod SJ (1991) Liming acid stream: aluminium toxicity to fish in mixing zones. Water Air Soil Pollut 55:345–353

Weigelt C, Saare O, Schwab L (1885) Die Schädigung von Fischerei und Fischzucht durch Industrie- und Haus Abwässer. Arch Hyg 3:39–117

Wicklander L (1975) The role of neutral salts in the ion exchange between acid precipitation and soil. Geoderma 14:93–105

Wilkinson KJ, Jones HG, Campbell PGC, Lachance M (1992) Estimating organic acid contributions to surface waters acidity in Quebec (Canada). Water Air Soil Pollut 61:57–74

Wilkinson KJ, Bertsch PM, Jagoe CH, Campbell PGC (1993) Surface complexation of aluminium on isolated fish gills. Environ Sci Technol 27:1132–1138

Witters HE, Van Puymbroeck S, Vangenechten JHD, Vanderborght OLJ (1990) The effect of humic substances on the toxicity of aluminium to adult rainbow trout, *Oncorhynchus mykiss* (Walbaum). J Fish Biol 37:43–53

Wold CM, Selset R (1977) Glycoproteins in the skin mucus of the char (*Salmo alpinus* L.) Comp Biochem Physiol [B] 56:215–218

Wright RF (1983) Predicting acidification of North American lakes. Acid Rain Research, report 4/1983. Norwegian Institute for Water Research, Oslo

Wright RF (1989) Rain project: role of organic acids in moderating pH change following reduction in acid deposition. Water Air Soil Pollut 46:251–259

4 Climatic and Hydrologic Control of DOM Concentration and Quality in Lakes

P. Jefferson Curtis

4.1 Introduction

Dissolved organic matter (DOM) is a poorly-defined mixture of mainly naturally occurring substances in water. However, it has operational and qualitative properties that have significant effects on aquatic ecosystems. Operationally, DOM passes through most common filters and is composed mainly of organic matter. DOM is a significant source of carbon fueling microbial metabolism (Wetzel et al. 1995) and attenuates significantly the penetration of photosynthetically active and damaging ultraviolet radiation (Scully & Lean 1994; Schindler et al. 1996). Climate affects DOM by altering the rates or quality of DOM loading, and the rates of in situ loss or transformation (Schindler et al. 1992; Curtis & Schindler 1997). Concentrations of DOM in lake waters depend on loading, in situ loss and production, and on dilution or concentration from exchange of water directly with the atmosphere (precipitation and evaporation). Thus, quality of organic matter in surface waters depends on biologic and hydrologic properties of the system and on regional climate.

The quality of DOM is typically characterized by chemical and related optical properties. For example, DOM is characterized by its affinity to adsorb to synthetic resins, its elemental composition, its isotopic composition and "age" (^{14}C activity), the density of different functional groups and double bonding, and spectral attenuation of light and subsequent fluorescence. These qualitative properties have been linked to different sources of DOM (allochthonous and autochthonous), and to in situ transformation. Mechanistic interactions between climate and hydrologic setting are complicated, but empirical relationships between hydrologic properties and DOM concentration are well documented within climatic regions. For a few lakes, DOM has been monitored during a period of climatic change.

In this chapter, the effects of climate on DOM in aquatic systems are examined in three ways. First, DOM is compared among climatic zones. Second, DOM is compared among hydrologically diverse systems. Third, the effect of recorded climatic change on DOM is compared with that predicted from regional hydrology. Finally, the author speculates on the relative sensitivity of

Ecological Studies, Vol. 133
Hessen, Tranvik (eds) Aquatic Humic Substances

different aquatic systems to climatic effects on DOM, and identifies areas for future research.

4.2 DOM Concentration

Broad patterns of DOM concentration across climatic zones have been reported (Meybeck 1982); however, within-zone variation is extremely large, probably because of differences in production, loading, and flushing rates. For rivers, Meybeck (1982) concluded that concentrations were greatest in tiaga rivers, followed by rivers in the wet tropics, temperate, and tundra zones, respectively. Within zones, hydrologic conditions were deemed important modifiers of DOM concentration. For example, wetlands typically have the highest concentration (10–50 mg L^{-1}; Curtis and Bayley, in prep.) and alpine lakes the lowest (0.05–3.0 mg L^{-1}, Baron et al. 1991). These large differences in DOM concentration are partly caused by differences in the yield of DOM (g m^{-2} $year^{-1}$) from catchments, and partly by differences in the relative amounts of water.

The yield of DOM appears to be less variable among catchments than is the concentration of DOM. This observation was originally drawn from analyses of four hydrologically and physiographically different catchments and subbasins (Moeller et al. 1979). Subsequently, DOM yield measurements and empirical analyses suggest an average yield from terrestrial catchments of between 3 and 4.5 g m^{-2} $year^{-1}$ (Schlesinger and Melak 1981; Rasmussen et al. 1989; Schindler et al. 1992; calculations from Dillon and Molot 1997). Exceptions may be found where the organic content of soils is very low and/or where the contact time with water is low. Examples of such systems include high elevation, high latitude, and desert ecosystems.

The single largest variable affecting the yield of DOM from most catchments is the proportion of the catchment that is wetland (Urban et al. 1989; Dillon and Molot 1997). Wetlands are formed in depressions when there is sufficient moisture to saturate the soils. Drainage from organic-rich wetlands tends to have the highest concentrations of DOM in freshwater (Meybeck 1982; Dillon and Molot 1997). In contrast, drainage from organic-poor alpine and montane catchments has the lowest concentrations (Meybeck 1982; Baron et al. 1991). Thus, the organic content of the catchment appears to limit the concentration of DOM of inputs to surface water.

DOM yield is also probably modified by the flow-path of water. Groundwater has significantly lower DOM concentration relative to surface runoff. For example, in North America concentrations of DOM in groundwater average only a few milligrams per liter (Leenheer et al. 1974), whereas surface runoff is often five times richer. However, DOM in groundwater discharge to surface water can be increased by passing through shallow soils before discharge (Hinton et al. submitted; Schiff et al. 1997). Systematic comparisons have not been made, but it seems likely that seepage lakes would have lower concentrations of DOM than would comparable drainage lakes. Furthermore, the DOM in seepage lakes may be of different quality to that in drainage lakes. For example, DOM from seepage to rivers was significantly less colored than the highly stained DOM characteristic of runoff (Larson 1978). Thus, topography, surficial geology, and climate interact to modify the yield of DOM to lakes.

The concentration of DOM from terrestrial catchments is related to climate by the balance of moisture in catchments, or precipitation minus evapotranspiration. Thus the concentration of DOM in inputs from terrestrial catchments is determined from Eq. (1):

$$DOM = Y_{DOM}/(P\text{-}E_T), \tag{1}$$

where DOM is organic matter concentration as carbon (g m^{-3}), Y_{DOM} is the yield of organic matter as dissolved organic carbon (g m^{-2} $year^{-1}$), and P-E_T is excess precipitation or precipitation minus evapotranspiration (m $year^{-1}$).

In lakes, DOM can be concentrated or diluted by exchange of water with the atmosphere [Eq. (2)]:

$$DOM = Y_{DOM} \times Ad/((Ad \times (P\text{-}E_T) + Ao \times (P\text{-}E)), \tag{2}$$

which reduces to

$$DOM = Y_{DOM}/((P\text{-}E_T) + (P\text{-}E) \times Ao/Ad) \tag{3}$$

where Ad is the drainage area, Ao is the lake area, P is precipitation, and E is lake evaporation (assuming for the moment that DOM is diluted or concentrated conservatively). Equation #3 illustrates how the moisture balance (P-E and P-E_T) of a climatic zone can regulate the concentration of DOM in lakewater. This is consistent with the pattern of increasing DOM concentration in freshwater lakes in Canada from humid southcentral Ontario, (1–4 mg L^{-1}; Dillon and Molot 1997), increasing to the west through northwestern Ontario (2–12 mg L^{-1}) to the semi-arid western great plains of Alberta (e.g. 20–40 mg L^{-1}; Curtis and Adams 1995). It is also consistent with observed decreases in DOM in response to increased precipitation (Tate and Meyer 1983), and increases in response to drought (Schindler et al. 1992). It is also apparent from Eq. (3) that properties of the catchment, including the relative sizes of lake (Ao) and catchment (Ad) area, can strongly modify DOM concentration within climatic zones.

Empirical relationships of increasing DOM with increasing catchment:lake area (Ad:Ao) have been reported for lakes in districts characterized by significant external loadings of DOM and humid climate (e.g. Fig. 4.1). The first series of observations were based on relationships between "organic color" and Ad:Ao (Schindler 1971; Gorham et al. 1983; Engstrom 1987; Rassmussen et al. 1989; Curtis and Schindler 1997). Similarly, the organic color of water has been related inversely to water residence time (Meili 1992; Curtis and Schindler 1997; Dillon and Molot 1997) and to mean depth (proportional to water resi-

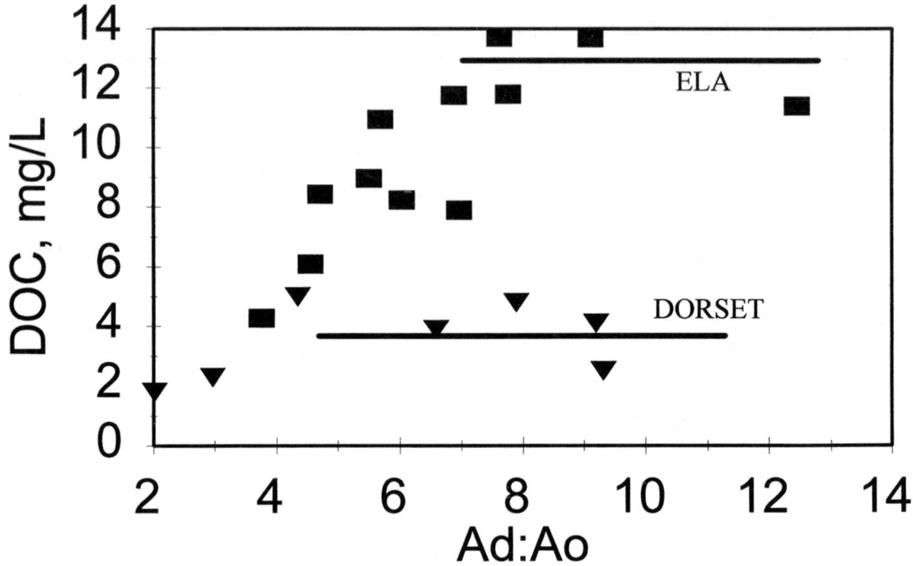

Fig. 4.1. The dependence of DOM concentration as carbon (DOC) on the relative drainage area (Ad:Ao) for lakes at the Experimental Lakes Area (ELA; *squares* Curtis & Schindler 1997) northwestern Ontario and for the Dorset region (*triangles*) of southcentral Ontario (Dillon & Molot 1997).

dence time; Gorham et al. 1983). Subsequently, DOM concentration has been related to Ad:Ao and water residence time (Curtis and Schindler 1997).

Dilution of DOM by precipitation is responsible for a small part of the decrease in DOM and organic color observed in humid climates (Fig. 4.2). Dilution of DOM or organic color is calculated by Eq.(4),

$$X_l = (X_t \times Q/Ao)/(P\text{-}E+Q/Ao), \qquad (4)$$

where X is DOM or organic color, Q is discharge from the terrestrial catchment to the lake, Ao is lake area, P is precipitation, E is evaporation, and the subscripts l and t indicate lake and terrestrial input, respectively. Concentrations of DOM and organic color in lake water from the Experimental Lakes Area (ELA) are all well below that for conservative dilution, and are evidence of in situ loss or transformation of DOM and organic color (Fig. 4.2).

Analyses of DOM mass balances for lakes of different water residence times indicate that DOM retention increases with increasing water residence time (Fig. 4.3). Retention rates vary between zero (no residence time) and about 75% following a rectangular hyperbola. Extrapolation of this trend suggests that the upper limit for DOM retention is about 80% at infinite residence time. It also indicates that the bulk quality of DOM is altered during its residence in lakes because some portion of the DOM is very refractory.

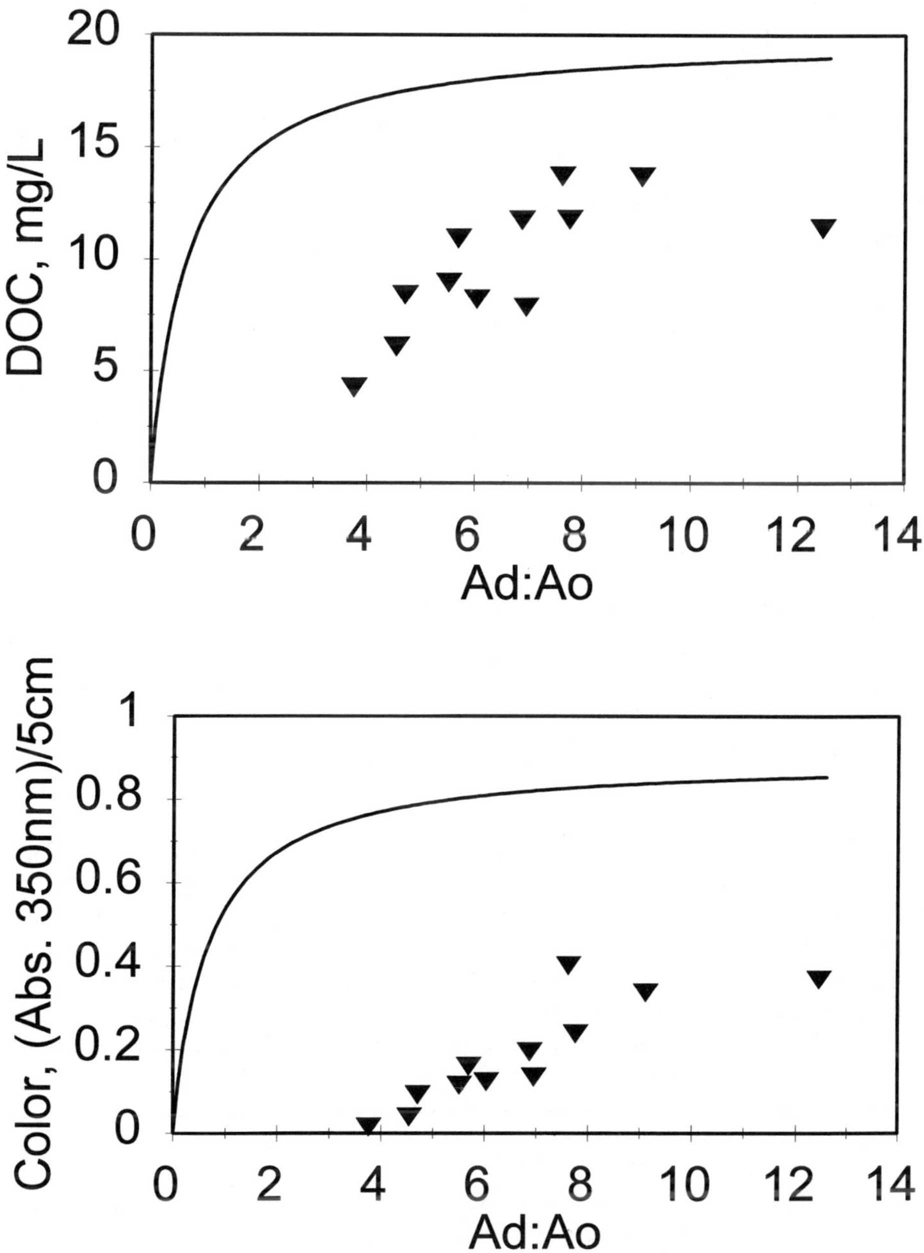

Fig. 4.2a. The dependence of DOM concentration (as carbon) on the relative drainage area (Ad:Ao) for the Experimental Lakes Area (ELA; *triangles*), northwestern Ontario. The *solid line* represents conservative dilution of DOM inputs. **b** Same as for a except for organic color as absorbance units at 350 nm with a 5-cm path length. (Adapted from Curtis & Schindler 1997)

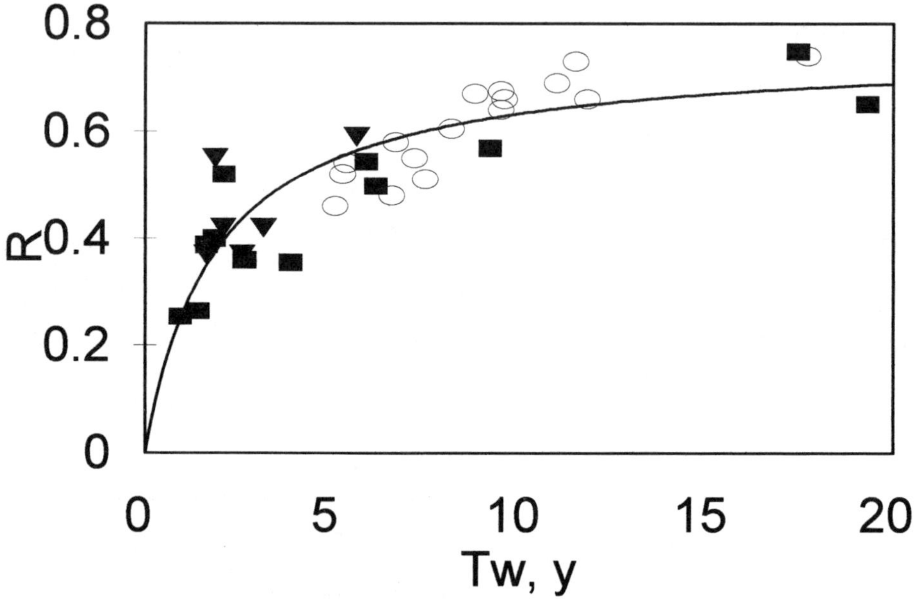

Fig. 4.3. Retention coefficients for DOM as carbon for Ontario Lakes as a function of water residence time. *Symbols* are as in Fig. 4.1 except *open symbols* are for Lake 239 (Experimental Lakes Area during 16 years of climatic warming of 2°C). The *curve* is the best fit line to the solid symbols by the equation $R = R_{max} \times T_w/(R_{0.5} + Tw)$, where R_{max} is the value of R at infinite water residence time (T_w), and $R_{0.5}$ is the value of T_w, where R is $0.5R_{max}$. Values (S.E. in parentheses) of R_{max} and $R_{0.5}$ were 0.754 (0.034) and 1.92 (0.342), respectively. Corrected R^2 was 0.793, with $P<0.01$. The curve explained a 60% of the variation if R in Lake 239 ($F=2.43$, $P<0.05$).

There are important regional/climatic differences in the empirical relationships among Ad:Ao, water residence time, and organic color and DOM that are related to a regional climate. Principally, dependencies are most strong in humid climates. This may be related to water residence times being sufficiently long for in situ processes to affect DOM concentration. Consistent with this interpretation, dependencies of DOM and organic color on Ad:Ao and water residence time are weaker for maritime climates (e.g. Engstrom 1987; Clair and Sayer 1997), probably because of rapid flushing.

For lakes having longer water residence times, DOM concentration can be modified within lakes by processes that consume or produce DOM. For example, concentrations of DOM may decrease by scavenging (loss to sediments) and mineralization (loss to the atmosphere). Alternatively, the concentrations may increase by in situ production of DOM (Schindler et al. 1992) or by evapoconcentration (Curtis and Adams 1995). High rates of nutrient loading appear to enhance DOM concentration by as much as a few milligrams per liter (Schindler et al. 1992; Curtis and Schindler, in prep).

In contrast, evapoconcentration is capable of enhancing significantly DOM concentration (Curtis and Adams 1995). In semi-arid western Canada, DOM concentrations increase with increasing salinity and inferred water residence times (Curtis and Adams 1995). By comparing the evapoconcentration of conservative ions relative to DOM, it appears that 10% of DOM inputs are concentrated. Thus, DOM retention in saline lakes is very efficient despite the high concentrations, and approximates the upper limit for DOM retention calculated for lakes in humid Ontario (Curtis and Schindler 1997).

If indeed DOM retention (as suggested above) and the yield of DOM from catchments were similar among climatic zones (Moeller et al. 1979), then predicting the concentration of DOM in headwater lakes is determined by the balance of water and its effect on water residence time. Concentrations of DOM in higher order lakes would be somewhat more difficult to predict for two reasons. First, a portion of the input water has prior residence in a lake with loss of DOM. Second, DOM loss rates among lakes suggest that DOM quality is altered with lake residence. Thus, loss rates of DOM in higher order lakes are likely slower than in headwater lakes.

4.3 DOM Quality

Variables affecting DOM quality are less well known than are those affecting concentration. Qualitative measures of bulk DOM include molecular weight, aromaticity and optical properties, elemental composition, and proportion of humic substances extractable with XAD resins. Most of the qualitative analyses of DOM have been performed to distinguish between DOM sources – primarily between allochthonous and autochthonous DOM.

In general, DOM from terrestrial sources is nitrogen-poor, optically dense, rich in aromatic structures, and contains a high but variable portion of substances that adsorb to XAD resins (McKnight et al. 1991; McKnight et al. 1994; Curtis and Adams 1995). In contrast, autochthonous DOM is nitrogen-rich, relatively transparent, with a moderate proportion of substances adsorbing to XAD resins (McKnight et al. 1994). Reported molecular weights for allochthonous and autochthonous DOM vary considerably between and less than 1000 and more than 10,000 Daltons, respectively (summarized in Curtis and Adams 1995).

There is some evidence that DOM quality is also affected by flowpath and in situ processes. For example, there is some suggestion that DOM from groundwater inputs may be more transparent to light relative to water draining through surficial materials (Larson 1978). However, the most striking change to DOM in surface water is probably photobleaching or loss of conjugated double bonds (aromaticity). Clearly, photochemical processes can disrupt conjugation of double bonds in DOM and break apart large molecules by photolysis (Kieber et al. 1990; Wetzel et al. 1995). Studies in lakes and enclosures reveal that the optical density of DOM decreases significantly over time (Curtis and Schindler, in prep.). Thus, hydrologic residence time can modify DOM optical quality.

The elemental composition and molecular weight of DOM can also apparently be modified by in situ processes. For example, nitrogen and phosphorus can be photolysed from DOM and possibly assimilated by microorganisms. However, there have apparently been no studies that unambiguously demonstrate that in situ processes alter significantly the elemental composition of bulk DOM in surface water.

DOM quality may also be affected by fractionation of the bulk DOM pool. This is consistent with variable portions of the allochthonous and autochthonous DOM that are resistant to degradation (or sedimentation) in lakes. For example, Schiff et al. (1997) suggested that the lower limit for DOM concentration in lakes might be determined by the loading of extremely refractory DOM present in groundwater. There is also some evidence from whole-lake radiocarbon experiments that a significant proportion of the autochthonous DOM production may also be very refractory (Hesslein et al. 1980).

In situ processes appear to affect qualitative properties of DOM similarly across large climatic gradients. In humid central Ontario and in semi-arid central Alberta, the optical density of DOM decreased with increasing water residence time (Curtis and Adams 1995; Curtis and Schindler 1997). Processes consistent with the observed pattern included photobleaching, or selective loss by scavenging or microbial degradation. The processes are not mutually exclusive and all may be operating simultaneously.

4.4 Effects of Climate Change on DOM in Lakes

Climate changes are generally of much longer duration than the typical scientific study. Climate changes may take decades or centuries to run their course. Thus, it is very fortunate that a series of experiments at the Experimental Lakes Area, northwestern Ontario, provided the sufficient data on reference lakes over a period of 20 years coincident with a climatic warming of 2°C. Detailed accounts of these studies are reported elsewhere (Schindler et al. 1992; Schindler et al. 1996; Schindler et al. 1997; Schindler and Curtis 1997).

Briefly, climatic warming was associated with a decrease in precipitation and an increase in evaporation and evapotranspiration. The yield of DOM [measured as dissolved organic carbon (DOC)] decreased slightly with increased warming; however, concentrations of DOM in stream runoff increased significantly because the yield of water decreased more than the yield of DOM. In contrast to streams, the concentration of DOM in lakes decreased even though concentrations of DOM in runoff increased. The decrease of DOM in lakewater was attributed to increased in situ retention at longer water residence times – retention coefficients were related directly to water residence times for Lake 239.

The relationship between DOM retention coefficients and water residence time in Lake 239 during a period of climatic warming was consistent with observed relationships between DOM retention coefficients and water residence time among lakes from climatically and hydrologically diverse regions (Curtis and Schindler 1997; Dillon and Molot 1997, Fig. 4.3). These findings are significant for two reasons. First, they demonstrate that space (among lakes and catchments) might be substituted for time in predicting the effects of climatic change on lakewater DOM. Second, they demonstrate that the relationship between DOM retention and water residence time is robust for water residence times from about 1 to 30+ years.

The most significant effects of climate change on DOM quality were on its optical properties. Specifically, attenuation of damaging UV radiation by DOM decreased with climatic warming and a subsequent increase in water residence time (Schindler et al. 1996; Schindler and Curtis 1997). The increase in UV transparency was further enhanced by acidification of lakewater with strong acids (Schindler et al. 1996; Yan et al. 1996). Also, there was a suggestion that the C:N ratio of DOM decreased during climatic warming, for which there was no clear mechanism.

Similar qualitative changes in the optical density of water were observed across a gradient of water residence times (Curtis and Schindler 1997). Specifically, the absorbance of blue light by DOM decreased with increasing water residence time in lakes of humid Ontario and semi-arid central Alberta (Curtis and Adams 1995; Curtis and Schindler 1997). Because measurements of optical properties were different for the long-term record, and the subsequent analyses of lakes spanning gradients of water residence time, it was not possible to compare the dependencies directly as was done for DOM concentration. Thus, space cannot yet be substituted for time in predicting the response of DOM quality to climatic change. In contrast to the response of C:N in DOM during climatic change, there was no apparent relationship between C:N in DOM with water residence for semi-arid lakes of central Alberta.

Moreover, it seems reasonable to speculate that processes and process rates affecting DOM are similar among lakes despite wide diversity of physical, biological, and chemical properties of lakes. This contention is supported by the relatively small variation in DOM yields from terrestrial catchments, the often dominant role of allochthonous DOM in the DOM pool, the robustness of the relationship between retention coefficients and water residence time, and the similarity of empirical relationships between DOM quality and water residence time. Mechanisms regulating DOM concentration and DOM quality require further study because they cannot be inferred unambiguously from rates.

4.5 Sensitivity of Aquatic Systems to Changes in DOM Quantity and Quality

The sensitivity of aquatic ecosystems to climatic perturbation of DOM concentration or quality clearly depends on the complex interaction among climate, surficial geology and topography. Nevertheless, the sensitivity to climatic perturbation can be estimated from loading and flushing rate relationships. In humid regions, concentrations of DOM will likely decrease with climatic warming. Lakes most sensitive to change are probably those having relatively water long residence time and DOM dominated by allochthonous inputs. In such lakes, relatively small climate changes can cause large changes in water residence time and changes in the relative importance of in situ processes in regulating DOM concentration and quality. For example, the water residence time in Lake 239 increased from a minimum of about 4 years to more than 15 years as the climate warmed. Over the same period of time, DOM retention (measured as DOC) increased from about 30% to almost 70% (Fig. 4.3).

Loss of DOM from long water-residence-time, freshwater lakes (+10 years) in response to climatic warming has been linked to enhanced penetration of underwater UV radiation. Such lakes are especially sensitive to elevated UV radiation because the penetration of UV radiation increases exponentially with decreasing DOM concentration (Scully and Lean 1994; Schindler et al. 1997).

In arid and semi-arid regions, climatic warming will probably increase the concentration of DOM by evapoconcentration. The sensitivity of these systems to UV radiation is not known. However, very high DOM concentrations appear to compensate in part for the specific UV transparency of the DOM (Curtis and Adams 1995). The effects of elevated DOM concentration in such lakes on aquatic food webs are poorly known; however, there is some suggestion that the evapoconcentrated DOM is highly refractory and poorly utilized by bacteria (Waiser and Robarts 1995).

Short water-residence-time lakes are probably insensitive to climate-dependent changes in DOM concentration and quality because DOM loading can overwhelm the relatively slow processes of retention and transformation (Curtis and Schindler 1997; Dillon and Molot 1997). A portion of these lakes with very high DOM loading may be dystrophic or allotrophic, and consequently the structure of the aquatic ecosystem may depend on very high DOM loading rates. Reductions in DOM loading from climatic warming could cause such lakes to shift from an allotrophic state to an autotrophic state.

There may also be large perturbations to DOM for which we have no contemporary or historical analogs. For example, in temperate and boreal regions climatic warming will likely reduce the proportion of catchment area occupied by wetlands. Consequently, DOM loading rates will likely decrease. In contrast, climatic warming may increase DOM loading to high latitude lakes by melting permafrost and increasing the proportion of catchment area occupied by wetlands.

In the years to come, it is clear that research on DOM and climate interactions should expand to wetland and catchment process, and into seepage, low latitude, and high elevation lakes. Finally, using the past as a guide for future study, it would seem expedient to pursue empirical and mechanistic/process lines of study simultaneously.

4.6 Summary

Dissolved organic matter (DOM) in lakes is derived from terrestrial and aquatic primary production. The terrestrial or allochthonous DOM delivered to lakes depends on climate and surficial geology because these variables determine capacity to produce DOM (soil productivity and moisture content), the concentration of DOM by dilution (with excess precipitation) and the contact time between organic matter and water in the catchment (slope and permeability). Yields of DOM from terrestrial catchments to surface waters are often within the range of 3 and 4.5 g m^{-2} y^{-1}. Concentrations in lakes vary widely from as little as 1 to as high as 400 mg L^{-1}, however typical values are between 2 and 12 mg L^{-1}.

Aquatic (autochthonous) DOM is less clearly dependent on climate except possibly where concentrations are diluted or concentrated by precipitation or evaporation, respectively. Whole-lake fertilization experiments indicate that DOM concentrations can increase by a few milligrams per liter. Thus, autochthonous production delete seldom dominates the DOM pool.

DOM in lakes is transformed and degraded within lakes by such processes as photolysis, microbial degradation and scavenging (roughly transformation to particulate form followed by sedimentation). The climatic dependence of the individual processes cannot be inferred from the current literature. However, there is clear evidence that loss of DOM within lakes, operationally defined as retention, is strongly dependent on climate.

Retention of bulk DOM increases with increasing water residence time within and among lakes. Similarly, the optical density of DOM decreases with increasing water residence time among lakes and probably within lakes (inferred from enclosure studies). Thus, climatic warming and consequent increases in water residence time have caused DOM concentration to decrease and UV transparency to increase.

References

Baron J, McKnight D, Denning AS (1991) Sources of dissolved and particulate organic material in Loch Vale Watershed, Rocky Mountain National Park, Colorado, USA. Biogeochemistry. 15:89–110

Clair TA, Sayer BG (1997) Environmental variability in the reactivity of freshwater dissolved organic carbon to UV-B. Biogeochemistry 36:89–97

Curtis PJ, Adams HE (1995) Dissolved organic matter quantity and quality from freshwater and saline lakes in eastcentral Alberta (Canada). Biogeochemistry 30:59–76

Curtis PJ, Schindler DW (1997) Hydrologic control of dissolved organic matter in low-order Precambrian Shield lakes. Biogeochemistry 36:125–138

Dillon, PJ, Molot, LA (1997) Dissolved organic and inorganic carbon mass balances in central Ontario lakes. Biogeochemistry 36:29–42

Engstrom DR (1987) Influence of vegetation and hydrology on the humus budgets of Labrador lakes. Can J Fish Aquat Sci 44:1306–1314.

Gorham E, Dean WE, Sanger JE (1983) The chemical composition of lakes in the north-central United States. Limnol Oceanogr 28:287–301

Hesslein RH, Broecker WS, Quay PD, Schindler DW (1980) Whole-lake radiocarbon experiment in an oligotrophic lake at the Experimental Lakes Area, northwestern Ontario. Can J Fish Aquat Sci 37:454–463

Kieber RJ, Zhou X, Mopper K (1990) Formation of carbonyl compounds from UV-induced photodegradation of humic substances in natural waters: fate of riverine carbon in the sea. Limnol Oceanogr 35:1505–1515

Larson RA (1978) Dissolved organic matter of a low-coloured stream. Freshwater Biol 8:91–104

Leenheer JA, Malcolm RL, McKinley PW, Eccles LA (1974) Occurrence of dissolved organic carbon in selected ground-water samples in the United States. J Res 2:361–369

McKnight DM, Aiken GR, Smith RL (1991) Aquatic fulvic acids in microbially based ecosystems: results from two desert lakes in Antarctica. Limnol Oceanogr 36:998–1006

McKnight DM, Andrews ED, Spaulding SA, Aiken GR (1994) Aquatic fulvic acids in algal-rich antarctic ponds. Limnol Oceanogr 39:1972–1979

Meili M (1992) Sources, concentrations and characteristics of organic matter in softwater lakes and streams of the Swedish forest region. Hydrobiologia 229:23–41

Meybeck M (1982) Carbon, nitrogen, and phosphorus transport by world rivers. Am J Sci 282:401–451

Meyer JL, Tate CM (1983) The effects of watershed disturbance on dissolved organic carbon dynamics of a stream. Ecology 64:33–44

Moeller RJ, Minshall GW, Cummins KW, Petersen RC, Cushing CE, Sedell JR, Larson RA, Vannote RL (1979) Transport of dissolved organic carbon in streams of differing physiographic characteristics. Organic Geochem 1:139–150

Rasmussen JB, Godbout L, Schallenberg M (1989) The humic content of lake water and its relationship to watershed and lake morphometry. Limnol Oceanogr 34:1336–1343

Schiff SL, Araveba R, Trumbore SE, Hinton MJ, Elgood R, Dillon PJ (1997) Export of DOC from forested catchments on the Precambrian Shield of central Ontario: Clues from ^{13}C and ^{14}C. Biogeochemistry 36:43–65

Schindler DW (1971) A hypothesis to explain similarities and differences among lakes in the Experimental Lakes Area, northwestern Ontario. J Fish Res Bd Can 28:295–301

Schindler DW, Bayley SE, Curtis PJ, Parker B, Stainton MP, and Kelly CA (1991) Natural and man-caused factors affecting the abundance and cycling of dissolved organic substances in precambrian shield lakes. Hydrobiologia 229:1–21

Schindler DW, Bayley SE, Parker BR Beaty KB Cruikshank DR, Fee EJ, Schindler EU, Stainton MP (1996) The effects of climatic warming on the properties of boreal lakes and streams at the Experimental Lakes Area, northwestern Ontario. Limnol Oceanogr 41:1004–1017

Schindler, DW, Curtis, PJ (1997) The role of DOC in protecting freshwaters subjected to climatic warming and acidification from UV exposure. Biogeochemistry 36:1–8

Schindler DW, Curtis PJ, Bayley SE, Parker B, Beaty KG, Stainton MP (1997) Climate-induced changes in the dissolved organic carbon budgets of boreal lakes. Biogeochemistry 36:9–28

Schlesinger WH, Melack JM (1981) Transport of organic carbon into the world's rivers. Tellus 33:172–187

Scully NM, Lean DRS (1994) The attenuation of UV radiation in temperate lakes. Arch Hydrobiol 43:135–144

Tate CM, Meyer JL (1983) The influence of hydrologic conditions and successional state on dissolved organic carbon export from forested watersheds. Ecology 64:25–32

Urban NR, Bayley SE, Eisenreich SJ (1989) Export of dissolved orgainc carbon and acidity from peatlands. Water Resour Res 25:1619–1628

Waiser MJ, Robarts RD (1995) Microbial nutrient limitation in prairie saline lakes with high sulfate concentration. Limnol Oceanogr 40:566–574

Wetzel RG, Hatcher, PG, Bianchi, TS (1995) Natural photolysis by ultraviolet irradiance of recalcitrant dissolved organic matter to simple substrates for rapid bacterial metabolism. Limnol Oceanogr 40:1369–1380

Yan ND, Keller W, Scully NM, Lean DRS, Dillon PJ (1996) Increased UV-B penetration in a lake owing to drought-induced acidification. Nature 381:141–143

II Humus, Light Regimes and Primary Production

5 Attenuation of Solar Radiation in Humic Waters

David Lean

5.1 Introduction

Underwater solar spectral distribution has become critically important in assessing not only the impact of increased solar UV-B (290–320 nm) radiation resulting from stratospheric ozone depletion (Kerr and McElroy 1993) but also the role of ambient levels of both UV-B, UV-A (320–400 nm) and photosynthetically active radiation (PAR) (400–750 nm). Furthermore, as noted in other chapters, dissolved organic carbon (DOC) compounds compete for photons of visable light with the phytoplankton, which lowers areal primary production in humic waters.

Most measurements of UV attenuation have been made in marine waters (reviewed by Kirk 1994) and the notion that suspended particulate materials were important was advanced. In contrast to oceans where DOC concentrations are low (< 1 mg C l^{-1}), lakes contain DOC levels generally range from 2 to 8 mg C l^{-1} and the role of particles in UV attenuation is generally not significant (Scully and Lean 1994; Morris et al. 1995; Laurion et al. 1997). This chapter attempts to extend our ability to predict underwater spectral irradiance in aquatic systems as DOC concentrations increase from 8 to 30 mg C l^{-1}.

5.2 Measurements of Attenuation Coefficients for UV-B and UV-A

The well-known Beer-Lambert equation states that the intensity of radiation at any depth (I_d) in aquatic systems is equal to the intensity just under the surface (I_o) times $e^{-K_d \times D}$ where K_d is the vertical attenuation coefficient and D is depth. In lakes with DOC levels less than 8 mg l^{-1}, the K_d for UV-B radiation was correlated with DOC concentration from the equation provided by Scully and Lean (1994):

$$K_d \, \text{UV-B} = 0.415\,(\text{DOC})^{1.86} \qquad r^2 = 0.97.$$

Ecological Studies, Vol. 133
Hessen, Tranvik (eds) Aquatic Humic Substances

The attenuation coefficient for UV-A, i.e. K_d UV-A, was found to be equal to 0.299 $(DOC)^{1.53}$. The r^2 was also excellent and equal to 0.95.

There are so few irradiance data that have been collected for humic lakes in a consistent manner that for this review the author used data obtained in 11 ponds and small humic lakes in south-central Ontario (Lean et al., submitted). Three sites were located on the Canadian Shield, near Anstruther Lake, four were on transitional bedrock near Jacks Lake and four on limestone, south of the Shield near Peterborough with DOC values from 8 to 20 mg l^{-1}. In these waters, spectral irradiance was measured using techniques similar to that in Scully and Lean (1994) every 2 nm between 280 and 800 nm using an OL 752 spectroradiometer (Optronics Inc.). The system was calibrated with a NIST (National Institute for Standards and Testing, Washington D.C., USA) traceable OL 752-10E spectral irradiance plug-in standard. Wavelength accuracy was checked using the mercury lines of the Optronics dual calibration module. A submersible 8-in. integrating sphere with a cosine receptor assembly was connected to the dual monochromator using a 3.5-m fibre optic cable. The system was powered with a 12-Volt Gates Cyclon gel cell battery. In coloured waters, one must be careful not to include values that are below the detection limit of the instrument and to measure depth precisely. We used a metal rod which was screwed into the sensor and pushed into the soft sediments to the exact depth.

Typical spectral irradiance data are shown in air and underwater at 25 and 50 cm for Ranger Lake, Ontario, Canada. This lake was used by Amyot et al. (1997) for their mercury experiments and would be considered a fairly clear humic lake with DOC values near 6 mg C l^{-1} (Fig. 5.1). As such, these data illustrate not only the lower end of the results presented below but also the dif-

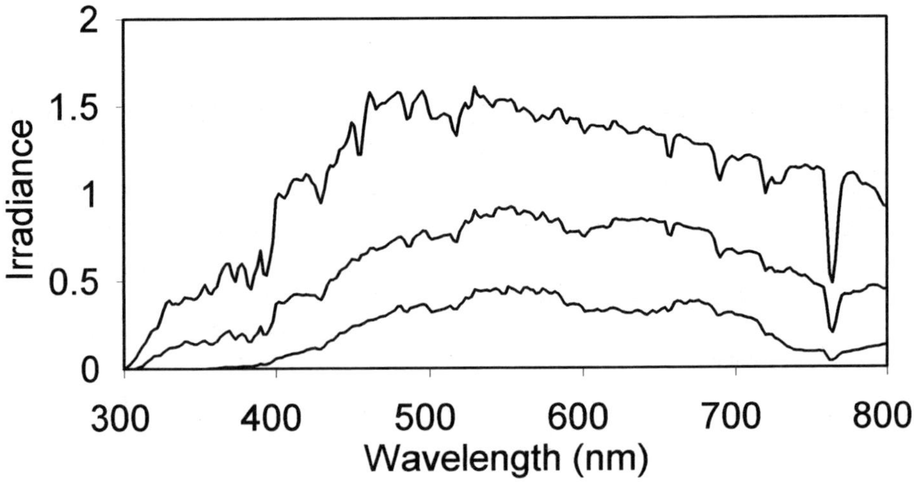

Fig. 5.1. Incident irradiance (Watts m^{-2} nm^{-1}) measured in air and underwater at depths of 25 and 50 cm for Ranger Lake, Ontario, Canada

ficulty in obtaining reliable information and the need to precisely measure the depth where measurements are being obtained. The jagged trace was not due to instrument noise. The peaks and valleys were reproduced at other times and at other depths and are a consequence of absorbance by chemicals in the atmosphere above the earth (Madronich 1993).

To quantify the attenuation of UV radiation, values were integrated over the UV-B range (290–320) from 320 to 400 for UV-A and 400 to 750 nm for PAR. On sunny days UV-B and UV-A values were about 0.6 and 10% of total irradiance (290–750) respectively. While absolute values were lower under cloudy conditions, the percent UV-B and UV-A of total irradiance was higher, reaching about 1 and 16% of total irradiance respectively. Rapid attenuation of the shorter wavelengths can be seen. At 25 cm almost all the UV-B is lost, and only a small amount of the UV-A remains at 50. Clearly, a consistent method to handle such information is required.

Integrated data from Fig. 5.1 along with other similar data collected at the same time at other depths were converted to logarithm values and plotted as a function of depth (Fig. 5.2). This example, which represents the lower end for humic waters, dramatically illustrates how abruptly UV-B and to a lesser extent UV-A is attenuated. The slope, generated by regression analysis, provides the vertical attenuation coefficient (K_d). Values for UV-B, UV-A and PAR were 14.3, 7.5 and 2.4 m^{-1} respectively. Rearranging the Beer-Lambert law above the depth where only 1% of each waveband remains is equal to 2.3 divided by K_d. This means that 99% of UV-B, UV-A and PAR is absorbed in 0.32, 0.61 and 1.9 m respectively.

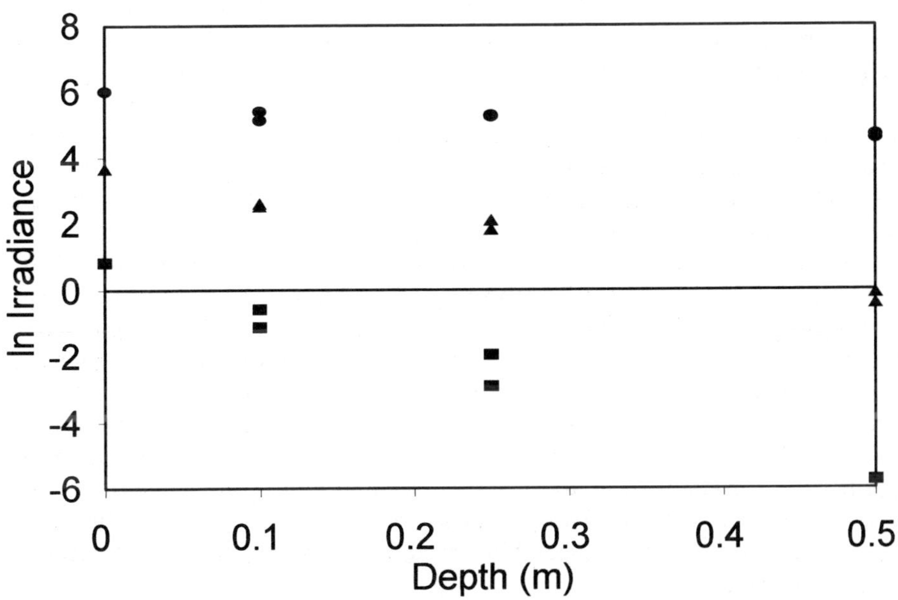

Fig. 5.2. Integrated values for UV-B (*squares*), UV-A (*triangles*), and PAR (*circles*) are converted to the natural logarithm and plotted as a function of depth

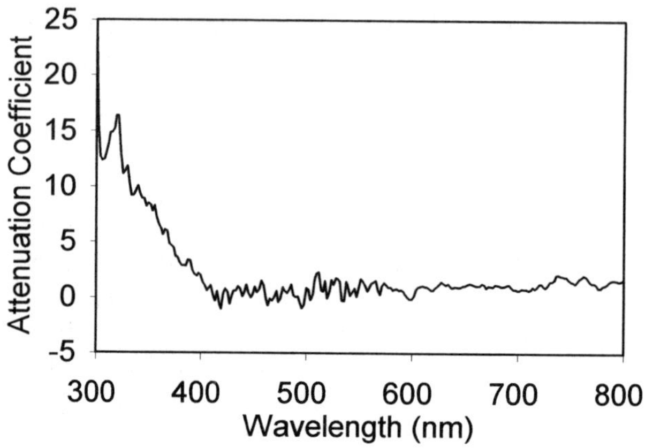

Fig. 5.3. Plot of attenuation coefficient (K_d) as a function of wavelength from 300 to 800 nm

The absolute amount of radiation increases with increasing UV-B wavelengths from 295 to 320 (Fig. 5.1). However, it is well known that the shorter wavelengths are the most damaging but are attenuated the most. It is therefore essential to know not only the attenuation coefficients for the UV-B waveband but also how the spectral composition is altered within each waveband. To illustrate this point, data from 10 and 25 cm were used to calculate the attenuation coefficient every 2 nm from 300 to 800 nm. Errors due to slight changes in the depth of the sensor or slight changes in cloud cover during the 8-min exposure required to measure spectral irradiance at each depth will cause errors in the relationship. Nevertheless, a consistent pattern emerges that shows (Fig. 5.3) that while the mean K_d values were 14.3, 7.5 and 2.4 m^{-1} for UV-B, UV-A, and PAR respectively, the K_d value at 300 nm was near 20, decreasing to 12 at 320 nm followed by further declines to near 2 at 400 nm (Fig. 5.3). Because the quantum yields, action spectra or biological weighting functions (see below) decrease exponentially with increasing wavelengths, such changes in the attenuation coefficients cannot be ignored.

One objective for future research is to predict the change in K_d with increasing wavelength. There is some hope that this can be achieved by comparing slopes of K_d as a function of wavelength with the slope of the absorbance curve measured on filtered water in a spectrophotometer in a laboratory. When these data were plotted on a natural logarithmic scale a straight line slope between 300 and 436 nm was near 0.015 but some examples ranged from 0.011 to 0.018. This observation is critical to the development of simple predictions for attenuation and will be discussed more completely below (see Fig. 3). When we simply integrate values over the UV-B and UV-A waveband, the calculations are more straightforward, but this obscures the differences across the wavebands.

It is an oversimplification to consider that all UV-B wavelengths have the same ability to cause damage or stimulate photochemical reactions. Some photochemical reactions are limited to the UV-B region alone while others extend into the blue region (photochemical iron reduction and hydrogen peroxide production). Appropriate weighting functions at all wavelengths are essential to quantify the integrated response. While UV-B is generally much more damaging than UV-A, there is more UV-A and it penetrates to much greater depths. Consequently, the role of UV-A and PAR cannot be overlooked.

While particulate materials are important in the attenuation of UV radiation in marine systems (Kirk 1994), they do not seem to be important in lakes. At first this does not seem to make sense until we make calculations of the relative importance of DOC. It represents about 40–45% of the total dissolved organic matter (DOM) and contains the chemical structures or chromophores for UV absorbance. The concentration of DOC in most non-humic lakes is generally 4–8 mg C l^{-1} compared with the concentration of particulate organic carbon which is generally 0.2–2 mg C l^{-1}. Furthermore, most of the particulate carbon is made up of carbon molecules such as protein, carbohydrates or lipids which do not absorb strongly in the UV range. The particulate carbon not only represents a much smaller amount of carbon but also lacks the appropriate structures for UV absorbance.

Even where there was a high level of suspended clay material (Scully and Lean 1994), UV attenuation was controlled by DOC concentration. In the Welland Canal, for example, the penetration of visible light was only a few centimetres, but with DOC levels similar to Lake Erie, the depth of UV-B penetration was still controlled by the DOC concentration. Only in deep ocean regions where DOC concentrations are low is there a significant attenuation due to particulate materials. Correlations with total chlorophyll or particulate organic carbon were poor especially when the influence of autocorrelation with DOC was removed. Although PAR is attenuated both by DOC and suspended particulates (algae, sediments, clay particles, etc.), the diffuse nature of UV radiation means that in lakes DOC controls the depth of penetration independent of the suspended particulates. With declining DOC levels the ratio of the depth of penetration of UV-B to PAR increases exponentally. In other words, the depth of damaging radiation to that required for photoplankton production reaches a critical ratio with DOC concentrations below 3 mg C l^{-1} (Laurion et al. 1997). These levels are typical in about a third of the lakes in Ontario.

5.3 Relationship Between Attenuation and DOC and DOC Fluorescence

In Fig. 5.4 attenuation coefficients for UV-B (top) and UV-A (bottom) measured in humic ponds and lakes (Lean et al. submitted) (closed symbols) were plotted along with data from Scully and Lean (1994) (open symbols) as a function of DOC (Fig. 5.4). DOC values were obtained using the UV oxidation

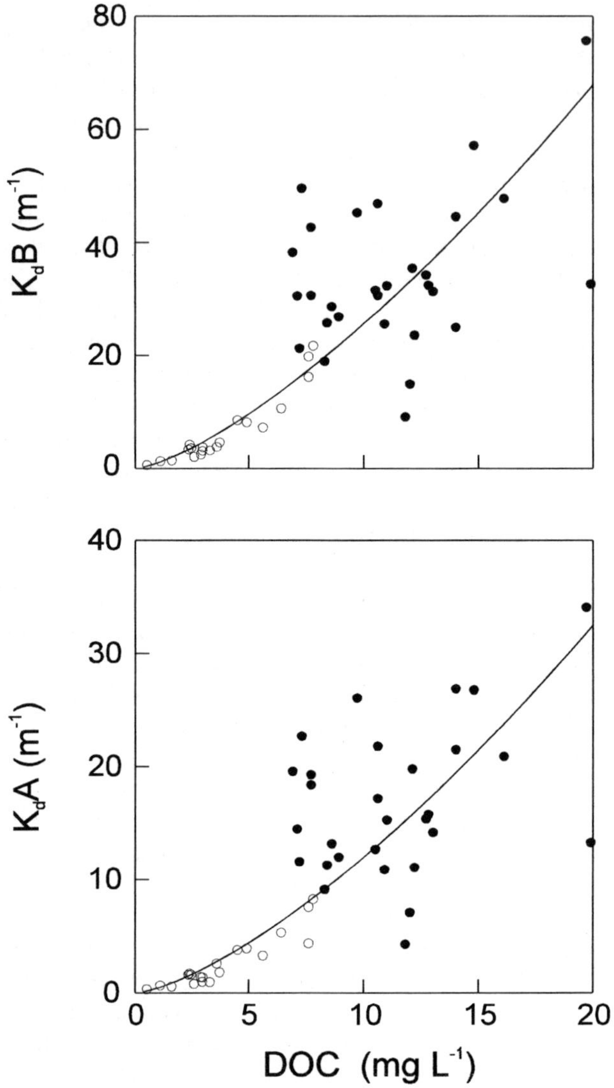

Fig. 5.4. Attenuation coefficients for UV-B (top) and UV-A (*bottom*) obtained from 11 humic ponds and lakes sampled on three occasions (*solid circles*) and lake data for less humic lakes (*open circles*), redrawn from Scully and Lean (1994), plotted as a function of DOC

method (Environment Canada 1994) but were highly correlated with values obtained using high temperature oxidation (Scully and Lean 1994). DOC was measured after samples were filtered through 0.45 μm Sartorius 11103 cellulose acetate filters. Samples were stored near 0 °C in glass bottles with Teflon tops. The sample was injected into a continuous flow system with nitric acid and potassium persulphate, passed over a UV source and purged with oxygen, and the gaseous CO_2 was measured using an infrared detector. Fluorescence of the filtered and unfiltered water samples (DOCFL) was also measured using Corning 7–60 and soft glass excitation filters to give a peak maximum at 365 nm. Fluorescence emission was measured at 437 nm using filters 2A and 47B from Turner. The reference filter was a 2A (Turner). Fluorescence values were standardized using quinine sulphate (QSU) units where 1 QSU = 1 $\mu g\ l^{-1}$ in 0.1 N H_2SO_4.

The r^2 values were 0.63 and 0.60 if the clear and humic lake data were used, but if only the high DOC water was used the r^2 values dropped to 0.15 and 0.11 for K_dB and K_dA respectively. When plotted against and DOCFL the r^2 values were 0.72 and 0.70 but dropped to 0.21 and 0.14 (without the lake data) for K_dB and K_dA respectively (Lean et al. submitted). Values were occasionally off the line of best fit (using all data) by ± 100%. Our attenuation coefficients were similar to the three K_dB values reported by Morris et al. (1995) and three K_dB values of Graneli et al. (1996) for high DOC lakes. Clearly, the factors which determine attenuation in humic waters are more complicated than those found in less coloured lakes.

Values for K_d were used to calculate the depth where 90% of the UV-B and UV-A was attenuated. These values were then plotted as a function of DOC (Fig. 5.5). The depth where 99% is attenuated is twice as great. The depth of

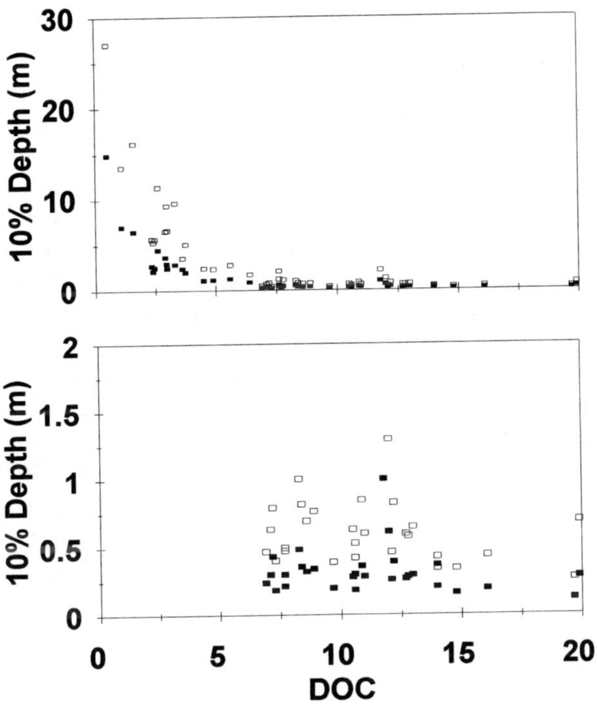

Fig. 5.5. Depth of 90% attenuation UV-B (*solid squares*) and UV-A (*open squares*) for data from Fig. 5.4 (*top*) and only highly humic ponds on an expanded scale (*bottom*)

90% UV-B attenuation in lakes (Scully and Lean 1994) ranged from less than 1 to 15 m and the penetration was 1–27 m for UV-A. In the humic systems studied here, 90% of the UV-B was attenuated in 12–60 cm while UV-A was about twice as great (see expanded scale on bottom of Fig. 5.5). It should be recalled that we are seeing as much attenuation in 10 cm of pond water as in 15 m in the Pacific or Antarctic Oceans (Kirk 1994). This means that all the photochemical activity due to UV radiation is confined to this shallow depth.

When the Beer-Lambert law is rearranged to calculate the depth of 10% penetration it takes the form $K_d \times D$ = constant. Since the K_d is also a function of DOC, it is not surprising to see this relationship take the shape of a rectangular hyperbola with a sharp transition near 2 mg C l^{-1}. We should also recall that lakes are losing their sun screen through warmer drier conditions (Schindler et al. 1996; Yan et al. 1996). This means that any decline in DOC through the 2 mg C l^{-1} level will result in rapid increases in UV-B penetration. This, however, is not likely to become a problem in humic waters where our measured DOC values were 7–20 mg C l^{-1}.

5.4 The Relationship Between DOC and DOC Fluorescence (DOCFL)

An unresolved problem for aquatic scientists is that DOC cannot be chemically classified (see Chap. 2, this Vol.). Aromatic groups tend to absorb UV radiation more strongly than aliphatic groups. Consequently, it would not be surprising to find different sources of DOC with different UV absorbing properties. Nevertheless, the relationship between DOC and UV attenuation is robust. Since fluorescence requires that UV-A is first absorbed, DOC fluorescence (DOCFL) should be a better measure of the aromatic proportion of DOC. When DOCFL values were plotted as a function of DOC concentration a power function resulted (Scully and Lean 1994). Several different patterns emerged from our humic lakes and ponds (Lean et al., submitted). Early in the year, corresponding to periods when attenuation was overestimated by DOC, the DOCFL values were high. During periods when DOC underestimated the attenuation coefficients, the DOCFL was lower. This happened later in the year when photobleaching may have occurred. Data were off the curve by +/– 50% when the maximum error in measurement was less than +/– 5%. These observations suggest that the DOC in humic waters may at times be more reactive when exposed to UV radiation but at other times may be photobleached and neither absorb nor stimulate photochemical reactions to the same extent.

5.5 Using Absorbance Values to Predict UV Attenuation in Lakes

Absorbance coefficients for specific wavelengths (e.g. K_a 310) were calculated by multiplying the absorbance value measured on water filtered through 0.45 μm Sartorius 11103 cellulose acetate filters by 2.303 divided by the path length of the cuvette in metres. Measurements were made using a 1-cm quartz cuvette for these highly coloured waters. A high correlation was found between the attenuation coefficient (K_d) measured in the ponds using a spectroradiometer and the absorbance coefficient (K_a) measured in the laboratory on filtered water using a spectrophotometer (Fig. 5.6). Here K_d (310) = 1.22 K_a (310) + 0.894 (r^2 = 0.94) and was similar to that found in Scully and Lean (1994) for low DOC lakes, i.e. K_d (310) = 1.,23 K_a (310) + 0.237. Integrated UV-B values can also be reliably predicted from the absorbance coefficient at 310 nm [K_d (BINT) = 1.02 K_a (310) + 1.22, (r^2 = 0.86].

It is clear that the absorbance of the water determines the attenuation coefficient and the percent DOC which acts to absorb UV-B radiation changes throughout the year. The highest attenuation and fluorescence per unit DOC occurred in May and to a lesser extent during September. This may reflect the recent input of fresh DOC that is far more photochemically active. In these shallow waters, the DOC becomes altered (photobleached) until it is more similar to that in lakewaters (June and August).

It is important to recall that the measurement of DOC fluorescence is conducted using excitation wavelengths at or near 360 nm. We are assuming that

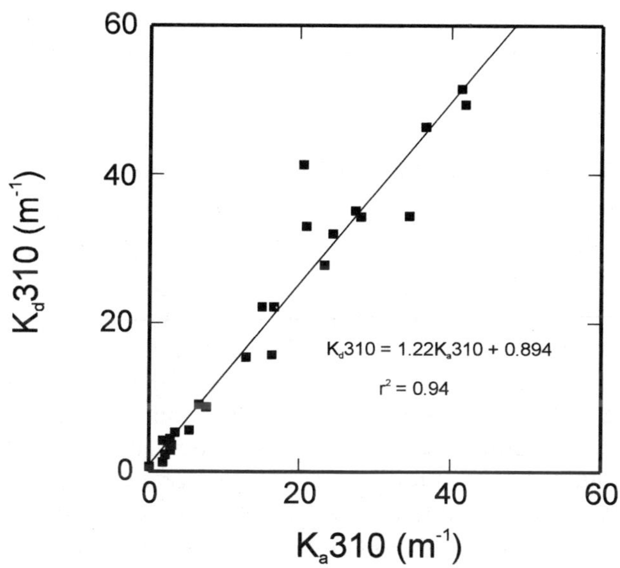

Fig. 5.6. Plot of K_d measured at 310 nm in the ponds with a spectroradiometer as a function of the absorbance coefficient (K_a) at 310 nm measured in the laboratory using a spectrophotometer

the slope of the absorbance and attenuation curves are constant, otherwise our predictions of K_d would be incorrect. Clearly, the extrapolation from 360 nm to the UV-B range requires a consistent slope. A main objective for future research should be to determine how much and why does the slope of the plots of K_d and K_a as a function of wavelength change from the 0.015 value. It may provide a useful fingerprint for the identification of the photoactivity of the DOC. At present we only know that when lakewater or seawater is placed in a quartz cuvette in a scanning spectrophotometer, the absorbance spectra shows a featureless exponential decrease from the lower end of the UV-B waveband (280 nm) into the visible portion. In other words, the absorbance increases exponentially with decreasing wavelength. This is consistent with the observation that the penetration of the shorter wavelengths is much less than that for the longer wavelengths.

This technique provides values which can be easily obtained in the laboratory that can be correlated to the labour-intensive and costly exercise of measuring the attenuation coefficients in the lake itself.

On a much larger data set, Morris et al. (1995) measured attenuation coefficients in lakes in Alaska, Colorado, Pennsylvania and the Bariloche region of Argentina. They also found K_d for UV-B and UV-A to be a simple function of DOC concentration but obtained a curve that was displaced about 2 mg C l^{-1} to the left of that published by Scully and Lean (1994). Reasons for this discrepancy have not been resolved. One explanation may be that Morris et al. froze their samples prior to DOC analysis. This may have resulted in some aggregation and loss of DOC. Alternatively, Scully and Lean may have had a consistent error of about 2 mg C l^{-1}. We know that both the instrument used by Morris et al. and the Optronics instrument used here are capable of providing similar values for K_d (Kirk et al. 1994).

5.6 Influence of Climate Change and Lake Acidification on DOC Levels

Since humic materials are the principal attenuators of UV-B radiation and affect the depth of visable light penetration declining concentrations of DOC will result in increased UV-B penetration (Schindler et al. 1996; Yan et al. 1996). During warmer-drier conditions, a complex series of events can occur which result in increased UV exposure for the underwater community. The export of phosphorus and DOC from drainage basins declines. Lakes are getting clearer and the mixing zone is becoming deeper. With longer water renewal times, there is a greater opportunity for DOC to be photodegraded. Exposed to direct sunlight, littoral zone reduced sulphur compounds are oxidized and during the next rainfall increased acidification occurs in poorley buffered systems. This, along with acid deposition, will cause further DOC declines. Since DOC is the sun screen which attenuates UV radiation in our lakes and rivers, the loss of DOC can result in a greater UV impact than that from stratospheric ozone depletion.

5.7 Surface Water Photochemistry

5.7.1 Interaction of UV Radiation andHumic Materials in Controlling Lake Mercury Levels

Photochemical reactions influence the transport rates for mercury in natural waters. Most of the soluble mercury in water is Hg^{2+}. It may be bound to various ligands, but when exposed to UV-B or UV-A radiation, it is reduced to Hg^0 which, due to its low solubility and favourable Henry's law constant, can volatilize from lakes (Amyot et al. 1997). Unfortunately, DOC seems to act as a competitive inhibitor for this reaction and rates are lower in coloured waters.

We also now know that in remote areas mercury loading is correlated with DOC loading (Mierle and Ingram 1991; Watras et al. 1995) and that wetlands export methyl-Hg (St. Louis et al. 1996; Krabbenhoft et al. 1997). Furthermore, it seems that methyl-Hg is also broken down by solar radiation (Sellers et al. 1996). This process is limited to the depth of penetration of radiation up to the blue region of the spectrum. Consequently, lower rates would be expected to occur in coloured waters. Such observations help to explain why fish in brown water lakes have greater methyl-Hg than fish in clear lakes. More mercury comes in but less gets out.

The overall influence of changing UV-B radiation on the mercury cycle cannot yet be determined since the wavelength-specific quantum yields for both photoreduction and photodemethylation have not been measured. Such information is required so that it can be integrated with spectral models for underwater irradiance to provide an overall mercury model for aquatic systems. We must conclude that UV radiation is particularly important in controlling the loss of methyl- and total mercury from aquatic systems, but we cannot yet predict the overall impact of increased UV radiation on mercury levels in aquatic systems. The reason for this is due to the relative contribution of UV radiation to the other side of the equation. The rate of input through wet deposition is also controlled by photochemical oxidation of mercury in the troposphere, is also driven by UV exposure and the oxidative capacity of the cloud cover. Some researchers have shown that the oxidizing capacity has already increased since pre-industrial times and that it will increase by at least another 50% in the next 50 years (Thompson et al. 1989; Thompson 1992; Willey et al. 1996). Few mercury researchers appreciate that mercury deposition rates may in fact be controlled by oxidation rates of the tropospheric Hg^0.

5.7.2 Interaction of UV Radiation andHumic Materials on Toxic Chemicals

While UV radiation is known to be involved in the breakdown of toxic chemicals including pesticides, rarely are all the breakdown products identified. Some are still toxic and persistent. Many of these reactions result from UV radiation being absorbed by humic substances or DOC with the liberation of an aqueous electron, and the formation of superoxide, hydrogen peroxide, and hydroxyl radicals. In addition, singlet oxygen, hydroperoxy, carbonate or nitrate radicals can be formed (Cooper 1989; Hoigne et al. 1989; Miller 1994; Mopper and Zhou 1990; Zafiriou et al. 1990). While some of these photochemical reactions have been identified, little reliable ecological interpretation has been made.

5.7.3 Photochemical Reactions

When solar radiation strikes DOC some photobleaching occurs (Miller 1994), lowering the ability of DOC to attenuate UV radiation. We have also observed a decline in fluorescence in surface waters in lakes over the season and as much as 30% is lost in a single day by direct exposure of a lakewater sample in a quartz tube to direct sunlight. The measured DOC does not decrease to the same extent, indicating change in the structure from UV-absorbing to non-absorbing carbon compounds. Thus, it is not unexpected that in some lakes (such as the saline lakes of the prairie regions) the correlation of K_d with DOC breaks down while the correlation of Kd and DOC fluorescence is more robust (Scully and Lean 1994).

For any photochemical reaction to occur, energy must be absorbed. In addition to energy released as fluorescence, the absorbance of sunlight by naturally occurring DOM generates a variety of photochemical transients that potentially could alter the metabolic processes which take place in aquatic systems. They include excited triplet state DOM, solvated electrons, organic radical cations, superoxide, singlet oxygen, hydroxyl radicals and peroxy radicals (Cooper 1989; Hoigne et al. 1989). Since these compounds may only last nannoseconds, molecular probes have been used to quantify the rate of production (see review by Zafiriou et al. 1990). The longer lived photoproducts that build to trace concentrations include hydrogen peroxide (H_2O_2). In lakes, concentrations are much higher than in marine systems and may reach in excess of 1000 nM during midday. Rainwater often contains more than 50000 nM H_2O_2 (Cooper et al. 1989; Cooper and Lean 1989, 1992; Cooper et al. 1994; Lean et al. 1993; Willey et al. 1996). H_2O_2 is thought to influence redox metal chemistry, membrane transport and other biogeochemical processes.

H_2O_2 is produced photochemically when UV radiation strikes DOM but its decay is principally biological in most systems. Exceptions have been recently found in coloured lakes with their higher iron concentration. Here, other photochemical reactions likely taking place with photochemically produced reduced iron are suspected and the formation of hydroxyl radicals is suspected (Toy 1997).

Such highly reactive reduced oxygen species are important in altering the persistence and structure of pesticides and other toxic organic compounds. Half-lives of such man-made compounds are often reported but generally the photochemical products are not identified. Many are toxic but only appear as unknown compounds during GC analysis, and may persist in the environment and bioconcentrate in aquatic food chains.

A strong correlation exists between the production of low-molecular-weight (LMW) carbon (carbonyl) compounds and photodegradation of DOM. Unlike that for the formation of hydrogen peroxide, and fluorescence bleaching which occurs to 400 nm and above, no carbonyls are produced through exposure to wavelengths greater than 320 nm (Kieber et al. 1989, 1990). The LMW photo products are also substrates which can stimulate microbial production and biomass. This has been measured in marine (Kieber et al. 1989, 1990; Mopper and Zhou 1990) and freshwater (Lindell and Tranvik 1995) ecosystems. Samples exposed to UV radiation even for short periods can result in stimulation of microbial activity by 2–6 times. The major photo-produced LMW carbon compounds are formaldehyde, acetone, glyoxal, methylglyoxal, glyoxylate and pyruvate. These compounds seem to be formed through hydroxyl radical substitution followed by rearrangement and cleavage of the carbonyl moity.

Carbon monoxide is also generated in both fresh and marine waters when DOM is exposed to sunlight. Wetlands with their high DOM absorbance produce CO at rates over 100 times greater than those reported for Sargasso Sea water (Jones 1991; Valentine and Zepp 1991). CO accounts for more than 50% of the carbonyl production (Jones 1991) with methane also a significant photo product. Field studies (Jones 1991) have shown that CO was supersaturated in surface waters with respect to atmospheric levels and exhibited a diel variation which correlated to light intensity. The dominant sink for CO in oligotrophic waters appears to be atmospheric venting.

Photoproduction of carbon dioxide was measured in filter-sterilized Suwannee River water (Miller 1994). In these acid waters, carbon dioxide production was about 20 times carbon monoxide production rates and both were linear with photo bleaching. Using a rate of 20 $nM^{-1}h^{-1}$ for CO, the production of CO_2 would be 4.8 $\mu gC\ l^{-1}h^{-1}$. In July 1996, we (Miller, Jones, Kieber, the author and his students) made similar measurements of CO, CO_2, hydrogen peroxide and primary production in waters from three lakes in central Ontario. We found that rates of CO production were generally about one-tenth of that for CO_2, and that the latter values were near 10% of maximal photosynthesis rates. The photo production of CO_2 from DOC is likely one reason why lakes are sources not sinks for atmospheric CO_2. With its low solubility and general slow rates of utilization, CO not only makes life difficult for organisms present but also is a principal loss of carbon from lakes and oceans. *To date,* *photochemical production of CO and CO_2 has been largely ignored in global carbon budgets but must represent a significant flux.*

Carbonyl sulphide (COS) is also produced photochemically when UV-B (only) radiation comes in contact with DOC. COS in a precursor for production of atmospheric sulphate aerosols which may serve as condensation nuclei for formation of clouds in remote locations (Hofmann 1990).

In our short history of UV radiation research we have learned some of the basics and we should now direct our attention to where problems will be the most significant. Critical processes are those where one pathway is controlled by UV and another controlled by some other physical and chemical factor. Speeding up photodegradation and flocculation by 10% is hardly perceptible on a time scale of days but over months and years could have drastic effects. Another example is the effect of UV on changing the bioavailability of iron. Since diatom production in the northern Pacific is thought to be limited by iron availability and UV radiation can increase iron availability it would seem that productivity will increase. Unfortunately, once used by phytoplankton the iron may be lost from the trophogenic zone and productivity will decline. Such hypotheses should not be left as speculation.

5.8 Summary

Shorter wavelengths of solar irradiance are attenuated more rapidly than longer wavelengths. As in clear lakes, the principal attenuating substance in humic waters is dissolved organic carbon (DOC). Consequently, 90% of UV-B (290–320 nm) radiation is attenuated in 10 to 20 cm in ponds and lakes with DOC levels above 10 mg C l^{-1}. Unfortunately, in contrast to that for clear lakes, DOC was not a useful predictor of the attenuation coefficients for UV-B or UV-A (320–400 nm). This results from changes in the quality of the DOC throughout the season perhaps due to the degree of photobleaching and alteration in the fluorescent properties of the DOC. Absorbance values measured in the laboratory with a spectrophotometer did reliably predict both UV-B and UV-A attenuation measured in the ponds with a spectroradiometer. While high DOC levels provide an effective sunscreen against the direct damage UV radiation, the high attenuation means that the photochemical activity and associated photochemical products which are normally spread over at least 15 m in the ocean is confined to only a few centimetres in humic waters. Reliable underwater spectral composition combined with quantum yields or biological weighting functions are both necessary to calculate the critical photochemical and biological responses to solar radiation.

Acknowledgements. This work was supported by a grant from the World Wildlife Toxicology Fund, co-sponsored by the World Wildlife Fund and the National Sciences and Engineering Research Council (NSERC), an NSERC Research Grant to D.L. and Environment Canada's Green Plan program.

References

Amyot M, Mierle G, Lean DRS, McQueen DJ (1997) Effect of solar radiation on the formation of dissolved gaseous mercury in temperate lakes. Geochim Cosmochim Acta 61:975–987

Cooper WJ (1989) Sunlight-induced photochemistry of humic substances in natural waters: major reactive species. In: Suffit ICH, MacCarthy P (eds) Aquatic humic substances: influence on fate and treatment of pollutants, Advances in Chemistry Series No. 219, American Chemical Society, Washington D.C., pp 333–140

Cooper WJ, Lean DRS (1992) Hydrogen peroxide dynamics in marine fresh water systems. In: Nierenberg WA (ed) Encyclopedia of earth system science, vol. 2. Academic Press, pp 527–535

Cooper WJ, Shao C, Lean DRS, Gordon AS, Scully FE (1994) The distribution of H_2O_2 in surface waters. Ch. 12. In: Baker LA (ed) Environmental chemistry of lakes and reservoirs. Advances in Chemistry Series # 237. ACS Pub Washington D.C., pp 391–422

Environment Canada (1994) Analytical methods manual. National Water Research Institute, Burlington, Ontario, Canada, 385 pp

Graneli W, Lindell M, Tranvik L (1996) Photo-oxidative production of dissolved inorganic carbon in lakes of different humic content. Limnol Oceanogr 41:698–706

Hofmann DJ (1990) Increase in stratospheric background sulfuric acid aerosol mas in the past 10 years. Science 249:996–1000

Hoigne J, Faust BC, Haag WR., Scully FE Jr, Zepp RG (1989) Aquatic humic substances as sources and sinks of photochemically produced transient reactants. In: Suffit ICH, MacCarthy P (eds) Aquatic humic substances: influence on fate and treatment of pollutants, Advances in Chemistry Series No. 219, American Chemical Society, Washington D.C., pp 329–332

Jones RD (1991) Carbon Monoxide and methane distribution and consumption in the photic zone of the Sargasso Sea. Deep Sea Res 38:625–632

Kerr JB, McElroy CT (1993) Evidence for large upward trends of UV-B radiation linked to ozone depletion. Science 262:1032–1034

Kieber DJ, McDaniel J, Mopper K (1989) Photochemical source of biological substrates in sea water: implications for carbon cycling. Nature 341:637–369

Kieber RJ, Zhou X, Mopper K (1990) Formation of carbonyl compounds from UV-induced photodegradation of humic substances in natural waters: fate of riverine carbon in the sea. Limnol Oceanogr 35:1503–1510

Kirk JTO (1994) Optics and UV-B radiation in natural waters. Arch Hydrobiol Beih Ergeb Limnol 43:1–16

Kirk JTO, Hargreaves BR, Morris DP et al. (1994) Measurements of UV-B radiation in two freshwater lakes: an instrument intercomparison. Arch Hydrobiol Beih Ergeb Limnol 43:71–99

Krabbenhoft DP, Hurley JP, Olson ML, Cleckner LB (1997) Diel variability of mercury phases and species distributions in the Florida Everglades. Biogeochem (in press)

Laurion I, Vincent WF, Lean DRS (1997) Underwater ultraviolet radiation: development of spectral models for northern high latitude lakes. Photochem Photobiol 65:107–114

Lean DRS, Cooper WJ, Pick FR (1994) H_2O_2 formation and decay in lake-waters, Ch. 12. In: Helz GR et al. (eds) Aquatic and surface photochemistry. Lewis Pub, pp 207–214

Lean DRS, Coulson D, Crump D, Berril M. Attenuation of ultraviolet radiation in humic waters: influence of dissolved organic carbon (submitted)

Lindell MJ, Graneli W, Tanvik LJ (1995) Enhanced bacterial growth in response to photochemical transformation of dissolved organic matter. Limnol Oceanogr 40:195–199

Madronich S (1993) UV radiation in the natural and perturbed atmosphere. In: Tevini, M. (Ed.): UV-B radiation and ozone depletion: effects on humans, animals, plants, microorganisms and materials. Lewis, pp 17–69

Mierle G, Ingram R (1991) The role of humic substances in the mobilization of mercury from watersheds. Water Air Soil Pollut 56:349–358

Miller WL (1994) Recent advances in the photochemistry of natural dissolved organic matter, Ch 7. In: Helz GR et al. (eds) Aquatic and surface photochemistry. Lewis Pub, pp 111–128

Mopper K, Zhou X (1990) Hydroxyl radical photoproduction in the sea and its potential impact on marine processes. Science 250:661–663

Morris DP, Zagarese H, Williamson CE et al. (1995) The attenuation of solar UV radiation in lakes and the role of dissolved organic carbon. Limnol Oceanogr 40:1381–1391

Schindler DW, Curtis PJ, Parker BR, Stainton MP (1996) Consequences of climate warming and lake acidification for UV-B penetration in North American boreal lakes. Nature 379:705–708

Scully NH, Lean DRS (1994) The attenuation of ultraviolet radiation in temperate lakes. Arch Hydrobiol Beih Ergebn der limnologie 43:135–144

Sellers P, Kelly CA, Rudd JWM, MacHutchon AR (1996) Photodegradation of methylmercury in lakes. Nature 380:694–697

St. Louis VL, Rudd JW M, Kelly CA et al (1996) Production and loss of methylmercury and loss of total mercury from boreal forest catchments containing different types of wetlands. Environ Sci Technol 30:2719–2729

Thompson AM, Owens MA, Stewart RW (1989) Sensitivity of tropospheric hydrogen peroxide to global chemical and climate change. Geophysical Res Letters 16:53–56

Thompson AM (1992) The oxidizing capacity of the earth's atmosphere: probable past and future changes. Science 256:1157–1159

Toy LJ (1998) Hydrogen peroxide production in humic ponds. M Sc Thesis, Trent University, 122 pp

Valentine RL and Zepp RG (1991) Formation of carbon monoxide from the photo-degradation of terrestrial dissolved organic carbon in natural waters. Environ Sci Technol 27:409–418

Watras CJ, Morrison KA, Host JS (1995) Concentration of mercury species in relationship to other site-specific factors in the surface waters of northern Wisconsin lakes. Limnol Oceanogr 40:556–565

Willey JD, Kieber RJ, Lancaster RD (1996). Coastal rainwater hydrogen peroxide: concentration and deposition. J Atmos Chem 25:149–165

Yan ND, Keller W, Scully NM et al. (1996) Increased UV-B penetration in a lake owing to drought-induced acidification. Nature 381:141–143

Zafiriou OC, Blough NV, Micinski E, Fister G, Kieber D, and Moffett J (1990) Molecular probe systems for reactive transients in natural waters. Marine Chem. 30:45–55

6 Effects of UV Radiation on Aquatic Humus: Photochemical Principles and Experimental Considerations

William L. Miller

6.1 Introduction

The field of natural water photochemistry has evolved over the last 20 years from a novelty to a necessity in the study of environmental chemistry in natural surface waters. As a consequence of this increasing awareness, there has been a sizable accumulation of reviews on the subject of marine and freshwater organic photochemistry. Previous publications have provided overviews, examples, and predictions of the potential significance of aquatic photochemistry (Zafiriou 1977, 1983; Zika 1981; Zafiriou et al. 1984). Other reviews have dealt with the photochemical processes specifically concerning aquatic humic substances (Choudhry 1981; Zepp 1988; Hoigné et al. 1989; Miller 1994) and the photochemical generation of reactive species (Cooper et al. 1989; Blough and Zepp 1995). A wealth of information is found in these papers and the reader is referred to them for additional discussions and references to topics not thoroughly covered here.

To summarize, sunlight-induced alteration of dissolved aquatic humic substances (DHS) results in reduction of dissolved organic carbon (DOC) average molecular weight, changes in water optical properties, and in the production of a complex mixture of reactive oxygen species (Blough and Zepp 1995) and carbon photoproducts (Miller 1994; Zepp et al. 1995). The largest of these photochemical carbon products results from the direct mineralization of DOC to CO_2 and/or inorganic carbonate species (Kuhnle et al. 1972; Chen et al. 1978; Miles and Brezonik 1981; Kotzias et al. 1986, 1987; Allard et al. 1994; Salonen and Vähätalo 1994; Miller and Zepp 1995; Granéli et al. 1996; Li et al. 1996; Lindell 1996). Both IR and NMR spectroscopy suggest that CO_2 production from photodecarboxylation is a dominant reaction in aquatic humic material (Chen et al. 1978; Li et al. 1996). Also among the carbon photoproducts are a suite of low molecular weight (LMW) compounds including carbonyls and organic acids.

Reports of these LMW carbon compounds originating from humic material have generated great interest in the significance of photochemistry to secondary biological production. Moran and Zepp (1997) recently reviewed

Ecological Studies, Vol. 133
Hessen, Tranvik (eds) Aquatic Humic Substances

the literature on biologically available compounds generated by photochemical reactions involving chromophoric dissolved organic material (CDOM). They list 13 different carbon substrates from 8 separate references and list 6 studies that noted growth enhancement of heterotrophic bacteria in natural samples exposed to sunlight. In contrast, Amon and Benner (1996) have noted the lack of bacterial growth stimulation by photochemistry in samples from the Amazon even though sunlight clearly altered their samples as evidenced by photochemical O_2 consumption and loss of DOC. Thomas and Lara (1995) reported that aged exudates from marine algae were resistant to photochemical degradation and saw no loss of DOC under natural levels of UV radiation. Naganuma et al. (1996) even showed a *reduction* in the ability of DOC (0.5% peptone solutions) to support bacterial growth after exposed to UV-B radiation.

In light of recent interest and conflicting results concerning the interaction between photochemical and biological processes in natural waters, a review of some fundamental photochemical concepts along with their potential impact on the interpretation of results from photochemical experiments with natural waters seems warranted. This Chapter will present an abbreviated version of some fundamentals for photochemistry and consider common processes which result during the course of experimental irradiations. While much of the following discussion will seem elementary to those familiar with photochemical methodologies, this Chapter should provide useful information for those new to aquatic photochemistry and consequently prevent erroneous interpretation of experimental results due to oversight of basic concepts.

6.2 Some Fundamentals of Aquatic Photochemistry

6.2.1 Absorption of Light

The first law of photochemistry, attributed to Grotthus and Draper in the early 1800s, states that only absorbed radiation is effective in producing photochemical changes. Stated another way, if the system (natural waters for this discussion) does not absorb light, photochemical changes will not occur. Consequently, absorption of incident radiation is *the* fundamental event behind all environmental photochemistry.

Quantifying the absorbance of radiation of a particular wavelength (λ) requires only a comparison of the intensity of the radiation entering a defined geometry of sample ($I_{o\lambda}$) with that exiting it (I_λ). Most spectrophotometers compare this ratio for dissolved compounds using a $\log_{10}$ scale which results in the measurement commonly called absorbance (A_λ):

$$A_\lambda = \log_{10} (I_{o\lambda}/I_\lambda). \quad (1)$$

The Lambert-Beer law describes the rate of light absorption in a homogeneous solution over pathlength l as being directly proportional to radiation intensity,

$$-(d\, I_\lambda / d\, l) = a_\lambda I_\lambda, \tag{2}$$

where a_λ is the absorptivity, a representation of the ability of the substance to absorb light at a given wavelength. Solving this differential yields

$$ln\,(I_\lambda / I_{o\lambda}) = -\, a_\lambda l. \tag{3}$$

Using the relationship $\alpha_\lambda = a_\lambda/2.303$, Eq. (1) can be written as

$$A_\lambda = \alpha_\lambda l. \tag{4}$$

When the molar concentration, [C], for the absorbing substance is known, the absorptivity is further defined as

$$\alpha_\lambda = \varepsilon_\lambda\, [C], \tag{5}$$

where ε_λ is the molar absorptivity.

The energy gained by a molecule from absorbed radiation (ΔE) is dependent on its wavelength as given by the Bohr equation:

$$\Delta E = h\nu = \frac{hc}{\lambda}, \tag{6}$$

where h is Planck's constant (6.63×10^{-34} J s), c is the speed of light (3×10^{8} m s^{-1}), and ν is the radiation frequency (s^{-1}). Accordingly, the energy contained in a mole of photons (1 einstein = 6.023×10^{23} photons) is inversely proportional to the wavelength of the radiation. Since atmospheric components such as ozone absorb virtually all solar radiation less than 290 nm before it reaches the earth's surface, UV-B (290–320 nm) and UV-A (320–400 nm) radiation provide the most energetic radiation available for photochemical reactions in surface waters.

A quick perusal of any one of the reviews listed above clearly shows that for most fresh and marine waters, CDOM (and the complex organic mixture called "humics" it contains) dominates the absorption of solar radiation in the energetic UV spectral region. It is in this role of "light gatherer" that humic material plays a pivotal role in environmental photochemistry. Its complexity, however, prevents precise definition of the molar concentration of chromophores present and restricts the use of "molar" absorptivities for CDOM. Consequently, light absorption by CDOM in natural waters is usually expressed by conversion of the absorbance measured in a spectrophotometer [Eq. (1)] to an absorptivity (m^{-1}):

$$a_\lambda = A_\lambda\, 2.303/l, \tag{7}$$

with spectrophotometric pathlength expressed in meters. This convention eliminates confusion involving reports of A_λ with no information on the pathlength used to make the measurement.

Optical analysis of natural surface waters from many different locations has resulted in a general description of the wavelength-dependent absorptivity. Because most natural waters exhibit a featureless exponential decrease in absorbance from the UV to the visible portion of the spectrum, the measured absorptivity can usually be fit to an equation of the form

$$a_\lambda = a_r 10^{[S(r-\lambda)]} \tag{8}$$

where a_λ and a_r are the absorptivities (m^{-1}) at wavelength λ (nm) and some reference wavelength (*r*). The parameter S is the slope of the log linearized spectrum, a measure of the rate at which absorptivity decreases at longer wavelengths. This general relationship has also been directly related to DOC for a variety of marine and freshwater samples by Zepp and Schlotzhauer (1981). Driven in part by efforts to quantify the CDOM signal present in remotely sensed optical data, various estimates of S have been published for natural waters. Typical values for S can be found in a number of references (Bricaud et al. 1981; Zepp and Schlotzhauer 1981; Carder et al. 1989; Blough et al. 1993; Green and Blough 1994). Initially, distinct differences between marine and freshwaters [$S_{(marine)} > S_{(fresh)}$] suggested that variations in S might be explained by simple mixing between specific end members with different values for S. Recent results, however, indicate this relationship may be more complex than originally thought, with photochemical conditioning also contributing to variations in S (Vodacek et al. 1997).

6.2.2 Photochemical Efficiency

Even though the first law of photochemistry states that light must be absorbed to affect photochemical change, all light that *is* absorbed does not contribute to photochemical reactions. When light energy is absorbed, an electronic transition occurs which elevates an electron from one orbital to another of higher energy. Once "energized," the molecule will rapidly return to its ground state by various paths. Generally, the pathways available to accomplish this *without* resulting in a chemical reaction include internal conversion (energy loss within singlet spin states), intersystem crossing (transition between singlet and triplet spin states), and emission of light energy in the form of fluorescence (transition from the lowest vibrational levels of the excited singlet to a ground-state singlet) or phosphorescence (transition from the lowest vibrational levels of the excited triplet to a ground-state singlet). A schematic diagram of these intramolecular photophysical processes is shown in Fig. 6.1 and additional details can be found in specialized texts such as Balzani and Carassiti (1970) or Horspool (1976).

Some portion of the absorbed energy which changes the molecular orbital energetics can also result in fundamental chemical changes in the molecule. Comparatively, energy transfers involving high energy excited states are much more rapid (deactivation $\leq 10^{-10}$ s) than those involving the lowest singlet and triplet states. Consequently, photophysical relaxation can result in

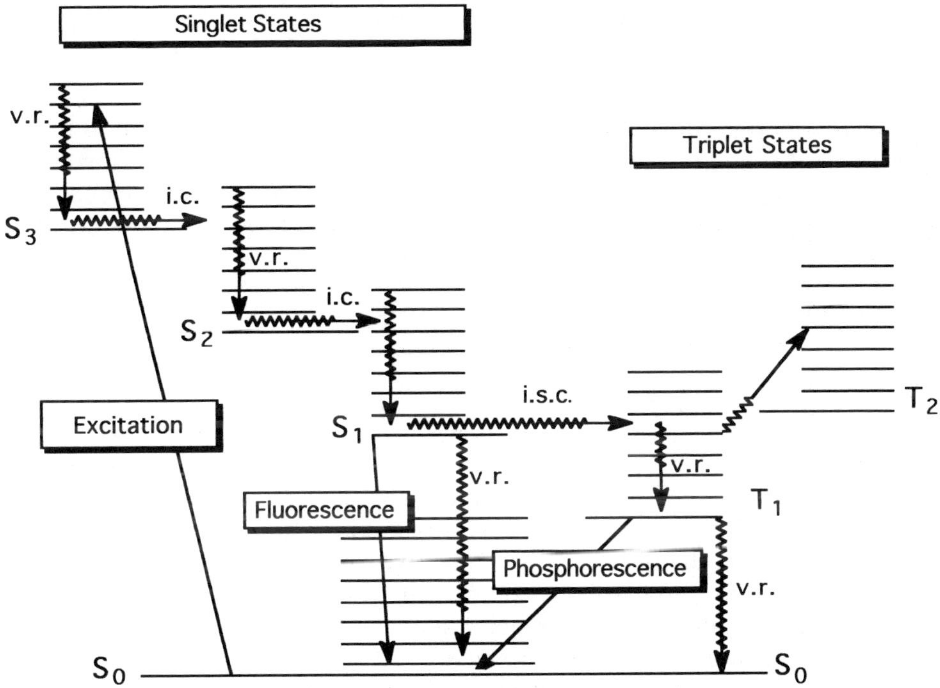

Fig. 6.1. Molecular electronic transitions involved in absorbance of light. *S* singlet spin state; *T* triplet spin state; *v.r.* vibrational relaxation; *i.c.* internal conversion; *i.s.c.* intersystem crossing

efficient S1 to T1 intersystem crossing, populating the lowest triplet state, T1, which can exhibit mean lifetimes of 10^{-3} s or longer. These longer-lived triplets are thought to be responsible for much of the primary photochemical behaviour exhibited by organic material in irradiated natural waters. Energized humic material can also serve as a sensitizer for secondary reactions, inducing photochemical reactions in compounds that do not absorb solar radiation directly. The progression of these types of reactions depends not only on humic reactions but also on the availability of various other compounds. Due to the potential complexity of photosensitized reactions, and because the primary intention of this chapter is to examine the photochemistry directly affecting aquatic humus, the following discussion will deal with primary reactions. Further information and references to papers on photosensitized reactions can be found in Zepp et al. (1985), Zepp (1988), and Brezonik (1994).

To approach both photophysical and photochemical primary processes in a quantitative manner, it is required that a relative efficiency be assigned to the process in question (e.g. DOM fluorescence, radical production, carbon product formation). In the most general terms, this efficiency, designated quantum yield (Φ), can be described as

$$\Phi = \frac{\text{Number of photoinduced events}}{\text{Number of photons absorbed}}. \tag{9}$$

For primary processes, Φ integrated for all events must equal unity. In other words, all of the energy gained by absorption of a photon must be equal to the sum of all of the subsequent energy transfers resulting in both photophysical and/or photochemical reactions. For quantitative consideration of photochemical reactions, a more specific definition is needed:

$$\Phi_\lambda = \frac{\text{Moles of product formed (or reactant lost)}}{\text{Moles of photons absorbed by recent}}. \tag{10}$$

The wavelength designation is required because the energy gained by absorption of any photon is related to wavelength [Eq. (6)]. Unlike the energetic balance for the total light absorbed by the system, photochemical reactions do not require Φ_λ values of 1. In many reactions, Φ_λ is less than 1 since some portion of the absorbed energy goes to processes other than photochemistry such as fluorescence. Also, any given photochemical reaction directly competes for absorbed energy with all other potential photochemical reactions; potentially a large number when considering a complex organic compound like humus. For certain reactions, it is also possible for Φ_λ to be greater than 1 due to secondary reactions involving radical transients or chain reactions. Many photochemical reactions exhibit a constant value for Φ_λ below a certain critical wavelength where the absorbed energy is sufficient to fully populate the T1 or S1 transitions. There is also a critical wavelength *above* which the energy gained from absorption of $I_{o\lambda}$ is not sufficient to promote photochemical change and the quantum yield is zero. The transition between $\Phi_\lambda = 0$ and its maximal value can span tens or hundreds of nanometers.

For many reactions involving humic substances, maximal Φ_λ requires energy transitions well above the level supplied by solar UV-B. Consequently, the quantum yield spectra applicable to sunlit natural waters usually have high values in the UV-B which tail to zero in the UV-A or visible depending on the particular reaction. Not surprisingly, these Φ_λ spectra are often similar in shape to the a_λ spectrum for CDOM. Two previously published Φ_λ spectra are shown in Fig. 6.2 (formaldehyde, Kieber et al. 1990; carbon monoxide, Valentine and Zepp 1990). It is clear from Fig. 6.2 that different photochemical reactions can have different wavelength efficiencies even though CDOM is likely to be the initiator of both of these reactions. This presumably reflects the fact that humic matter, and consequently the CDOM, consists of a complex mixture of potential electron transitions and chemical functionality.

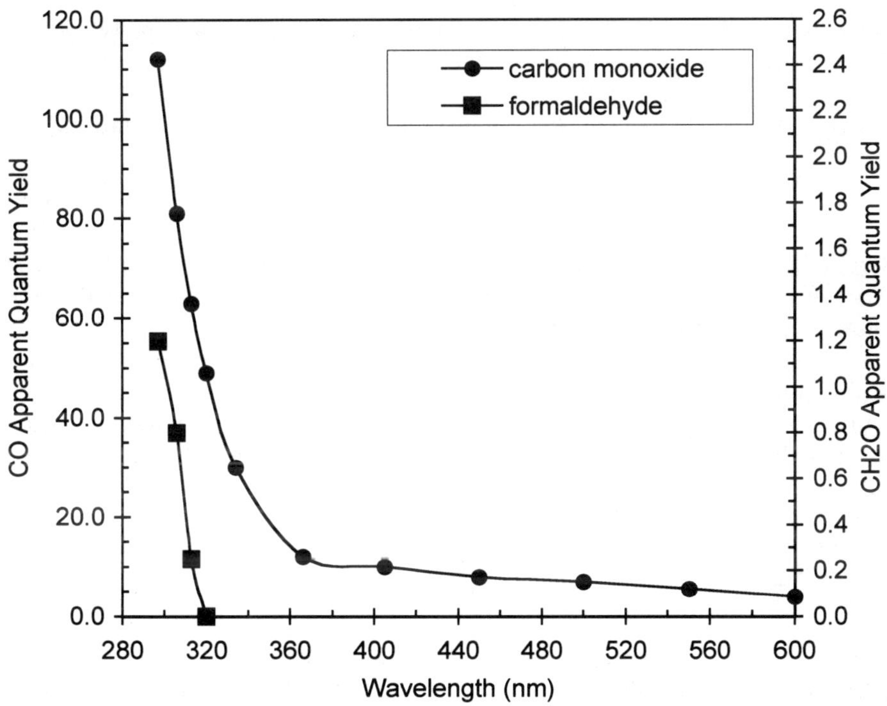

Fig. 6.2. Apparent quantum yield spectra for formaldehyde and carbon monoxide (Data from Kieber et al. 1990; Valentine and Zepp 1993)

6.2.3 Reaction Rates

Much of the interest in photochemical reactions associated with aquatic humic material concerns the rate at which carbon products are formed and/or the rate that chemical characteristics of the remaining humic molecule change. There is a growing body of evidence to suggest that photochemical reaction rates greatly affect the rate of biological consumption for humic substances, through either production of consumable LMW carbon compounds (see review by Moran and Zepp 1997) or chemical modification of the larger humic molecule (Miller and Moran 1997). Consequently, an understanding of the rate at which the microbial community oxidizes carbon in sunlit waters may require direct knowledge of photochemical kinetics.

In the simplest terms, the rate of a given photochemical reaction can be expressed by:

$$\frac{d[P]}{dt} = \sum_{\lambda} \phi_{\lambda} I_{a\lambda} \,, \tag{11}$$

where [P] is the measured molar concentration of the reactant or product and $I_{a\lambda}$ is the average light absorption rate (einsteins/l/s). Note that the quantity $I_{a\lambda}$ is not equivalent to incident light measured with a radiometer, I_o, which is usually reported in units of energy flux per area (e.g. W/cm^2 or J/cm^2). For the case where the photochemical reactant is the only chromophore in the system, Balzani and Carassiti (1970) note that $I_{a\lambda}$ and I_o (converted to einsteins / cm^2 / s) can be related by the following equation.

$$I_{a\lambda} = I_{o\lambda}(1-10^{-\varepsilon_\lambda [C]l})\frac{Area}{V}. \tag{12}$$

Here, [C] is the concentration of the chromophore, *Area* is the surface area (cm^2) of a columnated light beam passing through the sample, and *V* is the volume (l) of solution being irradiated.

In practical application to aquatic photochemistry, the reactant may not be the sole light absorber in the system. By expanding Eq. (11) to include light losses which do not lead to photochemical reactions and substituting Eq. (12), the reaction rate of a unimolecular, primary photochemical process can be described as follows

$$\frac{d[P]}{dt} = \sum_\lambda \Phi_\lambda I_{o\lambda}\left(1-10^{-(\alpha_\lambda+\varepsilon_\lambda [C])l}\right)\frac{Area}{V}F_\lambda \tag{13}$$

where α_λ is the absorptivity (cm^{-1}) of all components of the system other than the reactant and F_λ is the fraction of light absorbed by the reactant and is defined by

$$F_\lambda = \frac{\varepsilon_\lambda [C]}{\alpha_\lambda + \varepsilon_\lambda [C]}. \tag{14}$$

When the reactant under consideration accounts for the majority of light absorbance ($F_\lambda \cong 1$; $\alpha_\lambda \cong 0$), the following special relationships apply. For optically thick solutions [i.e., the reactant absorbs virtually all of the light within the system ($\varepsilon_\lambda[C]l > 2$)], Eq. (13) reduces to

$$\frac{d[P]}{dt} = \sum_\lambda \Phi_\lambda I_{o\lambda}\frac{Area}{V} \tag{15}$$

and the reaction is zero order. For an optically thin solution ($\varepsilon_\lambda[C]l < 0.05$), Eq. (13) reduces to

$$\frac{d[P]}{dt} = 2.303\sum_\lambda \Phi_\lambda I_{o\lambda}\frac{Area}{V}\varepsilon_\lambda [C]l \tag{16}$$

and the reaction is first order with respect to the concentration of the reactant.

From the brief presentation above, it is quite clear that exposure geometry and light attenuation must be precisely defined to accurately use photochemical rate equations. For additional reading and references on light attenuation and kinetics in photochemical experiments along with numerical models appropriate to surface water photochemistry, the reader is referred to Leifer (1988) and Brezonik (1994).

For natural waters where CDOM (and the humics which largely comprise CDOM) initiates most photochemical reactions, the molar concentration of photochemically reactive sites is ill-defined. Consequently, the term "$\varepsilon_\lambda[C]$" is not very useful for natural waters. By assuming a consistent irradiation geometry and a solution that is essentially clear, it is common practice to substitute the measured absorptivity, a_λ, into Eq. (13) and simplify to give the following.

$$\frac{d[P]}{dt} = \sum_{\lambda} {}_{a}\Phi_{\lambda} I_{o\lambda}\left(1 - 10^{-a_\lambda}\right) \tag{17}$$

Substitution of a_λ requires the use of an *apparent* quantum yield, ${}_{a}\Phi_{\lambda}$, to reflect the loss of a direct molar relationship between the primary absorber and the photochemical result. While the use of a_λ can simplify the description of light absorption when chromophore concentration is not known, examination of Eq. (13) reminds us that the photochemical rate remains very sensitive to irradiation geometry and exposure design: container characteristics and/or F_λ can have a dramatic effect on observed rates.

6.3 Evaluating Photochemical Rates for Natural Waters

There are two basic experimental approaches to evaluate the significance of photochemical reactions involving humics in natural waters. One is to measure chemical changes in irradiated surface waters directly in the field and correlate measurements with light intensity and/or integrated irradiation. This method gives a direct indication of the potential magnitude for a given photochemical process, thereby putting constraints on photochemical significance.

The second approach is to expose natural waters to irradiation inside a container. These experiments are common and are usually required to show conclusively that photochemistry, and not some other process which also varies with irradiation, accounts for the observed phenomenon. Contained irradiations allow precise knowledge of the composition of the sample and the possibility of experimental manipulation to eliminate other processes. In many cases, the trace amounts of product formed by photochemical reactions involving humic material can only be measured when allowed to accumulate in a container. Additionally, the samples can be subjected to a controlled light field which allows evaluation of the spectral efficiency of the photochemical process being studied.

Eventually, measurements made by these two methods must be justified with one another. Results from controlled irradiations should be comparable with observations made in the field. Perhaps the most difficult aspect of this extrapolation is the comparison between two light fields of differing geometry and intensity. This review will not present the complexities of a rigorous description of the in situ scalar irradiance in sunlit waters. The reader is encouraged to study any one of the available treatments of this subject (Jerlov 1976; Preisendorfer 1976; Kirk 1983; Mobley 1994; Spinrad et al. 1994; Ackleson 1995) for a comprehensive treatment. Quantifying the irradiation parameters related to contained samples, however, presents a more tractable problem and will be examined here.

Controlled irradiations are of little quantitative use if they cannot be related to real environmental processes and to similar research done by others. As evidenced by Eq. (11), the precise definition of the average light absorption rate, $I_{a\lambda}$, is required to extrapolate or compare rates. Failure to carefully consider $I_{a\lambda}$ and the parameters it contains (spectral absorbance, irradiation geometry and spectral distribution, and pathlength) can lead to erroneous interpretation of photochemical results and faulty comparisons with other research.

These considerations are particularly important when exposing samples to natural sunlight by suspending containers in the water column and/or in incubation baths. Reflection and refraction of light at container surfaces can alter the effective pathlength of exposure. It is quite possible that a sample in a round quartz flask will give a significantly different photochemical result from the same sample exposed beside it in a quartz test tube. Without careful attention to exposure geometry and light field, these effects can completely undermine careful analysis of photochemical products.

As shown in Eqs. (11) and (12), the three components critical in the description of photochemical rates for humic substances (Φ_λ, $I_{o\lambda}$, a_λ) all usually vary with wavelength. Consequently, the interpretation of observed photochemical rates can be wrong without consideration of the wavelength-dependence for each of these parameters. Different spectral transmissivity of container walls and wavelength differences in light sources (particularly in the UV where photon energy is high) can significantly alter photochemical results and make comparison to other experiments difficult, if not impossible.

6.3.1 Apparent Quantum Yield

As an example, we can examine the spectral dependence of carbon monoxide (CO) and formaldehyde photoproduction in several hypothetical exposures. Given a consistent radiation geometry, data for the three major variables contained in Eq. (17) ($I_{o\lambda}$, a_λ and ${}_a\Phi_\lambda$) allow comparison of relative photochemical rates. The quantum yield spectra for these two carbon products are shown in Fig. 6.2. Absorbance spectra for waters with two different DOC concentrations can be generated using Eq. (8) and data from Zepp and Schlotzhauer (1981). Fig. 6.3 shows the spectral overlap between these hypothetical absorbance spectra and modelled values for solar irradiance (Zepp and Cline 1977) as filtered though three different glasses (Pyrex®,

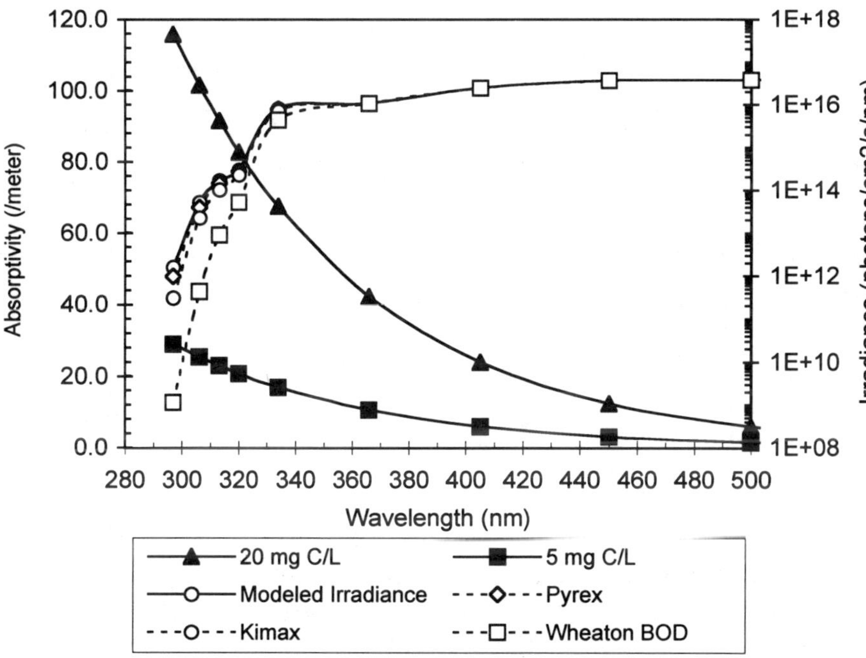

Fig. 6.3. Spectral distribution for modeled absorbance of natural waters with differing DOC (*dark symbols*) and for modeled irradiation intensity of sunlight passing through three types of glass

Kimax®, and a Wheaton BOD (biological oxygen demand) bottle). Multiplying $I_{o\lambda}$, a_λ and ${}_a\Phi_\lambda$ together at equivalent wavelengths gives the action spectra shown in Fig. 6.4. Using action spectra to calculate photochemical production rates requires great care in defining the geometry of spectral irradiance and attention to the units used in the calculations (see Leifer 1988 for detailed discussion). The action spectra as presented in Fig. 6.4 do not serve this purpose. Rather, they serve only to compare the expected spectral response for photochemical production rates.

In the case of formaldehyde and CO, a striking difference can be seen between the two spectral responses. Formaldehyde production depends almost entirely on UV-B radiation, with little or no photoproduction at longer wavelengths. Carbon monoxide, on the other hand, exhibits maximum production in the UV-A, with significant production in the visible. Fig. 6.4 shows clearly that exposing samples to sunlight without UV-B radiation may completely eliminate the photochemical production of formaldehyde while having very little effect on the production of CO.

Understanding the spectral response of discrete photochemical reactions can be critical to the accurate evaluation of experimental results. Comparing the photoproduction of CO with a suite of low molecular weight compounds including formaldehyde (CO_2 was not measured), Mopper et al. (1991) showed that CO accounts for about 60 to 80% of the total carbon photo-

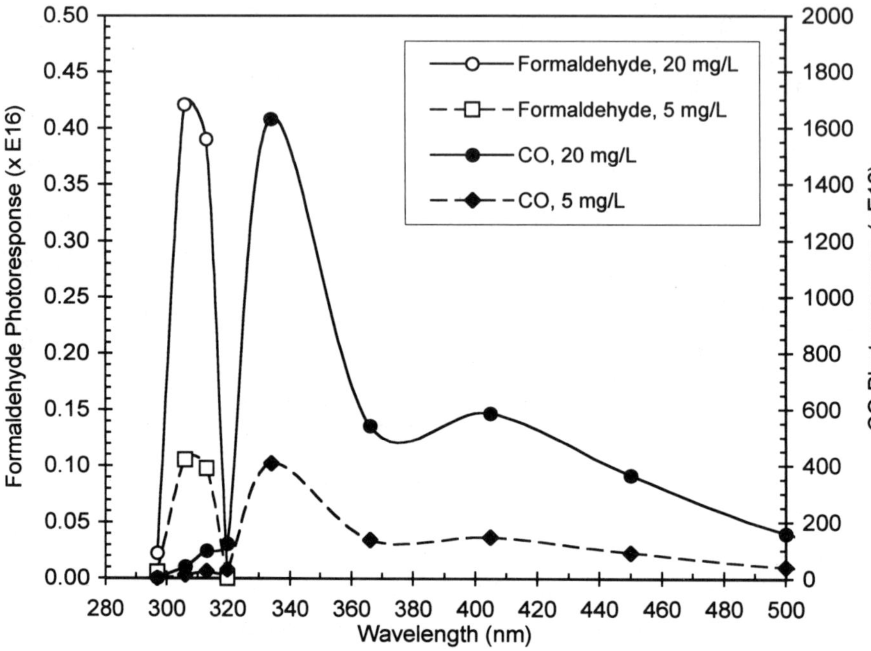

Fig. 6.4. Modeled photoresponse of carbon monoxide and formaldehyde production as a function of wavelength using the modeled absorbance spectra in Fig. 6.3

products measured. Limited data from both freshwaters and marine systems suggest that the photoproduction of dissolved inorganic carbon (DIC) from DOC is about 15 to 20 times larger than CO (Miller and Zepp 1995; Miller and Moran 1997). Even though few data currently exist, preliminary results suggest that DIC photoproduction exhibits a ${}_a\Phi_\lambda$ response similar in shape to CO. Consequently, spectral screening of UV-B by container walls and/or by in situ exposures at depths with little or no UV-B penetration would not necessarily prevent photochemical reactions that result in DOC loss from CO plus DIC formation. It could, however, prevent photoproduction of small carbonyl compounds. If these are the source of biological consumables which stimulate microbial growth, one might correctly observe strong evidence for photochemical reactivity of humics with little effect on their apparent microbial availability. Extension of this interpretation to the field where UV-B does impact the sample would be wrong.

It should be noted here that this example depends entirely on the accuracy of the quantum yield data since all other parameters for the calculation of CO and formaldehyde photoresponse are equivalent. Unfortunately, very few publications include quantum yield data for photochemical reactions involving humic material in natural waters. Spectral data for apparent quantum yields for CO (Valentine and Zepp 1993; Kettle 1994), formaldehyde (Kieber et al. 1990), and hydrogen peroxide (Cooper et al. 1988; Moore et al.

1993) have been published for a limited number of samples. While these studies suggest a general consistency for the ${}_a\Phi_\lambda$ of a given product within different water types, a complete understanding of how ${}_a\Phi_\lambda$ varies with differing CDOM source material, water chemistry, temperature, and photochemical exposure history still requires additional study.

6.3.2 Spectral Irradiance

As described by Eq. (17), the spectral quality of irradiance will also have a controlling effect on observed photochemical production rates. Using the case of formaldehyde photoproduction from the previous example, a closer examination of the potential effect of container composition is shown in Fig. 6.5. With a constant quantum yield and absorptivity, alteration of the sunlight irradiance spectrum by passing through three different glass containers can have large effects. Even high quality quartz glass will not transmit 100% of $I_{o\lambda}$ at its maximum transmissivity, the exact value depending on thickness, imperfections, and refractive index variations. Consequently, for Fig. 6.5, each glass was scaled to 100% transmission at wavelengths above 400 nm to isolate only those differences in the transmission of UV radiation.

Again, Fig. 6.5 makes it clear that changes in the UV-B portion of the solar spectrum could have a drastic effect on formaldehyde photoproduction. The requirement of precise knowledge of the quantity and quality of UV radiation present in photochemical experiments which employ broad-band irradiation can present a large problem. Ideally, a spectroradiometer and a competent operator can provide irradiance values for each wavelength throughout the spectrum. Unfortunately, these instruments are expensive and not yet easily accessible to many researchers. Radiometers with a spectral response spanning many wavelengths are less expensive and more common. When scaling photochemical results to radiation measured with a broad-band radiometer, however, much information is lost and faulty interpretations can occur.

For example, a radiometer which measures the total solar UV radiation between 290 and 400 nm would display a very small difference in total irradiance between unfiltered sunlight and sunlight which had passed through the BOD bottle shown in Fig. 6.3 (note the log scale for irradiance). Consequently, the radiometer measurement may suggest that there should be only a very small decrease in photochemical production. While this might actually be the case for CO production, Fig. 6.5 shows clearly that this conclusion is very wrong for formaldehyde production. Only when the spectral response of the detector exactly matches the spectral response for the photochemical reaction in question (${}_a\Phi_\lambda$) will the difference in measured irradiance exactly reflect photochemical changes.

Actinometry can provide precise knowledge of the total irradiance inside an experimental exposure. Aquatic actinometers (see Murov 1973; Leifer 1988), by nature of their well-defined reaction quantum yields and their re-

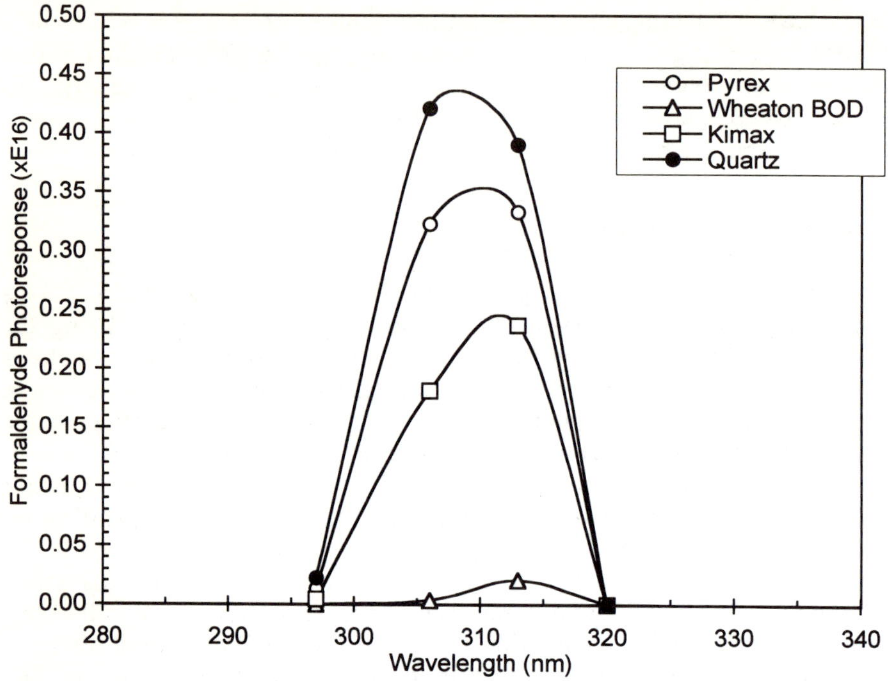

Fig. 6.5. Modeled spectral response for photoproduction of formaldehyde using sunlight filtered through various types of glass (spectra from Fig. 6.3)

fractive indices which closely match natural samples, can provide a direct measure of the integrated light intensity available for humic-mediated photochemical reactions. Any reflection or refraction affecting the light inside the container will be accounted for by actinometry. In monochromatic irradiations, actinometry is a very accurate measure of total irradiation relevant to photochemistry. For full spectral sunlight, however, actinometers will integrate irradiance over some known bandwidth specific for the particular actinometer. As in radiometric measurements, the spectral response of the actinometer may not match that of the photochemical reaction in question. Regardless of these limitations, it remains important to report any available measurement relative to the intensity and spectral quality of irradiation *inside* the container to allow some comparison with the work of others.

6.3.3 Absorptivity

Again from Eq. 17, the light absorbed by the sample (CDOM in natural waters) also exhibits direct control over observed production rates. As mentioned in section 6.2.1, a spectrophotometer can directly measure the sample absorptivity as a function of wavelength. Perhaps because of the straightforward nature of this measurement, the full impact of absorptivity variations in exposure experiments is sometimes ignored when evaluating photochemical results.

6.3.3.1 Self-Shading

Returning to Eq. (13), we can see that it is not only the light absorbance by the sample which controls photochemical rates, but also the ratio of sample to system absorbance. With an optically thin sample, a first order equation (Eq. 16) describes the photo-kinetics. A zero order equation (Eq. 15) can be used when essentially all of the light is absorbed by the sample. While these extreme situations are clear enough, samples falling somewhere in between are more difficult to compare. Most short pathlength containers used in photochemical exposures result in optically thin samples for natural waters. In other words, most of the light entering the cell will leave it without being absorbed and the intensity of light on the far side of the container is essentially the same as that in the portion of the solution nearest the light source. Using the simplifying assumptions in Eq. (16), we can define the photochemical reaction as first order.

When darker waters are evaluated, the sample may absorb enough light so that the solution closest to the light source experiences irradiance of higher total intensity and with different spectral quality than that on the far side of the container. This effect can result in an average photochemical rate than does not merely reflect the differences in absorptivity. In other words, for an equivalent incident irradiance, a solution with four times more absorptivity at any given wavelength would not simply produce four times more product, as might be expected from Eq. (13).

A typical field experiment might be to survey the photochemical reactivity of waters collected from different lakes exhibiting large differences in a_λ. Examining progressively darker samples under identical exposure conditions, one might observe progressively lower absorbance normalized photochemical production rates. Based on this result alone, it would be erroneous to conclude that dark waters are less photochemically efficient than their clear counterparts. Indeed the hypothetical dark water lake *could* have a lower photochemical efficiency than a similar clear lake, but the observed rate changes may also merely reflect self-shading within the container. Comparison of photochemical efficiency in samples with different humic content should only be done after accounting for variations in light absorption.

6.3.3.2 Photochemical Fading

It is well known that exposing natural waters and humic material to sunlight results in the eventual loss of colour across the entire absorbance spectrum (Zepp 1988; Kouassi and Zika 1992; Hongve 1994; Miller 1994), a process known as photochemical fading. This phenomenon has been noted in controlled exposures and recent evidence suggests that the effect is directly observable in surface waters (DeHaan 1993; Siegel and Michaels 1996; Vodacek et al. 1997). Obviously, when fading occurs during the course of exposure, observed photochemical rates will not reflect a_λ at the beginning of the exposure, but rather some average a_λ during the experiment. Without direct knowledge of the degree of fading, comparison of observed rates can be difficult.

Use of a single end point measurement in the estimation of photochemical product formation assumes that all variables controlling the rate ($I_{o\lambda}$, a_λ, and ${}_a\Phi_\lambda$) remain constant, or at least well-defined, during the course of exposure. The magnitude that fading effects this assumption will depend on the length and intensity of the experimental exposure. Miller and Zepp (1995) have shown that significant fading can occur during solar simulation experiments designed to examine DIC photoproduction. In fact, the fading rate of CDOM in salt marsh waters proceeded so rapidly that only about 15% of the DOC would be converted to DIC before the absorbance at 350 nm faded to essentially zero. For short exposure times and sensitive analytical methods, fading may not present a problem.

Similar to self-shading artifacts, photochemical fading presents the potential for misinterpretation of photochemical experiments involving natural waters. For example, monitoring an exposure over time may reveal a nonlinear accumulation of product with a decreasing rate of change as the exposure progresses. Without knowledge of the degree of photochemical fading, one might interpret these data with a scenario involving the initial consumption of the more photoreactive moieties followed by reactions involving less efficient sites, a reasonable explanation considering the complexity of humic material. In reality, an exponential loss of colour due to photochemical fading may produce the same result. Only after simple corrections for fading and/or self-shading have been performed can the more subtle aspects of photochemical reactions involving humic material be explored.

6.4 Summary

There has been a recent realization of the importance of organic photochemistry in the cycling of humic substances, and consequently DOC, in sunlit surface waters. Many new studies have evaluated the effect of photochemistry on the production of trace carbon compounds, DOC oxidation, O_2 consumption, CDOM optical properties, and the biological availability of humic substances. The number of recent publications related to the impact of organic photochemistry on ecological and biogeochemical cycles suggests that many new researchers are entering this promising field. With this in mind, one would expect that innovations and revelations are imminent.

It is critically important, however, that sound fundamentals be applied to this relatively new field of endeavour. This Chapter has attempted to demonstrate that three basic parameters contribute direct control over photochemical rates: (1) the spectral absorbance of light; (2) the intensity and spectral quality of the radiation involved; and (3) the spectral efficiency with which collected energy is transformed into a photochemical observation. Accurate experimental intercomparisons and environmental extrapolation requires careful attention to variations in these three parameters. Employment of a sound understanding of basic fundamental principles will ensure that the field progresses solidly toward a new understanding of the roles that photochemical reactions play in the biological and geochemical cycling of humic substances in sunlit waters.

References

Ackleson SG (ed) (1995) Advances in ocean optics: issues of closure. J Geophys Res Spec Issue 100(C7)

Allard B, Borén H, Pettersson C, Zhang G (1994) Degradation of humic substances by UV radiation. Environ Int 20:97–101

Amon RMW, Benner R (1996) Photochemical and microbial consumption of dissolved organic carbon and dissolved oxygen in the Amazon River system. Geochim Cosmochim Acta 60(10):1783–1792

Balzani V, Carassiti V (1970) Photochemistry of coordination compounds. Academic Press, New York

Blough NV, Green SA (1995) Spectroscopic characterization and remote sensing of nonliving organic matter. In: Zepp RG, Sonntag CH (eds) Roles of nonliving organic matter in the earth's carbon cycle. Wiley, New York, pp 23–45

Blough NV, Zepp RG (1995) Reactive oxygen species in natural waters. In: Foote CS, Valentine JS, Greenberg A, Liebman JF (eds) Active oxygen in chemistry. Chapman and Hall, New York, pp 280–333

Blough NV, Zafiriou OC, Bonilla J (1993) Optical absorption spectra of waters from the Orinoco River outflow: terrestrial input of colored organic matter to the Caribbean. J Geophys Res 98(C2):2271–2278

Brezonik PL (1994) Chemical kinetics and process dynamics in aquatic systems. Lewis, Boca Raton

Bricaud A, Morel A, Prieur L (1981) Absorption by dissolved organic matter of the sea (yellow substance) in the UV and visible domains. Limnol Oceanogr 26(1):43–53

Carder KL, Steward RG, Harvey GR,Ortner PB (1989) Marine humic and fulvic acids: their effects on remote sensing of ocean chlorophyll. Limnol Oceanogr 34(1):68–81

Chen Y, Khan SU, Schnitzer M (1978) Ultraviolet irradiation of dilute fulvic acid solutions. Soil Sci Soc Am J 42:292–296

Choudhry GG (1981) Humic substances. II. Photophysical, photochemical and free radical characteristics. Toxicol Environ Chem 4:261–295

Cooper WJ, Zika RG, Petasne RG, Plane JMC (1988) Photochemical formation of H_2O_2 in natural waters exposed to sunlight. Environ Sci Technol 22(10):1156–1160

Cooper WJ, Zika RG, Petasne RG, Fischer AM (1989) Sunlight induced photochemistry of humic substances in natural waters: major reactive species. Adv Chem Ser 219:333–362

DeHaan H (1993) UV-light penetration and photodegradation of humic substances in peaty lake water. Limnol Oceanogr 38(5):1072–1076

Granéli W, Lindell M, Tranvik L (1996) Photo-oxidative production of dissolved inorganic carbon in lakes of different humic content. Limnol Oceanogr 41(4):698–706

Green SA, Blough NV (1994) Optical absorption and fluorescence properties of chromophoric dissolved organic matter in natural waters. Limnol Oceanogr 39:1903–1916
Hoigné J, Faust BC, Haag WR, Scully FE Jr, Zepp RG (1989) Aquatic humic substances as sources and sinks of photochemically produced transient reactants. Adv Chem Ser 219:363–381
Hongve D (1994) Sunlight degradation of aquatic humic substances. Acta Hydrochim Hydrobiol 22:117–120
Horspool WM (1976) Aspects of organic chemistry. Academic Press, New York
Jerlov NG (1976) Marine optics. Elsevier, New York
Kettle AJ (1994) A model of the temporal and spatial distribution of carbon monoxide in the mixed layer. Master's Thesis, Woods Hole Oceanographic Institute, Massachusetts Institute of Technology
Kirk JTO (1983) Light and photosynthesis in aquatic ecosystems. Cambridge University Press, Cambridge
Kieber RJ, Zhou X, Mopper K (1990) Formation of carbonyl compounds from UV-induced photodegradation of humic substances in natural waters: fate of riverine carbon in the sea. Limnol Oceanogr 35:1503–1515
Kotzias D, Herrmann A, Zsolnay A, Russi H, Korte F (1986) Photochemical reactivity of humic materials. Naturwissenschaften 73:35–36
Kotzias D, Herrmann A, Zsolnay A, Beyerle-Pfnür R, Parlar H, Korte F (1987) Photochemical aging of humic substances. Chemosphere 16:1463–1468
Kouassi AM, Zika RG (1992) Light-induced destruction of the absorbance property of dissolved organic matter in seawater. Toxicol Environ Chem 35:1950–211
Kuhnle JA, Lundin RE, Waiss AC (1972) Photodecarboxylation of dicarboxylic acids in the presence of iron(III) chloride. J Am Chem Soc Chem Commun 1:287–288
Leifer A (1988) The kinetics of environmental aquatic photochemistry: theory and practice. ACS professional reference book. American Chemical Society, Washington DC
Li JW, Yu Z, Gao M, Zhang L, Cai X, Chao F (1996) Effect of ultraviolet irradiation on the characteristics and trihalomethanes formation potential of humic acid. Water Res 30:347–350
Lindell MJ (1996) Effects of sulight on organic matter and bacteria in lakes. PhD Dissertation, Lund University, Sweden
Miles CJ, Brezonik PL (1981) Oxygen consumption in humic-colored waters by a photochemical ferrous-ferric catalytic cycle. Environ Sci Technol 15(9):1089–1095
Miller WL (1994) Recent advances in the photochemistry of natural dissolved organic matter. In: Crosby D, Helz GR, Zepp RG (eds) Aquatic and surface photochemistry, ACS symposium series. Lewis, Boca Raton
Miller WL, Zepp RG (1995) Photochemical production of dissolved inorganic carbon from terrestrial organic matter: significance to the oceanic organic carbon cycle. Geophys Res Lett 22:417–420
Miller WL, Moran MA (1997) Interaction of photochemical and microbial processes in the degradation of refractory dissolved organic matter from a coastal marine environment. Limnol Oceanogr 42:1317–1324
Mobley CD (1994) Light and water, radiative transfer in natural waters. Academic Press, San Diego
Moore CA, Farmer CT, Zika RG (1993) Influence of the Orinoco River on hydrogen peroxide distribution and production in the eastern Caribbean. J Geophys Res 98(C2):2289–2298
Mopper K, Zhou X, Kieber RJ, Kieber DJ, Sikorski RJ, Jones RD (1991) Photochemical degradation of dissolved organic carbon and its impact on the oceanic carbon cycle. Nature 353:60–62
Moran MA, Zepp RG (1997) Role of photochemistry in the formation of biologically labile compounds from dissolved organic matter. Limnol Oceanogr 42:1307–1316
Murov SL (1973) Handbook of photochemistry. Marcel Dekker, New York
Naganuma T, Konishi S, Inoue T, Nakane T, Sukizaki S (1996) Photodegradation or photoalteration? Microbial assay of the effect of UV-B on dissolved organic matter. Mar Ecol Prog Ser 135:309–310
Preisendorfer RW (1976) Hydrologic optics. US Department of Commerce, Washington DC

Thomas DN, Lara RJ (1995) Photodegradation of algal derived dissolved organic carbon. Mar Ecol Prog Ser 116:309–310

Salonen K, Vähätalo A (1994) Photochemical mineralisation of dissolved organic matter in Lake Skjervatjern. Environ Int 20:307–312

Siegel DA, Michaels AF (1996) Quantification of non-algal light attenuation in the Sargasso Sea: implications for biogeochemistry and remote sensing. Deep Sea Res 43(2–3):321–345

Spinrad RW, Carder KL, Perry MJ (1994) Ocean optics. Oxford University Press, New York

Valentine RL, Zepp RG (1993) Formation of carbon monoxide from the photodegradation of terrestrial dissolved organic carbon in natural waters. Environ Sci Technol 27(2):409–412

Vodacek A, Blough NV, DeGrandpre MD, Peltzer ET, Nelson RK (1997) Seasonal variation of CDOM and DOC in the Middle Atlantic Bight: terrestrial inputs and photooxidation. Limnol Oceanogr 42:674–686

Zafiriou OC (1977) Marine organic photochemistry previewed. Mar Chem 5:497–522

Zafiriou OC (1983) Natural water photochemistry. In: Riley JP, Chester R (eds) Chemical oceanography, vol 8. Academic Press, New York, pp 339–379

Zafiriou OC, Joussot-Dubien J, Zepp RG, Zika RG (1984) Photochemistry of natural waters. Environ Sci Technol 18(12):358A–371A

Zepp RG (1988) Environmental photoprocesses involving natural organic matter. In: Frimmel FH, Christman RF (eds) Humic substances and their role in the environment. Wiley, New York, pp 193–214

Zepp RG, Cline D (1977) Rates of direct photolysis in aquatic environment. Environ Sci Technol 11:359–366

Zepp RG, Schlotzhauer PF (1981) Comparison of photochemical behavior of various humic substances in water. III. Spectroscopic properties of humic substances. Chemosphere 10(5): 479–486

Zepp RG, Schlotzhauer PF, Sink RM (1985) Photosensitized transformations involving electronic energy transfer in natural waters: role of humic substances. Environ Sci Technol 19(1): 74–81

Zepp RG, Callaghan TV, Erickson DJ (1995) Effects of increased solar ultraviolet radiation on biogeochemical cycles. Ambio 24:181–187

Zika RG (1981) Marine organic photochemistry. In: Duursma EK, Dawson R (eds) Marine organic chemistry: evolution, composition, interactions and chemistry of organic matter in seawater. Elsevier oceanography series, vol 31. Elsevier, Amsterdam, pp 299–325

7 Phytoplankton, Primary Production and Nutrient Cycling

Roger I. Jones

7.1 Introduction

The conception of this book arose from an awareness that there has been a surge of interest in the role of humic substances (HS) in the ecology of aquatic ecosystems. This increased interest is apparent in the case of phytoplankton. A search of literature published during the period 1981–1996 and using the linked keywords *phytoplankton* and *humic* yielded 104 references, of which the majority dated from after 1990. However, closer inspection of these articles revealed that most were primarily concerned with direct involvement of HS in carbon and nutrient cycles and only a few specifically addressed the impact of HS on phytoplankton production, growth and community composition. Hence, although phytoplankton are traditionally considered to be the base of the pelagic food web in aquatic systems, and although we now suspect that food webs may function differently in humic lakes compared with clearwater lakes, there is still little information on how phytoplankton performance may be modified in humic lakes. This chapter collates and assesses available information on this aspect of the ecological role of HS in lakes.

HS might influence lake phytoplankton in two essentially distinct ways: physical and chemical. Firstly, the presence of HS in lake water affects the penetration of solar radiation down the water column (Chapter 5) and in particular the penetration of potentially damaging UV-B radiation (Chapter 6). This effect of HS on light penetration could directly affect phytoplankton photosynthesis and growth, but could also affect phytoplankton performance indirectly through impacts on water column vertical structure and stability. Secondly, HS might critically affect the chemical environment of phytoplankton by altering the bioavailability both of essential nutrient elements and of potentially toxic chemicals. There is also the possibility that HS might be utilised directly by some phytoplankton as a source of carbon and energy. These potential impacts of HS on lake phytoplankton will be reviewed in turn.

Ecological Studies, Vol. 133
Hessen, Tranvik (eds) Aquatic Humic Substances

If the physical and chemical effects of HS on phytoplankton are sufficiently pronounced, it can be expected that phytoplankton in humic lakes should show distinctive physiological adaptations or behavioural strategies to enable them to cope effectively with the particular environmental conditions imposed by the presence of HS. These aspects of the influence of HS on the ecology of phytoplankton are also reviewed (see Section 7.4). Such adaptations or strategies would be expected to be most evident in lakes with high concentrations of HS. This Chapter finishes by considering whether such "humic lakes" represent a sufficiently distinctive environment for phytoplankton that they support characteristic phytoplankton species assemblages that can be consistently and clearly distinguished from the assemblages found in clearwater lakes.

7.2 Physical Effects

Dissolved HS profoundly affect the underwater light climate experienced by phytoplankton by altering the quality and quantity of solar radiation penetrating down through the water column. In pure water, blue and green light both penetrate deeply and to about the same extent, whereas red light is attenuated much more rapidly, being rather strongly absorbed by the water itself. This condition may be approximated in non-productive oceanic waters and in a few ultraoligotrophic, clearwater lakes (e.g. Crater Lake, Oregon: Smith et al. 1973). However, in most fresh waters the situation is altered by the presence of dissolved organic materials, often collectively referred to as "yellow substance", "gelbstoff" or "gilvin" (Kirk 1983) and mainly comprising dissolved HS. These absorb blue light strongly, so that in many lakes green wavelengths become the most penetrating. When the concentrations of HS are very high, red light may penetrate even further than green (e.g. Jones and Ilmavirta 1978a; see also Chapter 5). Thus the underwater spectral quality of irradiance experienced by phytoplankton in lakes can vary from blue/green-dominated in clear, ultraoligotrophic waters, through green-dominated in moderately coloured lakes to red-dominated in highly coloured lakes. This variability in spectral quality of underwater light in lakes does open the possibility that phytoplankton with particular pigmentation might be favoured in humic lakes with their predominantly orange/red light climate.

Marine biologists have long discussed the possibilty that the depth zonation of major groups of benthic algae in coastal waters is determined by the changing spectral distribution with depth due to selective absorption: the *chromatic adaptation* theory of Engelmann (1883, in Kirk 1983). In fact, this theory has been shown to be tenuous, with variation in intensity of irradiance probably a more influential factor (Dring 1981). Nevertheless, there is no doubt that the accessory pigments carried by certain algae in addition to chlorophyll *a* can help to fill the absorption window in between the strong chlorophyll *a* absorption in the blue and red regions of the spectrum. When phytoplankton biomass is sufficiently high, this absorption by accessory pigments, particularly the phycobiliproteins of cyanobacteria, can even be detected by in situ spectroradiometry (e.g. Jewson and Taylor 1978).

There is some laboratory evidence that certain types of phytoplankton with accessory pigments are better able than others to cope with chromatic fluctuations. For example, Ojala (1993) compared the growth in culture of three phytoplankters containing phycobiliproteins. She found that, whereas *Oscillatoria bourrellyi* (cyanobacteria) was able to sustain a high growth rate regardless of the light quality, two *Cryptomonas* strains (Cryptophyta) exhibited only weak chromatic adaptation. To date, the possible influence of chromatic adaptation in phytoplankton has received little field experimentation. Wall and Briand (1979) tested phytoplankton preferences for light intensity and colour in field experiments using coloured plexiglass cubes suspended at different depths in Heney Lake, Quebec. Their results were not clear-cut, but suggested that diatoms and green algae favoured higher intensities than dinoflagellates and other flagellates, and that red radiation increased the relative proportion of blue-greens, diatoms and green algae. Certainly the levels of phycoerythrin and phycocyanin in cyanobacteria can change drastically in response to growth in culture in light of differing spectral quality (e.g. Bennett and Bogorad 1973), although increased phycocyanin synthesis can also be induced at low irradiance even when no red light is present (Stramski and Morel 1990). Ganf et al. (1991) even reported apparent chromatic alteration in a slowly sinking population of *Tychonema bourelleyi* in Windermere, England, a lake in which green light is the most penetrating spectral block (Macan 1970).

Phycocyanin-dominated cyanobacterial cells absorb more strongly in the 600 to 650-nm region than do phycoerythrin-dominated cells, which in turn absorb more strongly in the 500 to 575-nm region. In principle, the former should be better adapted to the underwater light climate in humic lakes, with its red/orange character. Callieri et al. (1996) showed that a phycoerythrin-containing picocyanobacterium isolated from Lake Maggiore, Italy, (where the underwater light climate is green-dominated) grew best in green light, whereas a phycocyanin-containing picocyanobacterium isolated from Lake Balaton, Hungary, (where the underwater light climate is red-dominated) grew best in red light. As yet, few field data exist to ascertain the ecological importance of this in influencing the phytoplankton community in humic lakes. However, Pick (1991) reported that, in a survey of picocyanobacteria from 38 Canadian lakes, phycoerythrin-containing cells were particularly abundant in oligotrophic to mesotrophic hard-water lakes, while non-phycoerythrin types (presumably containing phycocyanin) were relatively more abundant in coloured lakes.

Of greater importance than the changes in spectral quality induced by HS in lakes is the much more rapid attenuation of total photosynthetically active radiation (PAR) in brown-water lakes. The vertical attenuation coefficient for downward irradiance of PAR (K_{PAR}) is strongly dependent on water colour

(Eloranta 1978; Jones and Arvola 1984), which is itself strongly dependent on dissolved organic carbon (DOC) (Jones and Arvola 1984). As a useful approximation (Kirk 1983; Jones 1992), K_{PAR} may be partitioned into a set of partial attenuation coefficients

$$K_{PAR} = K_W + K_G + K_{TR} + K_{PH} ,$$

where K_W, K_G, K_{TR} and K_{PH} are the partial attenuation coefficients for PAR due to water, gilvin (humic substances), tripton (non-living suspended particulate material) and phytoplankton, respectively. Kirk (1980) has proposed a crude optical classification for inland waters on the basis of the relative contribution of the different partial attenuation coefficients to total attenuation of PAR. The contribution of water is only rarely dominant (Type W) and in most lakes is around 20–30% (Kirk 1983); in humic lakes this proportion will be lower. The contribution of tripton has been little studied and is generally assumed to be negligible except in high turbidity rivers or shallow lakes with significant catchment erosion (Type T). In shallow lakes, resuspension of sediment can be important, particularly under windy conditions; Jewson (1993) reported a two- to three-fold fold increase in the tripton contribution to light interception in shallow Lough Neagh, Northern Ireland, during stormy weather. The remaining two components, gilvin and phytoplankton, can then be seen to be "competing" for available quanta within the water column.

This competition can be quantified if the assumption is made that the contribution of each component to K_{PAR} is linearly related to its concentration. If the contribution of water and tripton is assumed to be small and invariable, the proportion of total PAR captured by phytoplankton is approximated by (Jones 1992)

$$K_{PH}/K_{PAR} = B_c \cdot k_c /(B_c \cdot k_c + WC \cdot k_p)$$

where B_c is the phytoplankton biomass (milligrams of chlorophyll *a* m^{-3}), k_c is the specific vertical attenuation coefficient per unit of phytoplankton biomass (metres squared per milligram of chlorophyll *a*), WC is the water colour (milligrams of platinum per litre) and k_p is the specific vertical attenuation coefficient (metres squared per milligram of platinum). While k_c does vary between phytoplankton types, a typical mid-range value is 0.015 (see Table 9.1 in Kirk 1983). A value for k_p of 0.011 can be derived from Fig. 1a in Jones and Arvola (1984; see also Eloranta 1978). The proportion of total PAR captured by phytoplankton can then be calculated as a function of phytoplankton biomass under a range of water colour conditions (Fig. 7.1). At lower phytoplankton biomasses, increases in water colour greatly reduce the proportion of PAR which can be captured by phytoplankton, so that most most quanta are intercepted by gilvin (Type G waters, Kirk 1983). In highly coloured waters (> 100 mg Pt l^{-1}), phytoplankton only capture more than half the available quanta when the biomass reaches around 100 mg chlorophyll *a* m^{-3}, so that Type A waters (Kirk 1983), in which the algal biomass captures more PAR than the other components, are probably restricted to less coloured waters.

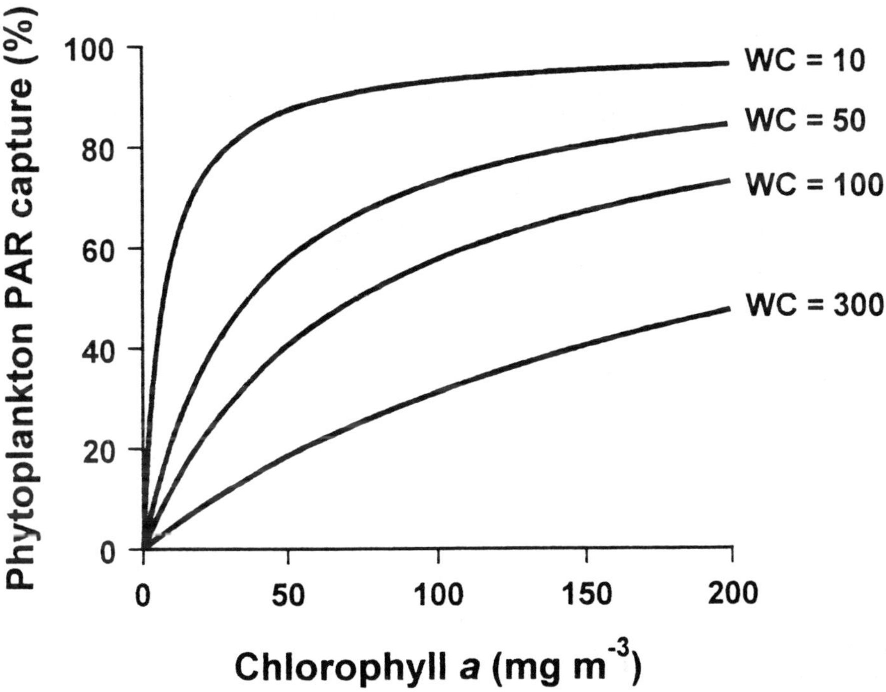

Fig. 7.1. Variation in proportion of PAR captured by phytoplankton from the total intercepted by phytoplankton and gilvin as a function of phytoplanktion biomass under different conditions of water colour (WC; mg Pt l^{-1})

Such calculations indicate how humic substances can compete with phytoplankton for available quanta and hence potentially restrict photosynthetic production within the water column. The importance of this effect will also depend on the distribution of phytoplankton within the water column and the depth to which cells are regularly circulated (Reynolds 1997); if the phytoplankton biomass is concentrated in the upper layers of the water column it will be able to compete more effectively for PAR. This can be achieved in two ways. Firstly, motile or buoyant algae may regulate their vertical position in the water column so as to maximise their photosynthetic efficiency; this strategy is discussed in Section 7.4. Secondly, if the mixed depth (Z_m) is reduced during the summer stratificaton period, phytoplankton cells will spend less time below the lower limit of the euphotic zone (Z_{eu}).

The euphotic zone depth is known to be strongly dependent on water colour (Eloranta 1978; Jones and Arvola 1984). However, in coloured, humic lakes the rapid absorption of incoming solar radiation within the surface layers also provides the potential for more abrupt thermal/density gradients to develop where wind-induced turbulence does not effectively disperse the accumulated heat energy (Bowling 1990). Jones (1992) showed how this effect

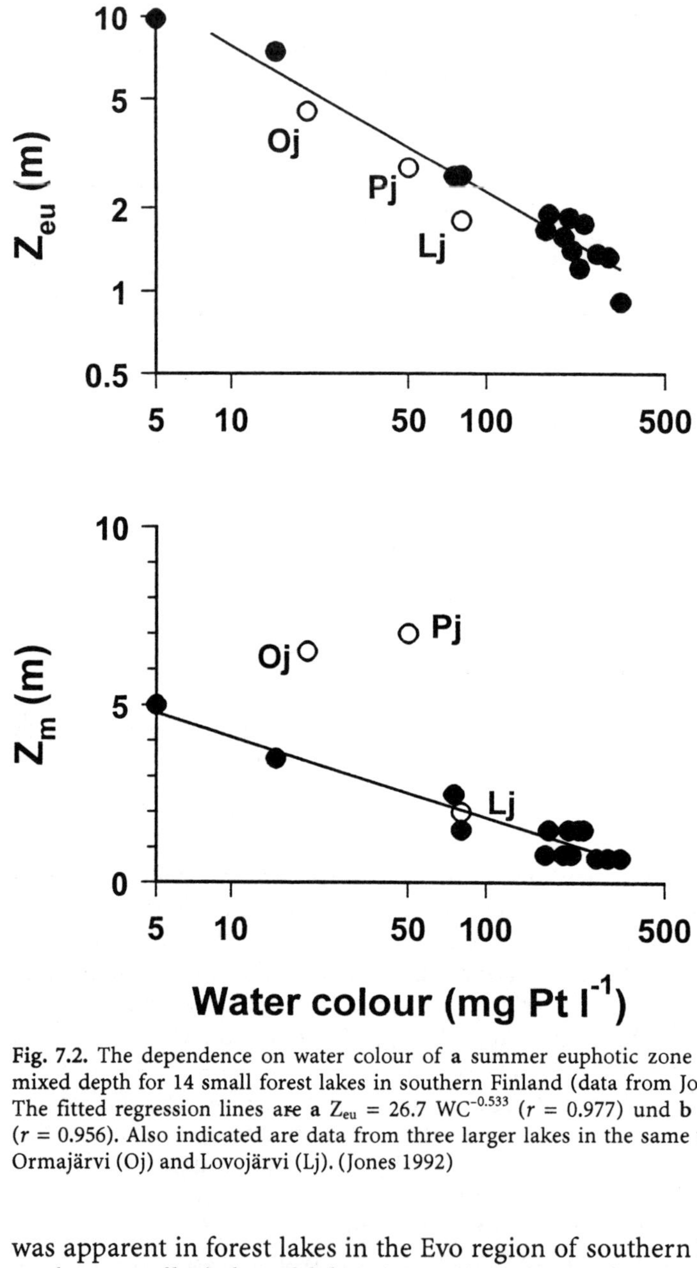

Fig. 7.2. The dependence on water colour of **a** summer euphotic zone depth and **b** summer mixed depth for 14 small forest lakes in southern Finland (data from Jones and Arvola 1984). The fitted regression lines are **a** $Z_{eu} = 26.7\ WC^{-0.533}$ ($r = 0.977$) und **b** $Z_m = 6.4–0.98 \ln WC$ ($r = 0.956$). Also indicated are data from three larger lakes in the same vicinity: Pääjärvi (Pj), Ormajärvi (Oj) and Lovojärvi (Lj). (Jones 1992)

was apparent in forest lakes in the Evo region of southern Finland (Fig. 7.2). In these small, sheltered lakes, increasing water colour reduces Z_{eu} but also leads to a decrease in Z_m, so that the effective light climate (Ramberg 1979) experienced by the phytoplankton in these lakes varies relatively little with water colour. However, in larger humic lakes in the same region, although the relation between Z_{eu} and water colour is consistent (Fig. 7.2a), the greater susceptibility to wind mixing destroys any correlation between Z_m and water colour (Fig. 7.2b). A similar relationship was reported for 21 Canadian Shield lakes by Fee et al. (1996). They found that over the full spectrum of lake sizes lake area was the primary determinant of the depth of the summer mixed layer. Transparency significantly modified this relationship only in small lakes (< 500 ha); this was in contrast to the conclusion of Mazumder and Taylor (1994) that transparency also affects mixing depth in large lakes (> 500 ha). In non-eutrophic shield lakes, transparency is controlled by the concentration of DOC, and Fee et al. (1996) concluded that lower DOC concentrations (a likely consequence of predicted climate change in the region) would cause transparency to increase, resulting in 1 to 2-m deeper epilimnia in small lakes, but with no equivalent impact on large lakes.

The interaction between euphotic zone depth and mixed depth in determining the underwater climate experienced by phytoplankton is further illustrated in Fig. 7.3, in which the effective light climate (normalised with respect to the surface irradiance) is plotted as a function of the ratio of mixed

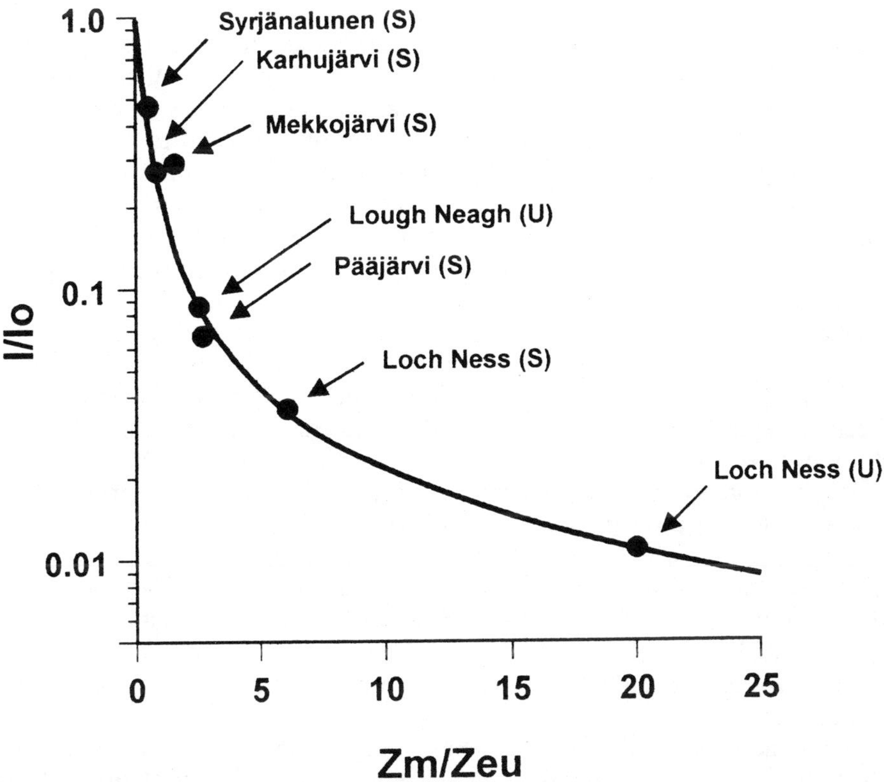

Fig. 7.3. Relationship between effective light climate ($\bar{I}/I_o$; normalised with respect to surface irradiance) and ratio of mixed depth to euphotic zone depth (Z_m/Z_{eu}). The curve follows the mathematically defined relationship, and the points indicate positions of a number of different lakes during summer-stratified (S) or unstratified (U) conditions. (Modified from Jones et al. 1996)

depth to euphotic zone depth. Small, sheltered forest lakes (data from Jones and Arvola 1984) cluster together with a relatively favourable light climate despite a wide range of water colour. Shallow, unstratified Lough Neagh, with high humic content and high phytoplankton biomass (Jewson 1993), and the mesohumic but mesotrophic Lake Pääjärvi, Finland, during summer stratification (Jones and Ilmavirta 1978a) lie together with a light climate typical of many larger lakes. By contrast, deep and turbulent Loch Ness, Scotland, has exceptionally deep mixing; even during summer stratification the mixed layer can still extend down to 40 m and during holomixis the mixed layer equates to the mean depth of 132 m. Thus in this lake even moderate colour, restricting the euphotic zone depth to around 6 m, when combined with the deep mixing produces an exceptionally unfavourable light climate for phytoplankton (Jones et al. 1996).

A consequence of the exceptionally unfavourable light climate in Loch Ness is that phytoplankton growth and production in this oligotrophic and moderately humic loch appears to be limited by light rather than nutrients (Jones et al. 1996), and the annual phytoplankton production of around 10 g C m^{-2} $year^{-1}$ is particularly low. Although it is often assumed that phytoplankton production will be depressed in humic lakes, the evidence for this is sparse and unsatisfactory, and Loch Ness may be an exceptional case. Beaver and Crisman (1991) reported that phytoplankton productivity in subtropical Florida lakes was more predictable from chlorophyll *a*, total phosphorus or total nitrogen in clear water (< 75 Pt units) lakes than in coloured (> 75 Pt units) lakes. While it is true that in humic lakes phytoplankton face increased competition for quanta from humic substances (see above), this will appreciably reduce production only if phytoplankton development is actually limited by light.

In a study of two small Swedish lakes, Ramberg (1979) found annual pelagial primary production in Botjärn to be about half that in the clearer and smaller Vitalampa. However, this may have been as much an effect of the greater circulation depth in the larger lake as of the difference in water colour. From a study of two small forest lakes in southern Finland, Arvola (1984) reported phytoplankton production to be only 25% higher in Alinen Mustajärvi (colour 50–75 mg Pt l^{-1}) than in nearby Horkkajärvi (colour 250–300 mg Pt l^{-1}), a finding consistent with the low variation in effective light climate in such small, sheltered forest lakes (see above). In contrast, Kankaala et al. (1984) found that annual phytoplankton production in the Kitka lakes in north-east Finland was about 50% higher in the moderately humic south-west basin than in the relatively clearwater northern basin. However, the phosphorus concentration was also about 50% higher in the south-west basin, so the differences in production were probably due to different nutrient loadings rather than water colour. This illustrates the difficulty of disentangling the effect of HS on phytoplankton from those of other environmental variables.

Some evidence concerning the effect of HS on phytoplankton production is available from experimental manipulations of humic content, mainly in various designs of mesocosms, but even this evidence is contradictory. Depression of primary productivity by humic matter in northern Manitoba lakes was reported by Jackson and Hecky (1980), who attributed this to reduced iron availability rather than to any effect on the light climate. In a subsequent limnocorral study in the same area, Guildford et al. (1987) found that addition of organic moss-peat material initially increased and then lowered primary productivity and biomass; the initial stimulation was traced to release of soluble nitrogen and phosphorus, while the subsequent depression was attributed to binding of iron or some other metal by dissolved humic material. Carlsson et al. (1995) also reported an initial stimulation of phytoplankton production and biomass during the first few days after addition of riverine humic substances to a coastal plankton community from the Skagerrak (west coast of Sweden) in enclosure experiments, and they attributed this stimulation to utilisation of nitrogen introduced with the HS (see also Carlsson and Granéli 1993). Arvola et al. (1996) used enclosures in Lake Pääjärvi to study the effect of enrichment with HS (derived from the outflow of a bog in the catchment area) on the plankton community of the lake. Phytoplankton production and biomass showed no response to the added humus in spring or autumn experiments and in the summer experiment only a small and transitory stimulation, which was probably due to the small increment of phosphorus added with the humus.

It should also be stressed that any possible depression of phytoplankton production by HS does not prevent many humic lakes from exhibiting all the characteristics of eutrophic lakes in terms of their phytoplankton biomass and species composition. Shallow humic lakes which receive heavy nutrient loadings, such as Lough Neagh in Northern Ireland and Tjeukemeer in The Netherlands, are amongst the most productive of studied temperate lakes and can produce phytoplankton biomass in excess of 100 mg chlorophyll a m^{-3} dominated in summer by filamentous cyanobacteria (Jewson 1976). Similarly, the deeper and partially meromictic lake, Lovojärvi, in southern Finland, which has high water colour (80 mg Pt l^{-1} in the epilimnion), also supports a high summer phytoplankton biomass (approaching 100 mg chlorophyll a m^{-3}) dominated by filamentous cyanobacteria (Keskitalo 1977; Jones and Ilmavirta 1978b). Therefore, although in principle the presence of high concentrations of HS in lakes will reduce the quanta available for phytoplankton photosynthesis, it appears that in the majority of lakes studied this has negligible impact because phytoplankton development is constrained by the availability of nutrients rather than quanta.

A final aspect of the light climate of humic lakes with respect to phytoplankton is the reduced penetration of UV-B radiation in humic waters compared with clear waters (Chapter 6). Since the impact of UV-B radiation on phytoplankton in freshwaters has recently been comprehensively reviewed by Karentz et al. (1994), it will be considered only briefly here. There is considerable evidence, especially from studies of marine phytoplankton, that photosynthesis, primary productivity and growth can all be depressed by UV-B.

However, because UV-B attenuation by dissolved organic carbon (DOC) in freshwaters is generally greater than in the oceans (Kirk et al. 1994), lake phytoplankton may be less susceptible to changes in incident UV-B than marine phytoplankton. This should be especially true in humic lakes, in which penetration of UV-B at potentially damaging intensities can be restricted to the top few centimetres of the water column. However, as Karentz et al. (1994) correctly caution, lakes are shallow relative to marine systems, so even small depths of penetration can influence a relatively large proportion of the water column. Besides any direct damaging effect on phytoplankton cells, UV-B can also affect motility of flagellates (Ekelund 1990, 1993), which is potentially important since flagellate vertical migrations are an important aspect of phytoplankton ecology in many humic lakes (Jones 1991). Moreover, even moderate UV irradiance of humic water can induce an algicidal effect which can persist for at least several days (Gjessing and Källqvist 1991; Hessen and Van Donk 1994).

7.3 Chemical Effects

As complex organic compounds, dissolved humic substances in lakes might affect phytoplankton either by serving as a carbon/energy source or by their ability to interact with and modify the properties of other chemicals in the water. In recent years there has been much interest in the direct utilisation of HS within the food chains of humic lakes (Jones 1992), but this has mainly concerned humus and secondary production. It has long been known that many algae, including planktonic forms, are facultatively heterotrophic, capable of uptake and metabolism of organic solutes (e.g. Sepers 1977; Sandgren 1988). However, in virtually all reported cases the compounds involved were low molecular weight (MW) sugars, organic acids and amino acids, which typically are present only in very low concentrations in lakes (Wetzel 1983), whereas gel filtration of lake humic substances suggests that the bulk of the HS carbon has a nominal molecular weight around 30 000 or more (e.g. De Haan et al. 1987). Few experiments have been reported which directly test the response of phytoplankton growth to HS. Tulonen et al. (1992) used ultrafiltration to fractionate the HS from the highly humic Lake Mekkojärvi, southern Finland, and then tested algal growth in the different fractions supplemented with inorganic nutrients and vitamins. No growth was observed at an irradiance of 2 $\mu E\ m^{-2}\ s^{-1}$ in any of the fractions, suggesting that the algae could not grow heterotrophically using the HS carbon. At higher irradiance good growth was observed in all HS fractions except that with the smallest MW (< 1000) in which algal growth ceased at low densities. Since all essential nutrients and vitamins were added separately, the low MW fraction presumably had some effect, either phototoxic or on nutrient availability, which was suppressed in the presence of higher MW fractions of HS. On the basis of the present, limited evidence, it appears reasonable to conclude that HS do not make a significant direct contribution to the carbon/energy budgets of phytoplankton.

One indirect effect of HS on phytoplankton is through a modification of the toxicity of metals. It is widely accepted that toxic metals have their greatest impact on phytoplankton and other organisms when present as free ions and that compounds which reduce the free ion activity, such as the chelator ethylene diamine tetra-acetic acid (EDTA), therefore reduce the toxicity of metals like copper (Anderson and Morel 1978). In natural waters, metal complexation appears to be mainly associated with dissolved organic matter. While this may include organic molecules released by algae, and especially cyanobacteria in eutrophic lakes (Simpson and Neilands 1976; McKnight and Morel 1980), in most lakes HS are more important complexation agents. Much work has been carried out on the effect of HS on copper toxicity to phytoplankton, presumably because of the past importance of copper as an algicide in water treatment. For example, Sunda and Lewis (1978) showed that the toxicity of copper to an estuarine diatom was reduced by dissolved HS in river water. An equivalent effect was found in the case of cadmium toxicity to the common test alga *Selenastrum capricornutum* (Sedlacek et al. 1983). Hongve et al. (1980) reported that humic bog waters detoxified zinc, lead, mercury and copper with respect to natural phytoplankton assemblages, although no significant effect was found with cadmium.

However, in most freshwaters metal toxicity is not an important aspect of phytoplankton ecology. Of greater importance is the potential of HS to modify the availability to and uptake by phytoplankton of essential metals such as iron and anionic nutrients (particularly phosphate) which may constrain phytoplankton growth. Early attempts to develop artificial growth media for algal culture quickly established that organic chelators (initially "soil extract" and later EDTA) are essential to prevent precipitation of iron from the medium and to maintain its availability to algae (Fogg and Thake 1987). In natural waters, humic substances are widely acknowledged to perform this function, and the enhancement of iron uptake in the presence of HS has been demonstrated in laboratory studies (e.g. Prakash et al. 1973). The precise mechanisms behind this enhanced uptake are still not really known despite more recent studies of aqueous iron chemistry and algal uptake (Anderson and Morel 1982). Nevertheless, the importance of chelated iron as a potential limiting nutrient for freshwater phytoplankton has often been indicated (see Sandgren 1988), and Van Donk (1983, in Sandgren 1988) suggested that the near absence of chrysophytes from hard-water, alkaline lakes might be explained by the low levels of dissolved organic carbon and hence of chelated iron in such lakes. Conversely, the frequent importance of chrysophytes in humic lakes (e.g. Eloranta 1986; Ilmavirta 1988) could be related to the greater availability of chelated iron or other essential micrometals in such lakes. It is also possible that in some circumstances humus-metal complexation may depress biological activity; for example, Jackson and Hecky (1980) attributed the reduced primary productivity they observed in humic waters to HS making iron less available to phytoplankton.

Although the C:N ratios of HS are high (typically around 50:1; Wetzel 1983) making them stoichiometrically unsuitable to support growth of organisms, photochemical release of nitrogen, mainly as ammonium, has been suggested to be a potentially important mechanism in nitrogen-limited aquatic ecosystems (Bushaw et al. 1996). This could account for the increased nitrogen availability and stimulation of primary production observed in Swedish coastal waters following addition of riverine humic substances (Carlsson and Granéli 1993; Carlsson et al. 1995).

In addition to a direct effect on iron (and possibly nitrogen) availability, there is now extensive evidence that HS can also affect the chemical speciation and bioavailability of phosphorus in freshwaters. This effect probably occurs mainly through interaction of phosphate with humus-iron complexes. It has long been known that humic lake waters tend to exhibit higher total phosphorus concentrations than otherwise comparable clearwater lakes (Hutchinson 1957; Wetzel 1983; Kämäri et al. 1990). Meili (1992), in a study of 18 Swedish forest lakes and streams, found that concentrations of total phosphorus (TP) tended to increase with water colour and total organic carbon (TOC). Despite these generally higher concentrations of TP in humic waters, other studies (e.g. Jackson and Hecky 1980; Chow-Fraser and Duthie 1983, 1987; Heyman 1983) have indicated that phytoplankton production and biomass development in humic lakes are frequently below that expected from widely used phosphorus loading models (Vollenweider 1968, 1975). These observations suggest that in humic lakes a higher proportion of TP is unavailable to phytoplankton, and the most obvious explanation is that a part of the phosphorus is complexed with the humic substances.

Progress in understanding the availability of phosphorus to phytoplankton in natural waters has been hindered by difficulties with the standard analytical techniques. It is generally acknowledged that free ionic orthophosphate is the form of phosphorus most readily exploited by phytoplankton (although organic phosphorus compounds may sometimes be taken up: Cembella et al. 1984; Bentzen et al. 1992; Cotner and Wetzel 1992). However, the standard acid-molybdate analytical technique for determining the concentration of dissolved inorganic phosphorus not only lacks sensitivity but also is known to measure a variable proportion of acid-hydrolysable phosphorus compounds as well as orthophosphate (e.g. Stainton 1980; Tarapchak et al. 1982). Ion chromatography can specifically detect orthophosphate, but still lacks sufficient sensitivity to be of use for water samples from most lakes. Nevertheless, there is some evidence that the concentration of dissolved humic material (DHM) may directly affect the rate at which ionic orthophosphate is taken up by plankton. Brassard and Auclair (1984) reported that planktonic orthophosphate uptake rates were mediated by the 1000 to 10 000 molecular weight fraction in Canadian Shield lakewater, while Francko (1986) found that *Typha* DHM, alone and with iron additions, could dramatically stimulate or depress uptake of $^{32}PO_4$ by plankton particles in some Oklahoma lakes. These rather conflicting reports highlight the need for more work on the possible influence of DHM on planktonic uptake of of free ionic orthophosphate.

Despite these problems, several studies have demonstrated that under various conditions DHM can associate with phosphate in the presence of iron (Francko and Heath 1979, 1982; Stevens and Stewart 1982; De Haan and De Boer 1986) or manganese (Steinberg and Baltes 1984). However, the difficulties with direct chemical analysis of the phosphorus component have led several researchers to use radioisotopes of phosphorus to help elucidate the processes involved. Early isotope studies of planktonic phosphorus cycling (e.g. Lean 1973; Lean and Rigler 1974) which were carried out in relatively clear water lakes concluded that abiotic complexing of phosphate was unimportant. Stewart and Wetzel (1981) first attempted to quantify interactions between $^{32}PO_4$ and DHM using gel filtration, but were unable to demonstrate any direct binding of phosphate to DHM. They attributed this "failure" to the high concentration of calcium carbonate in their sample water, since calcium carbonate colloids also bind phosphate, and speculated that in systems of lower alkalinity DHM-orthophosphate binding might be important in regulating bioavailability of orthophosphate.

Jones et al. (1988) reported detailed studies of abiotic movements of radiolabelled orthophosphate to and from higher molecular weight fractions, apparently through complexation with DHM-Fe. Water samples from Finnish forest lakes with various humic contents were filtered through 0.2-μm pore-size filters, then the filtrate was sterilized by addition of NaN_3 and spiked with $^{33}PO_4$. Following a period of incubation, the samples were run through a gel chromatography column containing Sephadex G-100 gel, and the radioactivity was determined in the eluted fractions. Three peaks of radioactivity were consistently identified (Fig. 7.4): peak I had a nominal molecular weight of > 100 000; peak II had a nominal molecular weight in the range 10 000–20 000; and peak III corresponded to free ionic orthophosphate. The movement of added ^{33}P to these peaks was greater when water samples contained more DHM and more iron. Addition of excess free $^{31}PO_4$ caused displacement of $^{33}PO_4$ from the high molecular weight complexes, indicating that an equilibrium exists between free PO_4 and the DHM-Fe-PO_4 complexes; however, this displacement was incomplete, suggesting that part of the PO_4 within the complex is more firmly bound. More direct evidence for the co-involvement of iron in the formation of complexes between DHM and phosphate was presented later by the same workers (De Haan et al. 1990) using double isotope labelling with ^{55}Fe and ^{32}P. During gel chromatography, co-elution of added ^{55}Fe and ^{32}P occurred within peaks I and II, but not peak III (Fig. 7.4). Moreover, particularly close co-elution of iron with DHM (estimated from UV absorbance) was observed within the peak-II band.

The importance of such DHM-Fe-P complexes in regulating the availability of P to phytoplankton may be considerable. In the absence of DHM,

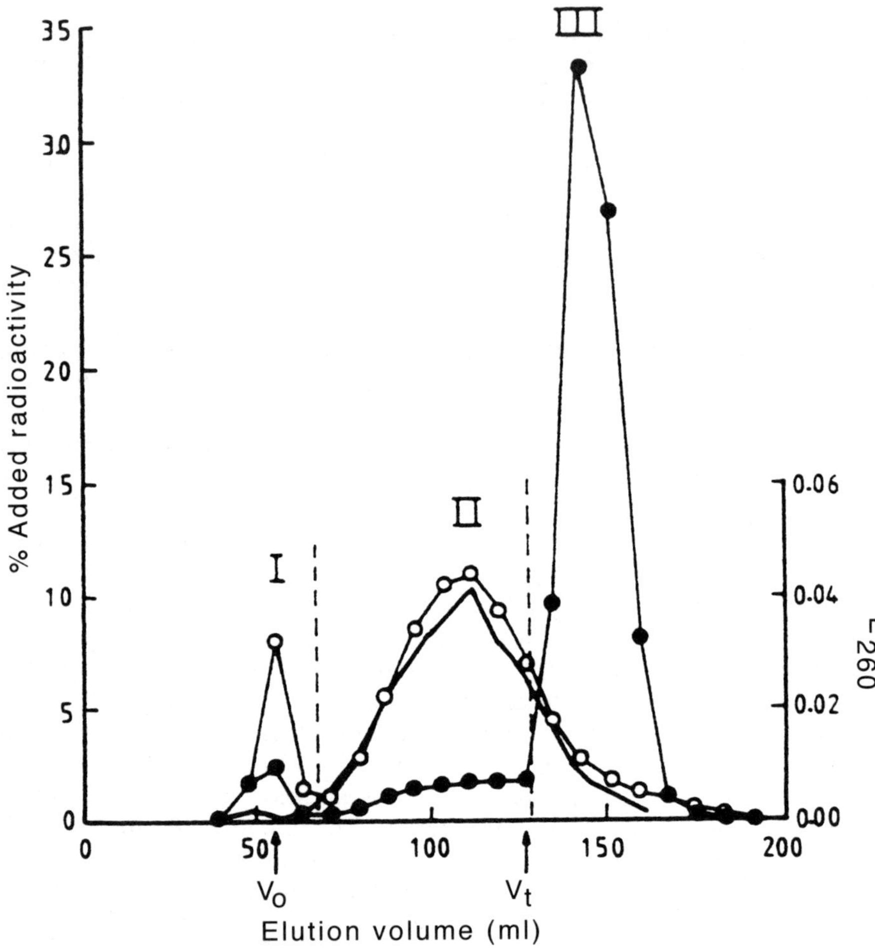

Fig. 7.4. Sephadex G-100 gel filtration in 0.02% NaN_3 of 2-ml < 0.2-μm filtered water from the polyhumic lake Mekkojärvi, southern Finland, containing 11.8 kBq $^{55}FeCl_3.6H_2O$ and 2.9 kBq $^{32}PO_4^{3-}$ after dark incubation at room temperature for 24 h. Elution pattern of ^{55}Fe (○–○); elution pattern of ^{32}P (●–●); elution pattern of DHM estimated as absorption at 260 nm (—). Elution profile is divided into three broad fractions corresponding to those identified by Jones et al. (1988). (De Haan et al. 1900)

phosphate can adsorb to ferric oxides and hydroxides (Lijklema 1980; Tipping 1981) and is effectively precipitated by ferric iron compounds (Hsu 1976; Stumm and Morgan 1981; Shang et al. 1992). The presence of even moderate amounts of DHM can substantially reduce such precipitation. For example, Jones et al. (1988) showed that $^{33}PO_4$ added to Milli-Q water remained in solution, but in the presence of added ferric chloride the ^{33}P was all retained by a 0.2 μm filter (Fig. 7.5). However, when the same procedure was applied to filtered water from lakes with even small concentrations of DHM, most of the added ^{33}P passed through the filter.

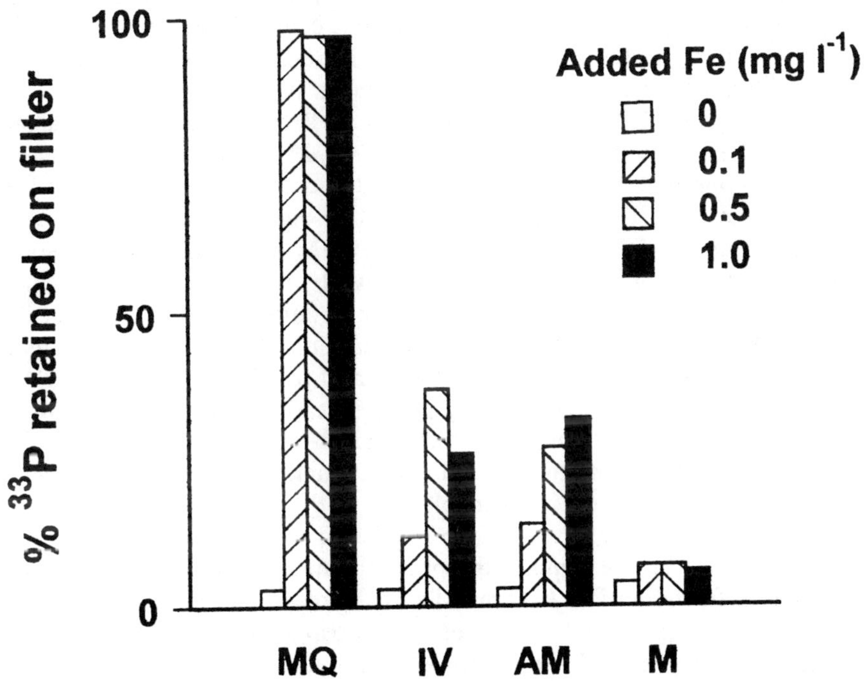

Fig. 7.5. Proportion of ^{33}P retained by 0.2-μm Nuclepore filter following addition of $^{33}PO_4$ to "Milli-Q" water (MQ) and water samples from three lakes of increasing humic content: Iso Valkjärvi (IV), Alinen Mustajärvi (AM) and Mekkojärvi (M). Iron (as ferric chloride) was first added to the samples in concentrations of 0, 0.1, 0.5 or 1 mg Fe l^{-1}. (Jones et al. 1988)

Shaw (1994) demonstrated a similar effect using water from a range of lakes of different ionic strength and DHM content. Humic waters of low and high ionic strength were able to maintain added $^{55}FeCl_3$ and $^{32}PO_4$ in "solution" (i.e. < 0.2 μm) under all pH conditions, whereas in samples from clear water lakes much of the added ^{55}Fe and, to a lesser extent, the added ^{33}P was in particulate form. The mechanism by which DHM maintains iron and phosphate in "solution" is still not fully understood. Shapiro (1967) proposed that inorganic colloids containing Fe (and by implication P) may be peptized by surface adsorption of DHM. Levesque and Schnitzer (1967) suggested that iron forms a "bridge" between HS and phosphate molecules to allow the creation of HS-Fe-PO_4 complexes. Probably both mechanisms can occur simultaneously (e.g. De Haan 1982; De Haan and De Boer 1986) and may account respectively for the peak I and peak II material identified in the gel chromatographs of Jones et al. (1988). De Haan et al. (1990) concluded, from refractionation of the peak II material, that Fe was bound more tightly than PO_4, suggesting that the material is a DHM-Fe complex to which PO_4 can become reversibly attached.

Although DHM can evidently prevent iron, and therefore also phosphate, from precipitating and being lost from the water column, phytoplankton still need to be able to access the nutrients within the DHM-Fe-PO_4 complex if these nutrients are to be considered bioavailable. It has been demonstrated that variable quantitites of high MW forms of P can be utilised by algae (Downes and Paerl 1978; White and Payne 1980; McGarrigle and Kilmartin 1992), so it is possible that algae may be able to access PO_4 directly from the complexes. However, it seems more probable that algae take up the PO_4 as it is released from the complex. As noted above, at least part of the PO_4 associated with these complexes is in equilibrium with free ionic PO_4 and hence can be released if the equilibrium conditions change (Jones et al. 1988; De Haan et al. 1990), for example by a fall in free ionic PO_4 concentration as a consequence of algal uptake. This regeneration of free ionic PO_4 from DHM-Fe-PO_4 complexes can be enhanced by photolysis, particularly by UV radiation apparently photoreducing iron to the ferrous state, thereby altering the ability of the DHM-Fe complex to bind PO_4 (Francko and Heath 1979, 1982). This photoreduction process is reversible with both UV-sensitive phosphorus and ferric iron being regenerated when irradiated samples were incubated in the dark, and may be the same as the photochemical ferrous-ferric catalytic cycle reported from humic waters by Miles and Brezonik (1981).

Phytoplankton can also gain access to phosphorus in high molecular weight compounds through enzyme-mediated hydrolysis of substrates such as phosphomonoesters (PME). Both membrane-bound and free phosphatase enzymes can be involved, and numerous studies have documented increased phosphatase activity under conditions of plankton phosphorus deficiency (e.g. Chróst 1990). Most studies have focused on alkaline phosphatases, but acid phosphatases may be more important in lakes with low pH (Olsson 1983), a category which includes many oligotrophic humic lakes. Münster et al. (1992) studied extracellular enzyme activity in Mekkojärvi in southern Finland, a polyhumic lake with mean pH 5.4, and found that phosphatases (as well as other enzymes) had their maximum activity at pH 5.4, with a smaller peak of activity at pH 8.5. Community alkaline phosphatase activity has been reported to be stimulated by additions of DHM (Stewart and Wetzel 1982; Auclair et al. 1985; Francko 1986), although Francko (1986) also found depression of enzyme activity by some size classes of DHM from certain lakes. Stewart and Wetzel (1982) hypothesised that DHM might also sequester PME compounds, thus affecting their capacity to yield phosphorus to phytoplankton; however, Francko (1986) reported that DHM addition to water samples from four Oklahoma lakes actually resulted in increased PME concentrations after 1 h of incubation, so this aspect of DHM involvement in phosphorus cycling remains highly questionable.

It is evident that phosphorus cycling in the plankton of humic lakes is likely to be more complex than in clearwater lakes lacking appreciable concentrations of DHM. As yet, the nature of the interactions of DHM with the various processes within the phosphorus cycle are incompletely understood, and further research is urgently needed. The capacity of DHM to reduce precipitation of phosphate with iron and other metals should help to maintain the pool of phosphorus in the water column, in which case the DHM-Fe-PO_4 complexes in humic lakes can be viewed as a reservoir of potential phosphorus for the plankton. However, since phosphate is the form of phosphorus most readily utilised by phytoplankton, any sequestration of phosphate by humic substances, in combination with iron, must reduce the instantaneous bioavailability of phosphorus. Availability of this phosphorus over time will then depend on the phosphorus-release kinetics, either through chemical equilibration with the changing free orthophosphate pool or mediated by UV radiation. The latter will be highly dependent on the depth of penetration of UV relative to the depth of the mixed layer. Radiotracer studies in a eutrophic lake (Levine et al. 1986) and in an enclosure within a small humic lake (Salonen et al. 1994) both indicated restricted labelling of the total dissolved phosphorus even after a period of 2–3 weeks, suggesting only very slow turnover of the larger part of the total dissolved phosphorus pool, particularly in humic lakes. Any reduction in immediate bioavailabilty of phosphate as a consequence of high concentrations of DHM might be expected to favour phytoplankton with either higher affinity for phosphate or alternative strategies for satisfying their phosphorus requirements, such as diurnal migration to phosphate-rich deeper water (e.g. Salonen et al. 1984) or ingestion of phosphorus-containing particulate material (e.g. Nygaard and Tobiesen 1993). These strategies are considered in the Section 7.4.

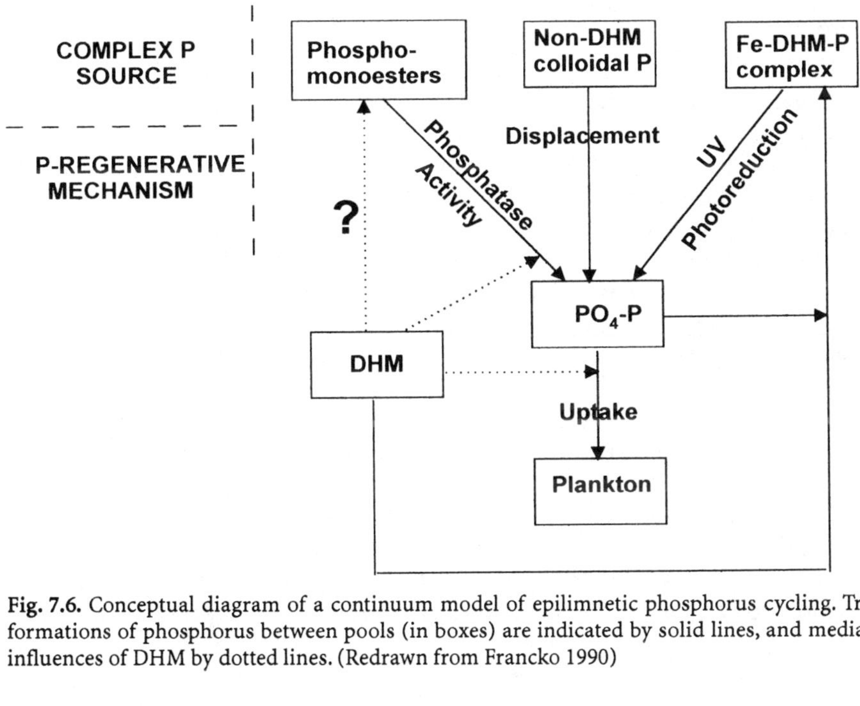

Fig. 7.6. Conceptual diagram of a continuum model of epilimnetic phosphorus cycling. Transformations of phosphorus between pools (in boxes) are indicated by solid lines, and mediating influences of DHM by dotted lines. (Redrawn from Francko 1990)

Francko (1986, 1990) has proposed a conceptual model of epilimnetic phosphorus cycling (Fig. 7.6). This model recognises three major sources of complexed phosphorus (PME, non-DHM colloidal phosphorus and DHM-Fe-PO_4 complexes), as well as three major orthophosphate regenerative mechanisms (phosphatase activity, colloidal displacement and UV photoreduction) which can co-occur in lakes. The relative importance of the major sources of complexed phosphorus and of the phosphorus regeneration mechanisms can vary between lakes. In clearwater or eutrophic lakes, enzymatic regeneration of phosphorus from phosphomonoesters may be most important, whereas in oligotrophic but polyhumic lakes the role of Fe-DHM-PO_4 complexes and the UV-promoted release from them of PO_4 should be much more significant. Francko emphasised that a given system could lie anywhere between two extreme ends of the spectrum represented by the model in terms of the relative importance of the phosphorus sources and regenerative mechanisms, and hence called his proposal a "continuum model of epilimnetic phosphorus cycling". An extremely important consequence of this concept is that, at least in terms of phytoplankton and nutrients, it is not possible to classify lakes definitively as "humic" or "clearwater"; all lakes have some content of DHM, and that DHM content will vary along a continuous gradient from scarcely detectable in some lakes to the high DHM content that gives the most humic lakes their rich-brown colour. The impact on phytoplankton communities can be expected to be equally various.

7.4 Adaptations by Phytoplankton in Humic Lakes

The influence of HS on the light climate experienced by phytoplankton has been evaluated earlier (Section 7.2). There it was shown that, although the spectral quality of underwater irradiance is clearly altered by the presence of HS, there is little real evidence that this has any marked effect on the pigmentation of phytoplankton in humic lakes and hence on the structure of the phytoplankton community. However, it is frequently assumed that the much more rapid attenuation of total PAR in humic lakes, with the consequent compression of the euphotic zone, will have a more profound impact on phytoplankton ecology.

Calculations presented in Section 7.2 showed how phytoplankton will compete more effectively with HS for available quanta when phytoplankton biomass is high. In eutrophic lakes with high phytoplankton biomass the loss of quanta to HS will have little impact on phytoplankton production, so that some humic lakes such as Lough Neagh (Jewson 1976) can be among the most productive known. In oligotrophic lakes the situation is different and there is the real possibility that photosynthetic production, and hence phytoplankton growth, might be seriously restricted by interception of available quanta by HS. This potential effect would be minimised if the phytoplankton biomass could be concentrated as near to the surface as possible.

To some extent this is achieved inadvertently, at least in small, sheltered lakes in which the depth of mixing and the depth of the euphotic zone both decrease as water colour increases (Fig. 7.2). In addition, their potential ability to regulate their vertical position in order to maximise their photosynthetic efficiency has led to the argument that motile or buoyant algae might be particularly favoured in humic lakes with poor light penetration (e.g. Ilmavirta 1983). Certainly several studies of groups of polyhumic lakes have emphasised the high proportion of flagellates in their phytoplankton communities (e.g. Croome and Tyler 1988; Ilmavirta 1988). However, clearwater lakes within the same group of lakes can also have a high proportion of flagellates, so that no clear relationship between water colour and the proportion of flagellates in the phytoplankton has been established (Jones and Arvola 1984; Ilmavirta 1988).

In fact, many of these studies have been of groups of small lakes, often in forested catchments providing shelter from wind, and characterised by reduced turbulence and a shallow epilimnion. Under these environmental conditions, the motility afforded to flagellate species of phytoplankton may be important as an adaptation to avoid settling out from a low-turbulence water column with a shallow mixed layer, rather than any response to the irradiance conditions (Jones 1991). In a survey of 54 small lakes in southern Finland, Arvola (1986) reported diatoms (which tend to have particularly high sinking rates because of their heavy silica cell walls) to be uncommon in the spring phytoplankton, but in greater abundance in the larger, less sheltered lakes. Interestingly, following clearance of the forest around one of the small lakes in 1982, the depth of spring mixing increased and the diatom *Asterionella formosa* became temporarily abundant in the phytoplankton (L. Arvola, pers. commun.). Larger humic lakes with greater turbulent mixing, although exhibiting strong representation of flagellates in their phytoplankton, typically show a more diverse phytoplankton community, including a high proportion of diatoms (e.g. Eloranta 1986; Jones et al. 1996).

In small, sheltered, humic lakes, nutrient depletion of the shallow, mixed layer can occur particularly rapidly in the spring. This can quickly lead to a particularly distinct spatial segregation of the key resources required by phytoplankton, with irradiance restricted to the shallow euphotic zone and adequate nutrient supplies only available below the thermocline. Although these conditions develop in many different kinds of lakes, their vertical compression in small humic lakes means that the distance separating the vertically segregated resources is usually very small, so that even small flagellates capable of vertical relocations of only some hundreds of centimetres can gain access to both adequate irradiance near the surface and nutrients deeper in the water column. Considerable evidence has accumulated over recent years that many flagellates in humic lakes undertake diurnal vertical migrations (DVM) which permit them to access irradiance near the surface during daylight hours and nutrients at or below the thermocline during the night (Jones 1991, 1993). Studies conducted over successive 24-h periods have demonstrated the regularity of these migrations (Arvola et al. 1987), while convincing evidence that this DVM behaviour allows flagellates to retrieve nutrients such as phosphorus from deep water reserves has been provided by experiments using radioisotope tracers (Salonen et al. 1984). In larger lakes, in which resources are segregated over greater distances, the possibilities for flagellates to circumvent this segregation by means of DVM are reduced because of the limited amplitude of DVM. Although the greatest amplitude of DVM by a freshwater flagellate reported to date appears to be the 18-m migrations of a *Volvox* sp. in tropical Lake Cahora Bassa, Mozambique (Sommer and Gliwicz 1986), most reported flagellate migrations have an amplitude around 5 m or less (Sommer 1988).

It is now well established that some photosynthetic phytoplankton are capable of ingesting and digesting bacteria to utilise the materials within the bacterial cell. Phytoplankton which combine autotrophic photosynthesis with ingestion of bacteria display a *mixotrophic* mode of nutrition. Mixotrophy is actually exhibited by a wide range of planktonic organisms (Jones 1994), but is particularly prevalent among the chrysophyte flagellates which are often an abundant component of the phytoplankton in humic lakes. Mixotrophy in photosynthetic cryptophyte flagellates, which are also widespread in humic lakes, appears to be of only limited importance (Tranvik et al. 1989). Early reports of mixotrophy as important in the ecology of phytoplankton emphasised the ingestion of bacteria as a means to supplement the carbon budget of algae growing under conditions of light limitation (Bird and Kalff 1986, 1989), which could be applicable to phytoplankton in humic lakes. Alternatively, ingestion of bacteria could be more advantageous in providing mixotrophic phytoplankton which are photosynthetically competent with an alternative source of nutrients such as phosphorus (Nygaard and Tobiesen 1993) or iron (Veen 1991) during times of nutrient limitation. Given the likely accentuated problem of availability of phosphorus in humic lakes due to the formation of complexes with HS and iron (Section 7.3), it may well be that in humic lakes mixotrophy by phytoplankton is primarily a nutrient acquisition strategy. The generally greater biomass of bacteria in humic lakes (Hessen 1985; Tranvik 1988) would facilitate this strategy, particularly since bacterial cells tend to have relatively high phosphorus:carbon ratios (e.g. Fagerbakke et al. 1996).

Experimental evidence from field studies is needed to clarify these relationships. One recent study by Jansson et al. (1996) using nutrient enrichment experiments in enclosures in humic Lake Örträsket in northern Sweden suggested that obligate autotrophic phytoplankton were phosphorus limited and relatively scarce, whereas the mixotrophic flagellates, which dominated the phytoplankton community, exhibited nitrogen limitation induced by grazing of phosphorus-rich bacteria. Nutrient limitation and interactions between bacteria and phytoplankton are considered further in Chapter 8.

However, it is worth emphasising that mixotrophic phytoplankton are by no means restricted to humic lakes (e.g. Sanders et al. 1989; Berninger et al. 1992), nor even to oligotrophic environments (Sanders 1991).

7.5 Phytoplankton Community Structure in Humic Lakes

Sections 7.2 and 7.3 have considered the ways in which humic substances can modify the physical and chemical environment experienced by phytoplankton in lakes, and some of the adaptations, including behavioural strategies such as vertical migrations and mixotrophy, which, although by no means restricted to phytoplankton in humic lakes, may be of particular value in helping to cope with this modified environment. Therefore it is logical to conclude by considering whether "humic lakes" do in fact represent a sufficiently distinctive environment for phytoplankton that they support characteristic phytoplankton species assemblages clearly distinguishable from the assemblages found in clearwater lakes.

Hutchinson (1967) reviewed phytoplankton associations in different lake types, including an historical perspective. He naturally emphasised the gradient from *oligotrophic* to *eutrophic* introduced by Naumann (1919), but noted that Thienemann (1925) had added a third lake category, *dystrophic*, "because the brown peaty waters of many of the lakes of the mountains of northern Europe seemed to support a different plankton association from that of the deep blue or blue-green oligotrophic lakes of central Europe". Hutchinson also cited the extensive study of Finnish lakes by Järnefelt (1956) in which the lakes were divided into two groups according to their productivity and again into two groups according to humic water colour (*chthoniotrophy*), resulting in four lake classes. This typology was discussed further by Järnefelt (1958). The phytoplankton associated with the four lake classes were clearly determined primarily by the gradient of productivity from oligotrophic to eutrophic; the only influences of water colour noted by Hutchinson were "a tendency for diatoms to become increasingly dominant with increasing chthoniotrophy" and that "chrysophycean plankton occurred in all kinds of lakes, but apparently more commonly in the more humic waters". As noted in Section 7.3, this latter feature may be related to the greater iron chelation in humic lakes.

More recent comprehensive surveys of phytoplankton assemblages in relation to water colour are rather few. Ilmavirta (1980, 1983) studied the phytoplankton in more than 30 brown-water lakes in four lake chains with different trophic state in southern Finland. Diatoms were abundant in all lake types. Chlorophytes and cyanobacteria were dominant in more eutrophic lakes, while cryptophytes and chrysophytes were more abundant in the oligotrophic brown waters. Eloranta (1986) examined the summer phytoplankton in 58 lakes in central Finland, covering a range from oligotrophic to eutrophic. Many of these lakes had high water colour due to humic substances, but there was no clear influence of water colour on phytoplankton biomass or species composition. In a later study, Eloranta (1995) reported a survey of the summer phytoplankton in 103 lakes in 14 national parks in central and southern Finland, most of which were oligotrophic. Surprisingly, phytoplankton biomass was *positively* correlated with water colour (but r^2 was only 0.08), although this was considered an artefact of the active concentration of flagellate algae in some of the highly humic lakes at the standard sampling depth of 0.5 m. Number of taxa was negatively correlated with water colour, but again the r^2 was very low (0.04). Cluster analysis grouped the phytoplankton from the lakes into 9 main community types which suggested that those lakes in which cryptophytes and chrysophytes were dominant in the summer phytoplankton had high average water colour, whereas those dominated by dinophytes and chlorophytes were clear water lakes.

Willén et al. (1990) surveyed the summer phytoplankton in 73 mainly oligotrophic lakes in Sweden. Lake types were identified using both TWINSPAN classification and CANOCO ordination. Humic lakes mainly fell within a group with the dinophytes *Peridinium* spp. as characteristic phytoplankton, but also with the large raphidophyte flagellate, *Gonyostomum semen*, very frequent. However, these were humic lakes in southern Sweden affected by anthropogenic acidification which has lowered the pH by 1–2 units since the 1930s. Indeed, since many humic lakes may be naturally acidic or affected by anthropogenic acidification, it can be difficult to distinguish the effects of HS and acidity on phytoplankton community structure.

Earle et al. (1986) analysed the phytoplankton composition in 95 Labrador lakes in north-east Canada. A complete-linkage cluster procedure was used to examine the phytoplankton data for evidence of species associations (i.e. groups of co-occurring species having similar reactions to properties of the environment). Eight different species associations were identified. Correlation of the algal taxa comprising these associations with 22 measured environmental variables identified six derived environmental factors which could explain a significant part of of the variability in the phytoplankton data. One of these six factors, which explained 25% of the variance, was highly correlated with water colour, DOC, total dissolved solids, turbidity and sulphate, and was considered by Earle et al. (1986) to be a "dystrophy" factor. However, they also pointed out that low pH is associated naturally with highly coloured dystrophic lakes, which again highlights the difficulty of distinguishing pH effects from those of dissolved HS per se in determining phytoplankton composition in such lake surveys. In fact, when Arvola et al. (1990) studied 32 small lakes in southern Finland with a range of water colour and pH, they found no clear evidence that humic content affects algal composition, but emphasised the importance of pH as a factor regulating the phytoplankton community.

Thus the limited available evidence does not provide strong support for the view that humic lakes have a characteristic phytoplankton assemblage, although chrysophyte and cryptophyte flagellates do seem to be particularly well represented in nutrient-poor humic lakes. The most comprehensive analysis of phytoplankton community structure and its relation to environmental variables is that developed by Reynolds (1980, 1984a, 1984b, 1997). The generalised scheme put forward by Reynolds emphasises the interaction between turbulent mixing (or the stability of the water column) and the concentrations of nutrients, especially phosphorus and nitrogen. While Reynolds has not explicitly considered coloured, humic lakes in his scheme, it appears to cope with their phytoplankton assemblages reasonably well.

Humic lakes subject to high nutrient loading appear to be largely indistinguishable from other eutrophic lakes with respect to their phytoplankton. Shallow, well-mixed, eutrophic humic lakes have a phytoplankton dominated by cyanobacteria (especially *Oscillatoria agardhii* and *Oscillatoria redekei*), with important contributions from diatoms (especially *Stephanodiscus* spp.) in spring (e.g. Lough Neagh, Northern Ireland: Gibson 1993). This phytoplankton assemblage is equally characteristic of similar lakes which lack a high humic content (e.g. Loch Leven, Scotland: Bailey-Watts 1978). Eutrophic lakes that stratify in summer typically then develop a phytoplankton characterised by cyanobacteria such as *Anabaena*, *Aphanizomenon* and *Microcystis* and, in a later phase, the dinoflagellate *Ceratium*. This phytoplankton community can equally be found during summer stratification in eutrophic humic lakes (e.g. Lovojärvi, southern Finland: Jones and Ilmavirta 1978; Ilmavirta 1982; see also Eloranta 1986).

According to the scheme developed by Reynolds, nutrient-impoverished water columns of low stability will be dominated by diatoms: *Cyclotella* spp. and *Rhizosolenia* spp. in oligotrophic conditions with *Asterionella formosa*, *Aulacoseira* (*Melosira*) *italica* and *Synedra* spp. appearing under more mesotrophic conditions. These species are equally characteristic of oligotrophic, humic lakes under well-mixed conditions (e.g. Pääjärvi, Finland: Ilmavirta and Kotimaa 1974; Loch Ness, Scotland: Jones et al. 1996). When the water column stability increases during summer stratification, Reynolds' scheme predicts that the phytoplankton of these nutrient poor lakes should be characterised by chrysophytes or volvocales, with possible contributions from dinoflagellates and desmids. Again, humic lakes broadly fit within this scheme (Ilmavirta 1982), the main anomaly being the great importance in humic lakes of cryptophytes, a group which Reynolds is unable to link to any particular environmental conditions and considers ubiquitous. The observation that small, sheltered lakes can be dominated at all times by small flagellates, mainly cryptophytes and chrysophytes (Jones and Arvola 1984; Croome and Tyler 1988; Ilmavirta 1988), is accounted for under Reynolds's scheme by the combination of high water column stability and nutrient shortage which characterises these lakes through most of the year. This can apply irrespective of the humic content of the lakes, although both water column stability and nutrient unavailability may be accentuated in lakes with a high content of HS.

In conclusion, there appears to be no reason to suppose that "humic lakes" will consistently exhibit a particular and characteristic phytoplankton community. The phytoplankton composition in any lake is the outcome of competition between species for the resources (light and nutrients) needed to support growth, modified by their ability to resist losses from the water column (due to sinking and grazing). Water column stability, nutrient availability and grazing pressure are the key environmental factors for the phytoplankton. High concentrations of HS in lakes may modify the impact of these key factors but do not override their influence. However, HS can have a direct impact on other components of the plankton community in lakes, most notably the bacterioplankton, and this in turn may indirectly influence the function and importance of phytoplankton in the food chains of some humic lakes. These aspects of the ecology of humic lakes are addressed in Chapters 8 and 10.

7.6 Summary

It is important to emphasise that all lake waters contain some HS of allochthonous origin. Therefore, "humic lakes" with high water colour must be seen as lying towards one end of a continuum between lakes with low concentrations of HS and lakes with high concentrations. From this perspective, it is unlikely that humic lakes will exhibit completely distinctive characteristics; rather it may be supposed that the balance between various physical, chemical and biological processes will progressively shift as the content of HS in lakewater increases. This concept is well illustrated by the impact of HS on the ecology of phytoplankton.

The physical environment experienced by phytoplankton is influenced by the presence of HS. Both the spectral quality and the total available PAR from incoming solar radiation are affected by HS, but the latter appears to have the more important effect on phytoplankton. Absorption of PAR by HS can be regarded as a form of competition with phytoplankton for available quanta. When phytoplankton biomass is high, the proportion of available quanta "lost" to HS is small. However, when phytoplankton biomass is low, even moderate concentrations of HS can capture an appreciable proportion of total available quanta and, particularly under conditions of deep mixing, this can place phytoplankton growth under light limitation. This problem is offset to some extent in small lakes by an inverse relationship between mixing depth and water colour.

The chemical environment experienced by phytoplankton is also influenced by the presence of HS. There is no evidence that direct uptake of HS makes a significant contribution to the carbon/energy budgets of phytoplankton. Although HS are known to modify the toxicity of heavy metals, and possibly other pollutants, in most freshwaters metal toxicity is not an important aspect of phytoplankton ecology. Of greater importance is the potential of HS to modify the availability to phytoplankton of essential nutrients such as iron and phosphate. Chelation of iron by HS may be an important factor contributing to the relative importance of chrysophyte algae in humic lakes and their scarcity in hard-water, alkaline lakes. There is extensive evidence that HS affect the chemical speciation and bioavailability of phosphorus in freshwaters. This effect probably occurs mainly through interaction of phosphate with HS-iron complexes. Release of complexed phosphate is promoted by UV radiation.

Surveys of phytoplankton communities from a wide range of lakes, including lakes of differing HS content, highlight the difficulties in such surveys of disentangling the possible effects of HS from those of other environmental factors such as pH. However, the limited available evidence does not provide strong support for the view that humic lakes have a characteristic phytoplankton assemblage, although chrysophyte and cryptophyte flagellates do seem to be particularly well represented in nutrient-poor humic lakes. Generalised schemes which attempt to explain variations in phytoplankton community structure by reference to water column stability, nutrient concentrations and irradiance on the whole appear to cope adequately with the phytoplankton assemblages of humic lakes.

References

Anderson DM, Morel FMM (1978) Copper sensitivity of *Gonyaulax tamarensis*. Limnol Oceanogr 23:283–295

Anderson MA, Morel FMM (1982) The influence of aqueous iron chemistry on the uptake of iron by the coastal diatom *Thalassiosira weissflogii*. Limnol Oceanogr 27:789–813

Arvola L (1984) Vertical distribution of primary production and phytoplankton in two small lakes with different humus concentrations in southern Finland. Holarct Ecol 7:390–398

Arvola L (1986) Spring phytoplankton of 54 small lakes in southern Finland. Hydrobiologia 137:125–134

Arvola L, Salonen K, Jones RI, Heinänen A, Bergström I (1987) A three day study of the diel behaviour of plankton in a highly humic and steeply stratified lake. Arch Hydrobiol 109:89–106

Arvola L, Metsälä T-R, Similä A, Rask M (1990) Phyto- and zooplankton in relation to water pH and humic content in small lakes in southern Finland. Verh Int Verein Limnol 24:688–692

Arvola L, Kankaala P, Tulonen T, Ojala A (1996) Effects of phosphorus and allochthonous humic matter enrichment on the metabolic processes and community structure of plankton in a boreal lake (Lake Pääjärvi). Can J Fish Aquat Sci 53:1646–1662

Auclair JC, Brassard P, Couture P (1985) Effects of two molecular weight fractions on phosphorus cycling in natural phytoplankton communities. Water Res 19:1447–1453

Bailey-Watts AE (1978) A nine-year study of the phytoplankton of the eutrophic and non-stratifying Loch Leven (Kinross, Scotland). J Ecol 66:741–771

Beaver JR, Crisman TL (1991) Importance of latitude and organic color on phytoplankton productivity in Florida lakes. Can J Fish Aquat Sci.48:1145–1150

Bennett A, Bogorad L (1973) Complementary chromatic adaptation in a filamentous blue-green alga. J Cell Biol 58:419–435

Bentzen E, Taylor WD, Millard ES (1992) The importance of dissolved organic phosphorus to phosphorus uptake by limnetic plankton. Limnol Oceanogr 37:217–231

Berninger U-G, Caron DA, Sanders RW (1992) Mixotrophic algae in three ice-covered lakes of the Pocono Mountains, USA. Freshw Biol 28:263–272

Bird DF, Kalff J (1986) Bacterial grazing by planktonic lake algae. Science 231:493–495

Bird DF, Kalff J (1989) Phagotrophic sustenance of a metalimnetic phytoplankton peak. Limnol Oceanogr 34:155–162

Bowling LC (1990) Heat contents, thermal stabilities and Birgean wind work in dystrophic Tasmanian lakes and reservoirs. Aust J Mar Freshw Res 41:429–441

Brassard P, Auclair JC (1984) Orthophosphate uptake rate constants are mediated by the 10^3–10^4 molecular weight fraction in Shield lakewater. Can J Fish Aquat Sci 41:166–173

Bushaw KL, Zepp RG, Tarr MA, Schulz-Janders D, Bourbonniere RA, Hodson RA, Miller WL, Bronk DA, Moran MA (1996) Photochemical release of biologically available nitrogen from aquatic dissolved organic matter. Nature 381:404–407

Callieri C, Amicucci E, Bertoni R, Vörös L (1996) Fluorometric characterization of two picocyanobacteri strains from lakes of different underwater light quality. Int Rev Gesamten Hydrobiol 81:13–23

Carlsson P, Granéli E (1993) Availability of humic bound nitrogen for coastal phytoplankton. Estuarine Coastal Shelf Sci 36:433–447

Carlsson P, Granéli E, Tester P, Boni L (1995) Influences of riverine humic substances on bacteria, protozoa, phytoplankton, and copepods in a coastal plankton community. Mar Ecol Prog Ser 127:213–221

Cembella AD, Antia NJ, Harrison PJ (1984) The utilization of inorganic and organic phosphorus compounds as nutrients by eukaryotic microalgae: a multidisciplinary perspective. Part 1. CRC Crit Rev Microbiol 10:317–391

Chow-Fraser P, Duthie H (1983) An interpretation of phosphorus loadings in dystrophic lakes. Arch Hydrobiol 97:109–121

Chow-Fraser P, Duthie H (1987) Response of the phytoplankton community to weekly additions of monoammonium phosphate in a dystrophic lake. Arch Hydrobiol 110:67–82

Chróst RJ (1990) Ectoenzymes in aquatic environments: origin, activity and ecological significance. In: Overbeck J, Chróst RJ (eds) Advanced biochemical and molecular approaches to aquatic microbial ecology. Springer, Berlin, Heidelberg, New York, pp 47–48

Cotner JB, Wetzel RG (1992) Uptake of dissolved inorganic and organic phosphorus compounds by phytoplankton and bacterioplankton. Limnol Oceanogr 37:232–243

Croome RL, Tyler PA (1988) Phytoflagellates and their ecology in Tasmanian polyhumic lakes. Hydrobiologia 161:245–253

De Haan H (1982) Physico-chemical environment in Tjeukemeer with special reference to speciation of algal nutrients. Hydrobiologia 95:205–221

De Haan H. De Boer T (1986) Geochemical aspects of aqueous iron, phosphorus and dissolved organic carbon in the humic Lake Tjeukemeer, The Netherlands. Freshw Biol 16:661–672

De Haan H, Jones RI, Salonen K (1987) Does ionic strength affect the configuration of aquatic humic substances, as indicated by gel filtration? Freshw Biol 17:453–459

De Haan, Jones RI, Salonen K (1990) Abiotic transformations of iron and phosphate in humic lake water, revealed by double isotope labelling and gel filtration. Limnol Oceanogr 35: 491–497

Downes MT, Paerl HW (1978) Separation of two dissolved reactive phosphorus fractions in lakewater. J Fish Res Board Can 35:1636–1639

Dring MJ (1981) Chromatic adaptation of photosynthesis in benthic marine algae: an examination of its ecological significance using a theoretical model. Limnol Oceanogr 26:271–284

Earle JC, Duthie HC, Scruton DA (1986) Analysis of the phytoplankton composition of 95 Labrador lakes, with special reference to natural and anthropogenic acidification. Can J Fish Aquat Sci 43:1804–1811

Ekelund NGA (1990) Effects of UV-B radiation on growth and motility of four phytoplankton species. Physiol Plant 78:590–594

Ekelund NGA (1993) The effect of UV-B radiation and humic substances on growth and motility of the flagellate, *Euglena gracilis*. J Plankton Res 15:715–722
Eloranta P (1978) Light penetration in different types of lakes in central Finland. Holarct Ecol 1: 362–366
Eloranta P (1986) Phytoplankton structure in different lake types in central Finland. Holarct Ecol 9:214–224
Eloranta P (1995) Phytoplankton of the national park lakes in central and southern Finland. Ann Bot Fenn 32:193–209
Engelmann TW (1883) Farbe und Assimilation. Bot Z 41:1–13
Fagerbakke KM, Heldal M, Norland S (1996) Content of carbon, nitrogen, oxygen, sulfur and phosphorus in native aquatic and cultured bacteria. Aquat Microb Ecol 10:15–27
Fee EJ, Hecky RE, Kasian SEM, Cruikshank DR (1996) Effects of lake size, water clarity, and climatic variability on mixing depths in Canadian Shield lakes. Limnol Oceanogr 41:912–920
Fogg GE, Thake B (1987) Algal cultures and phytoplankton ecology, 3rd edn. University of Wisconsin Press, Madison
Francko DA (1986) Epilimnetic phosphorus cycling: influence of humic materials and iron coexisting major mechanisms. Can J Fish Aquat Sci 43:302–310
Francko DA (1990) Alteration of bioavailability and toxicity by phototransformation of organic acids. In: Perdue EM, Gjessing ET (eds) Organic acids in aquatic ecosystems. Wiley, New York, pp 167–177
Francko DA, Heath RT (1979) Functionally distinct classes of complex phosphorus compounds in lake water. Limnol Oceanogr 24:463–473
Francko DA, Heath RT (1982) UV-sensistive complex phosphorus: association with dissolved humic material and iron in a bog lake. Limnol Oceanogr 27:564–569
Ganf GG, Heaney SI, Corry J (1991) Light absorption and pigment content in natural populations and cultures of a non-gas-vacuolate cyanobacterium, *Oscillatoria bourrelleyi* (= *Tychonema bourrelleyi*). J Plankton Res 13:1101–1121
Gibson CE (1993) The phytoplankton populations of Lough Neagh. In: Wood RB, Smith RV (eds) Lough Neagh: the ecology of a multipurpose water resource. Kluwer, Dordrecht, pp 59–73
Gjessing ET, Källqvist T (1991) Algicidal and chemical effect of UV-radiation of water containing humic substances. Water Res 25: 491–494
Guildford SJ, Healey FP, Hecky RE (1987) Depression of primary production by humic matter and suspended sediment in limnocorral experiments at Southern Indian Lake, Northern Manitoba. Can J Fish Aquat Sci 45:1408–1417
Hessen DO (1985) The relation between bacterial carbon and dissolved humic compounds in oligotrophic lakes. FEMS Microbiol Ecol 31:215–223
Hessen DO, Van Donk E (1994) Effects of UV-radiation of humic water on primary and secondary production. Water Air Soil Pollution 75:325–338
Heyman U (1983) Relations between production and biomass of phytoplankton in four Swedish lakes of different trophic status and humic content. Hydrobiologia 101:89–104
Hongve D, Skogheim OK, Hindar A, Abrahamsen H (1980) Effects of heavy metals in combination with NTA, humic acid, and suspended sediment on natural phytoplankton photosynthesis. Bull Environ Contam Toxicol 25:594–600
Hsu PH (1976) Comparison of iron(III) and aluminium in precipitation of phosphate from solution. Water Res 10:903–907
Hutchinson GE (1957) A treatise on limnology. I Geography, physics and chemistry. Wiley, New York
Hutchinson GE (1967) A treatise on limnology. II Introduction to lake biology and the limnoplankton. Wiley, New York
Ilmavirta K, Kotimaa A-L (1974) Spatial and seasonal variations in phytoplanktonic primary production and biomass in the oligotrophic lake Pääjärvi, southern Finland. Ann Bot Fenn 11:112–120

Ilmavirta V (1980) Phytoplankton in 35 Finnish brown-water lakes of different trophic status. Dev Hydrobiol 3:121–130

Ilmavirta V (1982) Dynamics of phytoplankton in Finnish lakes. Hydrobiologia 86:11–20

Ilmavirta V (1983) The role of flagellated phytoplankton in chains of small brown-water lakes in southern Finland. Ann Bot Fenn 20:187–195

Ilmavirta V (1988) Phytoflagellates and their ecology in Finnish brown-water lakes. Hydrobiologia 161:255–270

Jackson TA, Hecky RE (1980) Depression of primary productivity by humic matter in lake and reservoir waters of the boreal forest zone. Can J Fish Aquat Sci 37:2300–2317

Jansson M, Blomqvist P, Jonsson A, Bergström A-K (1996) Nutrient limitation of bacterioplankton, autotrophic and mixotrophic phytoplankton, and heterotrophic nanoflagellates in Lake Örträsket. Limnol Oceanogr 41:1552–1559

Järnefelt H (1956) Zur Limnologie einiger Gewässer Finnlands. XVI. Mit besonderer Berücksichtigung des Planktons. Ann Zool Soc Vanamo 17:201 pp

Järnefelt H (1958) On the typology of the northern lakes. Verh Int Verein Limnol 13:228–235

Jewson DH (1976) The interaction of components controlling net phytoplankton photosynthesis in a well-mixed lake. Freshw Biol 6:551–576

Jewson DH (1993) The optical properties of Lough Neagh. In: Wood RB, Smith RV (eds) Lough Neagh: the ecology of a multipurpose water resource. Kluwer, Dordrecht, pp 59–73

Jewson DH, Taylor JA (1978) The influence of turbidity on net phytoplankton photosynthesis in some Irish lakes. Freshw Biol 8:573–584

Jones RI (1991) Advantages of diurnal vertical migrations to phytoplankton in sharply stratified, humic forest lakes. Arch Hydrobiol 120:257–266

Jones RI (1992) The influence of humic substances on lacustrine planktonic food chains. Hydrobiologia 229:73–91

Jones RI (1993) Phytoplankton migrations: patterns, processes and profits. Arch Hydrobiol Beih Ergebn Limnol 39:66–77

Jones RI (1994) Mixotrophy in planktonic protists as a spectrum of nutritional strategies. Mar Microb Food Webs 8:87–96

Jones RI, Arvola L (1984) Light penetration and some related characteristics in small forest lakes in southern Finland. Verh Int Verein Limnol 22:811–816

Jones RI, Ilmavirta V (1978a) Vertical and seasonal variation of phytoplankton photosynthesis in a brown-water lake with winter ice cover. Freshw Biol 8:561–572

Jones RI, Ilmavirta V (1978b) A diurnal study of the phytoplankton in the eutrophic lake Lovojärvi, southern Finland. Arch Hydrobiol 83:494–514

Jones RI, Salonen K, De Haan H (1988) Phosphorus transformations in the epilimnion of humic lakes: abiotic interactions between dissolved humic materials and phosphate. Freshw Biol 19: 357–369

Jones RI, Young JM, Hartley AM, Bailey-Watts AE (1996) Light limitation of phytoplankton development in an oligotrophic lake – Loch Ness, Scotland. Freshw Biol 35:533–543

Kämäri J, Forsius M, Kauppi L (1990) Statistically based lake survey: a representative picture of nutrient status in Finland. Verh Int Verein Limnol 24:663–666

Kankaala P, Hellsten S, Alasaarela E (1984) Primary production of phytoplankton in the oligohumic Kitka lakes in northern Finland. Aqua Fenn 14:65–78

Karentz D, Bothwell ML, Coffin RB et al. (1994) Impact of UV-B radiation on pelagic freshwater ecosystems: report of working group on bacteria and phytoplankton. Arch Hydrobiol Beih Ergebn Limnol 43:31–69

Keskitalo J (1977) The species composition and biomass of phytoplankton in the eutrophic Lake Lovojärvi, southern Finland. Ann Bot Fenn 14:71–81

Kirk JTO (1980) Spectral absorption properties of natural waters: contribution of the soluble and particulate fractions to light absorption in some inland waters in of southeastern Australia. Aust J Mar Freshw Res 31:287–296

Kirk JTO (1983) Light and photosynthesis in aquatic ecosystems. Cambridge University Press, Cambridge

Kirk JTO, Hargreaves BR, Morris DP et al. (1994) Measurements of UV-B radiation in two freshwater lakes: an instrument comparison. Arch Hydrobiol Beih Ergebn Limnol 43:71–99

Lean DRS (1973) Movements of phosphorus between its biologically important forms in lake water. J Fish Res Board Can 30:1525–1536

Lean DRS, Rigler FH (1974) A test of the hypothesis that abiotic phosphate complexing influences phosphorus kinetics in epilimnetic lake water. Limnol Oceanogr 19:784–788

Levesque M, Schnitzer M (1967) Organo-metallic interactions in soils: 6. Preparation and properties of fulvic acid-metal phosphates. Soil Sci 103:183–190

Levine SN, Stainton MP, Schindler DW (1986) A radiotracer study of phosphorus cycling in a eutrophic Canadian Shield lake, Lake 227, northwestern Ontario. Can J Fish Aquat Sci 43: 366–378

Lijklema L (1980) Interaction of orthophosphate with iron (III) and aluminium hydroxides. Environ Sci Technol 14:537–541

Macan TT (1970) Biological studies of the English Lakes. Longman, London

Mazumder A, Taylor WD (1994) Thermal structure of lakes varying in size and water clarity. Limnol Oceanogr 39:968–976

McGarrigle ML, Kilmartin L (1992) UV-sensitive phosphorus in Irish peaty waters: a study of potential effects on freshwater ecosystems. Environmental Unit Research Report, Department of the Environment, Dublin

McKnight DM, Morel FMM (1980) Copper complexation by siderophores from filamentous blue-green algae. Limnol Oceanogr 24:62–71

Meili M (1992) Sources, concentrations and characteristics of organic matter in softwater lakes and streams of the Swedish forest region. Hydrobiologia 229:23–41

Miles CJ, Brezonik PL (1981) Oxygen consumption in humic-colored waters by a photochemical ferrous-ferric catalytic cycle. Environ Sci Technol 15:1089–1095

Münster U, Einiö P, Nurminen J, Overbeck J (1992) Extracellular enzymes in a polyhumic lake: important regulators in detritus processing. Hydrobiologia 229:225–238

Naumann E (1919) Några synpunkter angående planktons ökologi. Med särskild hänsyn till fytoplankton. Svensk Bot Tidskr 13:129–163

Nygaard K, Tobiesen A (1993) Bacterivory in algae: a survival strategy during nutrient limitation. Limnol Oceanogr 38:273–279

Ojala A (1993) The influence of light quality on growth and phycobiliprotein/chlorophyll *a* fluorescence quotients of some species of freshwater algae in culture. Phycologia 32:22–28

Olsson H (1983) Origin and production of phosphatases in the acid Lake Gårdsjön. Hydrobiologia 101:49–58

Pick FR (1991) The abundance and composition of freshwater cyanobacteria in relation to light penetration. Limnol Oceanogr 36:1457–1462

Prakash A, Rashid MA, Jensen A, Subba Rao DV (1973) Influence of humic substances on the growth of marine phytoplankton diatoms. Limnol Oceanogr 18:516–524

Ramberg L (1979) Relations between phytoplankton and light climate in two Swedish forest lakes. Int Revue Gesamten Hydrobiol 64:749–782

Reynolds CS (1980) Phytoplankton assemblages and their periodicity in stratifying lake systems. Holarct Ecol 3:141–159

Reynolds CS (1984a) The ecology of freshwater phytoplankton. Cambridge University Press, Cambridge

Reynolds CS (1984b) Phytoplankton periodicity: the interactions of form, function and environmental variability. Freshw Biol 14:111–142

Reynolds CS (1997) Vegetation processes in the pelagic: a model for ecosystem theory. Excellence in Ecology 9, Ecology Institute, Oldendorf/Luhe

Salonen K, Jones RI, Arvola L (1984) Hypolimnetic phosphorus retrieval by diel vertical migrations of lake phytoplankton. Freshw Biol 14:431–438

Salonen K, Jones RI, De Haan H, James M (1994) Radiotracer study of phosphorus uptake by plankton and redistribution in the water column of a small humic lake. Limnol Oceanogr 39:69–83

Sanders RW (1991) Mixotrophic protists in marine and freshwater ecosystems. J Protozool 38: 76–81

Sanders RW, Porter KG, Bennett SJ, DeBiase AE (1989) Seasonal patterns of bacterivory by flagellates, ciliates, rotifers, and cladocerans in a freshwater planktonic community. Limnol Oceanogr 34:673–687

Sandgren CD (1988) The ecology of chrysophyte flagellates: their growth and perennation strategies as freshwater phytoplankton. In: Sandgren CD (ed) Growth and reproductive strategies of freshwater phytoplankton. Cambridge University Press, Cambridge, pp 9–104

Sedlacek J, Källqvist T, Gjessing E (1983) Effect of aquatic humus on uptake and toxicity of cadmium to *Selenastrum capricornutum* Prints. In: Christman RF, Gjessing ET (eds) Aquatic and terrestrial humic materials, Ann Arbor Science, Ann Arbor, pp 495–516

Sepers ABJ (1977) The utilization of dissolved organic compounds in aquatic environments. Hydrobiologia 52:39–54

Shang C, Stewart JWB, Huang PM (1992) pH effect on kinetics of adsorption of organic and inorganic phosphates by short-range ordered aluminium and iron precipitates. Geoderma 53: 1–14

Shapiro J (1967) Yellow organic acids of lake water: differences in their composition and behaviour. In: Golterman HL, Clymo RS (eds) Chemical environment in the aquatic habitat, North Holland, Amsterdam, pp 202–216

Shaw PJ (1994) The effect of pH, dissolved humic substances, and ionic composition on the transfer of iron and phosphate to particulate size fractions in epilimnetic lake water. Limnol Oceanogr 39:1734–1743

Simpson FB, Neilands JB (1976) Siderochromes in Cyanophyceae: isolation and characterization of schizokinen from *Anabaena* sp. J Phycol 12:44–48

Smith RC, Tyler JE, Goldman CR (1973) Optical properties and colour of Lake Tahoe and Crater Lake. Limnol Oceanogr 18:189–199

Sommer U (1988) Some size relationships in phytoflagellate motility. Hydrobiologia 161: 125–131

Sommer U, Gliwicz ZM (1986) Long range vertical migration of *Volvox* in tropical lake Cahora Bassa (Mozambique). Limnol Oceanogr 31:650–653

Stainton MP (1980) Errors in molybdenum blue methods for determining orthophosphate in freshwater. Can J Fish. Aquat Sci 37:472–478

Steinberg C, Baltes GF (1984) Influence of metal compounds on fulvic acid/molybdenum blue reactive phosphate associations. Arch Hydrobiol 100:61–71

Stevens RJ, Stewart BM (1982) Concentration, fractionation and characterization of soluble organic phosphorus in river water entering Lough Neagh. Water Res 16:1507–1519

Stewart AJ, Wetzel RG (1981) Dissolved humic materials: photodegradation, sediment effects, and reactivity with phosphate and calcium carbonate precipitation. Arch Hydrobiol 92: 265–286

Stewart AJ, Wetzel RG (1982) Influence of dissolved humic materials on carbon assimilation and alkaline phosphatase activity in natural algal-bacterial assemblages. Freshw Biol 12:369–380

Stramski D, Morel A (1990) Optical properties of photosynthetic picoplankton in different physiological states as affected by growth irradiance. Deep Sea Res 37:245–266

Stumm W, Morgan JJ (1981) Aquatic chemistry, 2nd edn. Wiley, New York

Sunda W, Lewis JAM (1978) Effect of complexation by natural organic ligands on the toxicity of copper to a unicellular alga, *Monochrysis lutheri*. Limnol Oceanogr 23:870–876

Tarapchak SJ, Bigelow SM, Rubitschun C (1982) Soluble reactive phosphorus measurements in Lake Michigan: filtration artefacts. J Great Lakes Res 8:550–557

Thienemann A (1925) Die Binnengewässer Mitteleuropas. Binnengewässer 1:255 pp

Tipping E (1981) The adsorption of aquatic humic substances by iron oxides. Geochim Cosmochim Acta 45:191–199

Tranvik LJ (1988) Availability of dissolved organic carbon for planktonic bacteria in oligotrophic lakes of differing humic content. Microb Ecol 16:311–322

Tranvik LJ, Porter KG, Sieburth JM (1989) Occurrence of bacterivory in *Cryptomonas*, a common freshwater phytoplankter. Oecologia 78:473–476

Tulonen T, Salonen K, Arvola L (1992) Effects of different molecular weight fractions of dissolved organic matter on the growth of bacteria, algae and protozoa from a highly humic lake. Hydrobiologia 229:239–252

Van Donk E (1983) Factors affecting phytoplankton growth and succession in Lake Maarsseveen (I). PhD Thesis, University of Amsterdam

Veen A (1991) Ecophysiological studies on the phagotrophic phytoflagellate *Dinobryon divergens* Imhof. PhD Thesis, University of Amsterdam

Vollenweider RA (1968) Scientific fundamentals of the eutrophication of lakes and flowing waters with particular reference to nitrogen and phosphorus as factors in eutrophication. OECD DAS/CSI/68.27, Paris

Vollenweider RA (1975) Input-output models with special reference to the phosphorus loading concept in limnology. Schweiz Z Hydrol 37:53–84

Wall D, Briand F (1979) Response of lake phytoplankton communities to in situ manipulations of light intensity and colour. J Plankton Res 1:103–112

Wetzel RG (1983) Limnology, 2nd edn. Saunders, Philadelphia

White E, Payne G (1980) Distribution and biological availability of reactive high molecular weight phosphorus in natural waters in New Zealand. Can J Fish Aquat Sci 37:664–669

Willén E, Hajdu S, Pejler Y (1990) Summer phytoplankton in 73 nutrient-poor Swedish lakes. Classification, ordination and choice of long-term monitoring objects. Limnologica 20: 217–227

8 Nutrient Limitation and Bacteria – Phytoplankton Interactions in Humic Lakes

Mats Jansson

8.1 Nutrient Limitation in Lakes

In most lakes, nutrient limitation is synonymous with P-limitation. Low concentrations of P restrict the productivity of freshwater organisms and freshwater ecosystems, a fact that has been stressed over the years by algal physiologists (Beijerinck 1890; Pringsheim 1912; Tilman et al. 1982) and limnologists (Rodhe 1948; Vollenweider 1968; Sakamoto 1966; Schindler 1977). Although other elements than P sometimes may limit the growth of certain members of the phytoplankton, i.e. Si for diatoms (Tilman et al.1982), Se for *Peridinium cinctum* (Lindström and Rodhe 1978) and N for cyanobacteria (Schindler 1977), the role of P as the key limiting inorganic nutrient for the productivity of freshwater ecosystems is today accepted and little questioned (Shapiro 1988). The convincing results of P addition to whole lakes in the Experimental Lakes Area in Canada in the 1970s (Schindler 1977) put an end to the discussion about nutrient limitation in lakes for some decades.

Early studies treated limiting nutrients in terms of nutrient access to phytoplankton, since phytoplankton, at least in systems dominated by pelagial production, were considered to be the base for the total production of lakes. After the introduction of the microbial loop concept (Azam et al. 1983), and the realization that bacteria are superior competitors for low P concentrations (Currie and Kalff 1984a–c; Jansson 1988; Vadstein et al. 1988), much of the concern for limiting nutrients shifted from the question of which nutrients are limiting, to studies on how the limiting nutrient, i.e. P, is shared between bacterioplankton and phytoplankton and other components of the food web. The most commonly accepted model predicts that bacteria win the competition for low P concentrations as long as their energy supply is sufficient (Currie and Kalff 1984c; Jansson 1993; Kirchman 1994). Since the major energy source for pelagic bacteria is carbon from the photosynthesis, phytoplankton can control the energy supply to the bacteria. The energy supply for bacteria is therefore restricted if the P supply to the algae decreases too much. The bacterial production then decreases and more P is allocated to phytoplankton, that start to produce more carbon for the bacteria,

Ecological Studies, Vol. 133
Hessen, Tranvik (eds) Aquatic Humic Substances

and so on (Currie 1990) The model is consistent with the view that P is the limiting nutrient, since both bacteria and phytoplankton production will ultimately be determined by P availability.

The nutrient supply and availability of limiting nutrients is not only dependent on the amount and quality of external loading. Interactions within the food web, e.g. grazing and excretion of dissolved nutrients from flagellates and zooplankton, may increase the turnover and supply of bioavailable nutrient fractions. Grazing on bacteria by flagellates can be a significant nutrient source for phytoplankton by excretion of dissolved nutrients (Rothaupt 1992). The quality and N:P ratio of excreted nutrients depend on grazer species composition (Sterner 1989). Moreover, different phytoplankton may have different requirements of nutrients as indicated by the highly variable N:P ratios in different phytoplankton species (Cembella et al. 1984). The limiting nutrient concept on the organism level is thus more complicated than on the system level. It is likely that the species composition of planktonic communities is largely influenced by interactions within the food web which more or less continuously alter the supply of available nutrients in different ways. However, it is not possible to dismiss the P limitation concept for whole lake ecosystems. The empirical evidence for that increased P supply increases the productivity of lakes is too strong (Schindler 1977).

Very few studies on nutrient limitation have been performed in humic lakes and the experiences from oligotrophic and eutrophic lakes are not necessarily applicable to brown water, dystrophic lakes. On the contrary, it seems as if humic lakes do not follow the relationships between nutrient concentration and biological responses established with data from other lake types. Primary production and phytoplankton biomass are often lower in humic lakes than what theoretically can be predicted from the P loading (Arvola 1984; Guildford et al. 1987). Heyman and Lundgren (1988; Fig. 8.1) showed that phytoplankton biomass and lake carrying capacity of phytoplankton in relation to total P concentration was inversely related to water colour, i.e. humic lakes produce less phytoplankton per P unit than non-humic lakes. Fig. 8.1 indicates low availability of P to phytoplankton in humic lakes or that some other factor than the P concentration limits the buildup of phytoplankton biomass. Ramberg (1979) argued that the biomass of phytoplankton was regulated by P supply and the light climate. Blomqvist et al. (1981) concluded that the primary production in Lake Siggeforasjön, Sweden was not regulated by P concentration and that the content of inorganic N in the epilimnion was too low to sustain the observed plankton production. Experiments in humic Finnish lakes have suggested that there was a balance between N and P limitation (Jones 1990).

In short, it appears as if a smaller share of the total P concentration is allocated to phytoplankton biomass in humic lakes than in non-humic lakes. On the other hand, there is no consensus on limiting nutrients in brown water lakes. Biological production in humic lakes can be P-limited, N-limited or N+P-limited according to the literature. However, none of the possibilities is supported by strong empirical or experimental evidence.

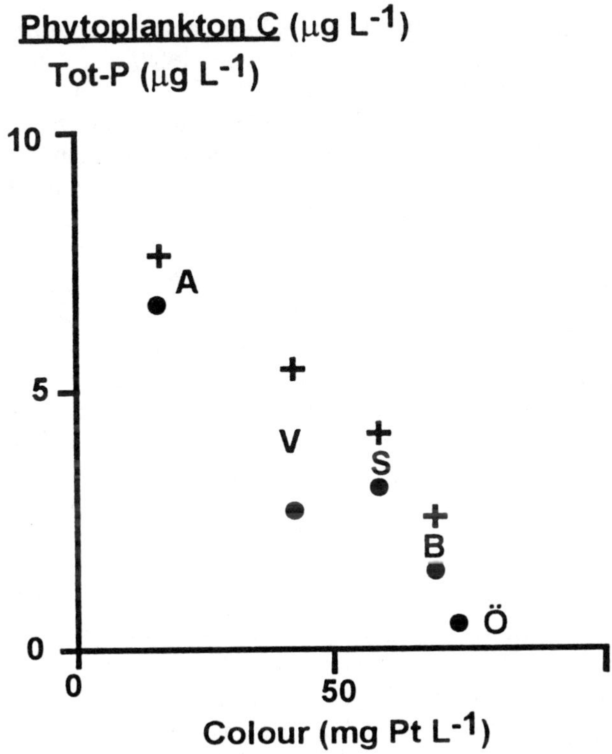

Fig. 8.1. The relation between summer means of phytoplankton biomass (*filled circles*) and lake carrying capacity for phytoplankton (*crosses*) per unit of total P and water colour in five Swedish lakes. A Lake Erken, V Lake Vitalampa, S Lake Siggeforasjön, B Lake Botjärn, Ö Lake Örträsket. (Modified from Heyman and Lundgren 1988; all data except for Lake Örträsket from Heyman and Lundgren 1988)

8.2 Influence of Humic Substances on Growth and Nutrient Demand of Bacterioplankton and Phytoplankton

Humic substances affect the prerequisites for nutrient uptake and planktonic production in several ways. First, humic substances absorb light and restrict the penetration of photosynthetic active radiation (PAR). The trophogenic layer is considerably thinner in humic lakes than in clearwater and moderately eutrophic lakes. In typical humic lakes with total organic carbon (TOC) concentrations > 5 mg l^{-1} and chlorophyll *a* concentrations of a few μg L^{-1} only a few percent of the PAR are captured by phytoplankton, which is about one magnitude less than in clearwater lakes (Jones 1992). Light adsorption by humic substances thus restricts the effective light climate in the epilimnion (Ramberg 1979; Heyman 1983). On the other hand, the heat adsorption of humic material means that humic water warms up quickly and summer stratification in temperate lakes gives thin epilimnia; the circulating layer is often only 1–2 m thick, sometimes even less (Jones 1992; Jones, this Vol.). Planktonic particles, randomly distributed in the epilimnion, may therefore be continuously dispersed within the light-saturated zone. However, the net effect of humus on phytoplankton production is that photosynthesis is aggravated. Humic compounds restrict the lake water volume available for photosynthesis and thereby also the amount of nutrients that can be extracted for phytoplankton production.

Second, humic substances serve as an energy source for bacteria which leads to considerably higher bacterial production than what can be postulated from primary production (Tranvik 1988,1989; del Giorgio and Peters 1993; Hessen et al. 1994). Bacterial biomass shows a positive correlation with humic content (Johansson 1983; Arvola 1984; Jones 1992) and bacterial production may be several times higher than phytoplankton production in humic lakes (Tranvik 1989; Jansson et al. 1996). Bacterioplankton appear to be able to use less than 10% of the total allochthonous organic carbon (Moran and Hodson 1990), but allochthonous organic carbon is, nevertheless, sufficient to sustain up to 90 % of the bacterial production in humic lakes (Hessen 1992). The bacterial degradation of humic compounds is probably enhanced by photo-oxidation, whereby complex recalcitrant compounds are degraded to more available smaller organic molecules (Lindell et al. 1995). The conditions in humic lakes, dictated by the characteristics of allochthonous carbon compounds, thus favour bacterial production and aggravate phytoplankton production. The flux of nutrients between bacteria and phytoplankton is, accordingly, different from that in other types of lakes.

8.3 Nutrients in Humic Lakes

Humic material is the main carrier of N and P in brown water lakes. Nitrogen is the most common limiting nutrient in terrestrial forest ecosystems in temperate regions. The export of inorganic N to groundwater and surface water is therefore small, and the bulk of dissolved N is bound in humic structures. Total N concentrations in temperate brown water lakes are typically 300–500 $\mu g\, l^{-1}$ while the inorganic N fractions are close to detection limits (Ramberg 1979; Jansson et al. 1996). The bioavailability of humus N is little known.

P concentrations may be rather high in humic lakes. Concentrations of between 10 and 25 $\mu g\, P\, l^{-1}$ are typical (Jansson et al. 1996). Most of the P is associated with humus colloids in iron-phosphorus-humus complexes (Ohle 1935; Tipping 1981). The negative surface charge of the complexes (H^-, COO^- and other functional groups of humic compounds) gives the complexes colloidal properties. The colloidal complexes have low settling rates and retention of P is probably lower in brown water lakes than in comparable non-humic lakes. Thereby, the P that enters humic lakes is exposed to uptake in plankton for a longer time than in non-humic lakes of comparable size and water renewal time.

The biovailability of humus-bound P is open to question. Allochthonous dissolved organic material (DOM) is poor in P which has been suggested to favour efficient P scavengers such as bacteria (Stewart and Wetzel 1982). De-Haan et al. (1990) demonstrated that humic complexes released orthophoshate to the ambient medium when orthophosphate was depleted, and suggested that humic complexes were a buffer system for bioavailable P. However, the phosphate release was slow and if it is a significant source for biological uptake it may cause long P turnover times in biota. Interactions between humus-bound P and biota deserve more attention in order to find out to what extent humus material serves as a reservoir for bioavailable P, and if bioavailable P is "exchanged" between abiotic organic colloids and plankton.

8.4 Energy Supply and Food Web Interactions in Humic Lakes

The generally accepted model for interactions in the planktonic food web in lakes assumes that photosynthesis is the major input of energy for biota and that energy mobilized by photosynthesis in phytoplankton is used directly in a straight food chain, but also via bacteria in the microbial loop. The model assumes that phytoplankton are nutrient- or light-limited and that bacteria are energy-limited and controlled by the energy mobilized by phytoplankton (Currie 1990, Kirchman 1994).

It is doubtful if this model is applicable in humic lakes. The role of humic substances as an energy source for bacteria means that there are two independent energy sources in humic lakes; light and allochthonous carbon, which are exploited by phytoplankton and bacterioplankton, respectively. Both these groups of organisms serve as energy mobilizers, making energy available to higher trophic levels in humic lakes (Jones 1992). Therefore, bacteria and algae should be able to exist independently of each other and it should be possible to find ecosystems including all trophic levels based either mainly on photosynthesis or mainly on bacterial transformation of chemical energy. The introduction of an external energy source, exclusive for bacteria, creates problems when explaining the coexistence of algae and bacteria and interactions within the microbial food web. The model on algal-bacterial relationships described above presupposes that bacteria are energy-limited and that the phytoplankton controls the energy mobilization. The critical question when allochthonous carbon is introduced as an energy source for bacteria is whether the bacteria become relieved of their dependence on phytoplankton as a source of energy. If they do, the bacteria may become nutrient-limited instead of energy-limited. Some evidence exists from recent research that bacteria are essentially nutrient-limited in humic lakes. Hessen et al.

Table 8.1. The most limiting nutrient (or combination of nutrients) for growth of bacterioplankton and phytoplankton in the surface (0 to 1m) water layer in Lake Örträsket during the summer 1995. (from experiments described and reported by Jansson et al. 1996)

	3 June	17 July	17 August	31 August
Bacterioplankton	N + P	P	P	P
Phytoplankton	N + P	N	N	N

(1994) demonstrated that N+P additions stimulated bacterial growth, while glucose did not, in humic lake water in Norway. Nutrient enrichment in mesocosms in the humic Lake Örträsket, Sweden, showed that P stimulated bacterial production (Table 8.1) and increased the bacterial biomass throughout the summer period (Jansson et al. 1996). N never stimulated bacterial growth, while N+P gave stimulance equal to, or slightly higher than, that of P alone.

It seems possible to postulate that bacterioplankton in humic lakes may be limited by P. If bacteria are P-limited, it is obvious that bacteria and phytoplankton must compete for available P. Nearly all experience from uptake studies in mixed cultures of algae and bacteria, or in whole lake water, indicates that bacteria are the winners of this competition (Currie and Kalff 1984; Jansson 1988; Vadstein et al. 1988; Rothaupt and Güde 1992, Kirchman 1994). Considering the overwhelming evidence for bacteria as superior competitors for low concentrations of inorganic P, it is therefore logical that bacterioplankton/phytoplankton production ratios are higher in humic lakes than in non-humic lakes. Conditions that favour bacterial growth and incorporation of P in bacterial biomass may also be part of the explanation of the relationship in Fig. 8.1, which shows that the phytoplankton yield per P unit is small in coloured lakes.

A critical question is how phytoplankton retrieve P in competition with P-limited bacteria that have an excess of energy. Humic lake ecosystems are not based entirely on bacteria but the phytoplankton production often equals or exceeds the bacterial production in the trophogenic layer (Salonen and Jokinen 1988; Tranvik 1989). There must, therefore, be an alternative (to the energy limitation model) explanation of how phytoplankton and bacterioplankton share a mutual limiting nutrient. There are several possibilities:

(1) phytoplankton in humic lakes may be more competitive with bacteria for low P concentrations than what is generally believed; (2) phytoplankton are not P-limited.

The possibility that phytoplankton P uptake can be competitive with bacterial uptake of P was given some support by the discovery of a certain high affinity P transport system in *Scenedesmus quadricauda* which allowed this algae to retrieve P from low concentrations of orthophosphate also in competition with bacteria (Jansson 1993). Similar uptake has not been described for other algae which may depend on the fact that conventional methods for uptake studies will fail to discover the P transport system described in

Scenedesmus. On the other hand, the many studies demonstrating that P added to mixed cultures of P-limited algae and bacteria or to whole lake water nearly always is rapidly incorporated in bacteria or particles of bacterial size are a strong argument against phytoplankton being competitive with bacteria. Another way to compete for low nutrient concentrations is by being small (high surface to volume ratio). Phytopicoplankton in the 0.2 to 2-μm size range are often dominant phytoplankters in oligotrophic clearwater lakes (Stockner 1991). Their distribution in brown water lakes is more or less unknown. The obvious advantage of being small in nutrient-poor media should encourage studies on the occurrence of phytopicoplankton in humic lakes.

The second possibility, i.e. that phytoplankton in humic lakes are not P-limited but restricted by some other nutrient or environmental factor, is in accordance with the basic resource competition theory (Tilman 1977) which postulates that organisms can exist together if they are not limited by the same nutrient. The theory was developed for different species of phytoplankton, but should, in principle, be relevant also for different functional groups of microplankton provided that they exploit mutual nutrient sources. That phytoplankton in humic lakes can largely be limited by N, instead of P, has recently been experimentally demonstrated (Jansson et al. 1996). As will be discussed in the following section, this result and its interpretation may help to explain the structure of microplankton communities in humic lakes and how nutrients are shared between bacterioplankton and phytoplankton.

8.5 Nutrient Limitation of Bacterioplankton and Phytoplankton in Humic Lakes

As mentioned in Section 8.4, very few studies have been carried out where the effect of different possible limiting nutrients have been systematically assessed in humic lakes. Jansson et al. (1996) tested the response to N and P enrichment in terms of chlorophyll *a* yield in nine humic lakes in northern Sweden (Fig. 8.2). P did not increase the chlorophyll concentration, while N gave higher concentrations in all lakes. N+P addition gave the highest chlorophyll concentrations. The same type of experiment (for experimental detail, see Fig. 8.2) was carried out on 4 occasions during the ice-free period in 1994 and five occasions in 1995 in Lake Örträsket (Jansson et al 1996). Lake Örträsket in northern Sweden is a large (7 km^2) and deep (mean depth 23 m) humic lake (TOC concentration of approximately 10 mg L^{-1}). The response to nutrient additions was followed by analysing phytoplankton production, biomass and species composition and bacterial production and biomass. The results demonstrated that phytoplankton (all response parameters) was N-limited during a large part of the summer. N-limitation was particularly pronounced at the summer phytoplankton biomass maximum which occurred

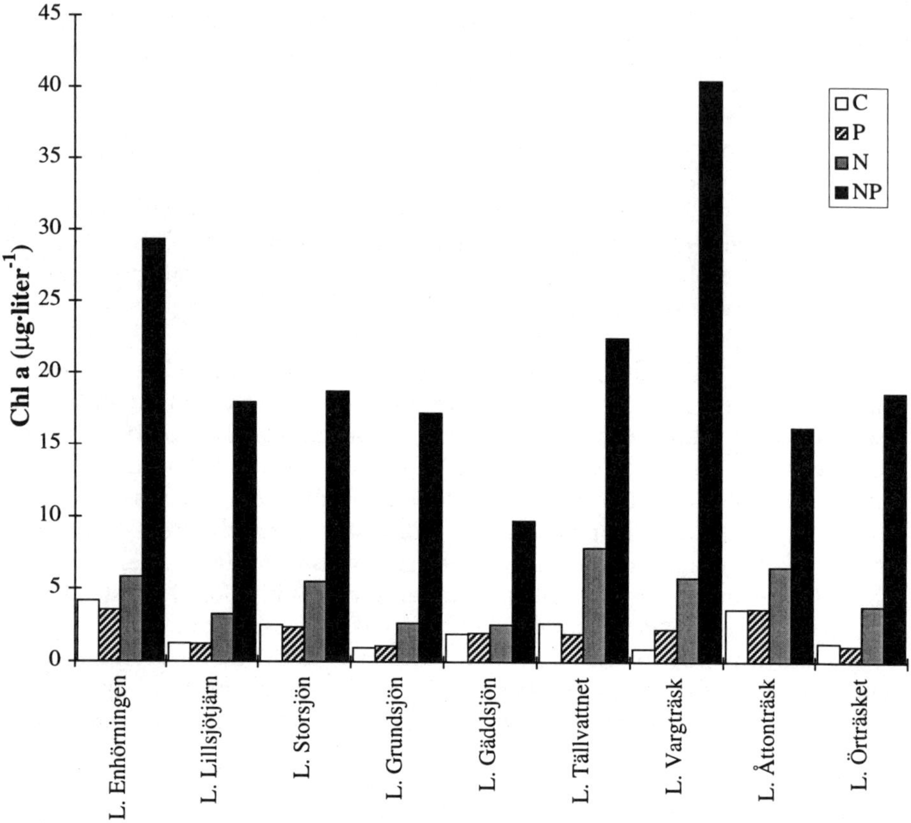

Fig. 8.2. Phytoplankton biomass measured as *Chl a* in enclosures from nine humic lakes in northern Sweden enriched with N, P and N+P. Twenty litres of surface lake water was pumped through a net (200 µm) into 20l polyethylene containers. P (KH_2PO_4) was added to a final concentration of 100 µg l^{-1} and N (NH_4NO_3) was added to a final concentration of 2 mg L^{-1}. Incubation was made in the lake for 4 days immediately below the surface. For statistical evolution of the response of different treatments versus the control (C), the nine lakes were considered as replicates of the same population (humic lakes in the region). One-tailed t-test showed that P did not increase phytoplankton biomass, while both N and the N+P treatments gave higher values than the control. (Modified from Jansson et al. 1996)

in late June and early August. The bacterioplankton were P-limited throughout the season (Table 8.1). These results demonstrate that phytoplankton in humic lakes can be N-limited while bacterioplankton at the same time is P-limited. To what extent the results from northern Sweden are applicable to "humic lakes" in general is hard to judge, but the lakes studied are typical brown water lakes of the boreal forest region. They are not to any significant extent affected by anthropogenic stress and thus represent a close-to-natural state.

The results from Lake Örträsket and the other lakes in Table 8.1 and Fig. 8.2 may seem confusing. That bacterioplankton were P-limited throughout the summer demonstrates that nutrient deficiency rather than exhaustion of available energy sources limits the bacterial production. In an epilimnion with P-limited bacterioplankton, phytoplankton should be extremely P-limited, since P-limited bacteria are more efficient in retrieving P from low medium P concentrations than phytoplankton (Currie and Kalff 1984a, b; Jansson 1993). That phytoplankton in such an environment are N-limited can be explained if they exploit mutual available P sources better than bacterioplankton, which is very unlikely, or if bacteria exploit available N sources better than phytoplankton. Since organic N is by far the largest N pool in humic lakes it is possible that bacteria, if they use humic compounds as a nitrogen source, become P-limited, while at the same time algal growth is depressed due to absence of available N. However, this explanation is not satisfactory. It has been shown that addition of humus to N-limited marine plankton stimulates the growth of phytoplankton (Carlsson et al. 1993) due to bacterial degradation of humus and subsequent liberation of inorganic N. In humic lakes with P-limited bacteria the bacterial exploitation of dissolved organic nitrogen (DON) should then be a net N source to the algae. However, the most serious criticism against the explanation that bacteria "take all the N and make algae N-limited" is that the bacteria are P-limited and it therefore still remains to explain how phytoplankton can obtain enough P to become N-limited. An alternative explanation which was supported by experimental results in Lake Örträsket is that phytoplankton in humic lakes exploit another P source than the bacteria and thereby more than satisfy their need for P.

Phytoplankton in humic lakes is often dominated by pigmented flagellates, mainly chrysophyceans and cryptophytes (Ramberg 1979; Salonen and Jokinen 1988; Jansson et al. 1996). Many of these algae are mixotrophic, i.e. they are capable of both heterotrophic and autotrophic growth. Heterotrophic growth is maintained by phagotrophy (intake of organic particles) or osmotrophy (intake of dissolved organic compounds) (Jones 1994). Phagotrophy by grazing on, e.g. bacteria or other picoplankton is the most frequently reported mixotrophic strategy among phytoflagellates and has been documented for many years (Skuja 1948, 1956; Pringsheim 1963). The ecological role of phagotrophy was not discussed very intensively until the mid-1980s in connection with the interest in flagellates which followed upon the introduction of the microbial loop concept (Azam et al. 1983). Several reviews on mixotrophic flagellates have been published since then (Boraas et al. 1988; Porter 1988; Sanders and Porter 1988; Jones 1994).

Beside their dominance in dystrophic and oligotrophic lakes, mixotrophic phytoplankton also occur episodically in eutrophic lakes, a typical situation being in connection with the bacteria maximum which often follows the crash of the spring diatom bloom (Blomqvist et al.1994). The dominance of mixotrophic flagellates in nutrient-poor lakes or nutrient-poor situations has

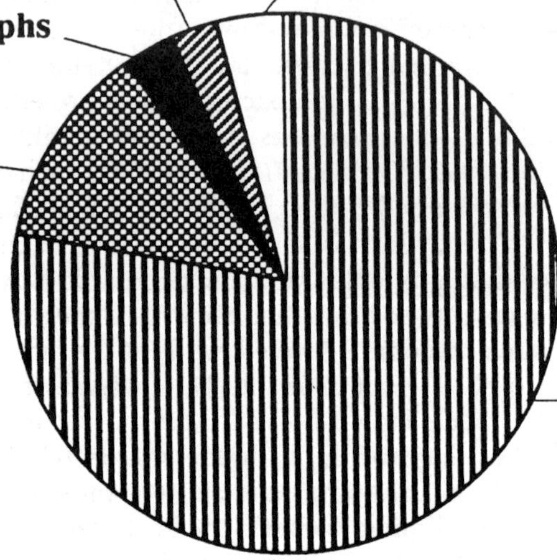

Fig. 8.3. Relative composition of microplankton in Lake Örträsket (mean composition for the period June–September 1993). For definition of species included in the categories mixotrophs and potential mixotrophs see Jansson et al. (1996)

led to speculations that phagotrophy is not only a means to obtain energy for heterotrophic growth but also a means to retrieve nutrients (Tranvik et al. 1989; Caron et al. 1990; Jansson et al. 1996). Mixotrophic phytoplankton have also been shown to have a high N- and especially P-incorporation efficiency from grazing on bacteria (Caron et al. 1990).

Mixotrophy must be considered as an advantageous strategy in brown water lakes where access to nutrients, especially P, is very restricted due to the nutrient competition with bacterioplankton. Mixotrophy thus offers the phytoplankton an alternative nutrient source besides dissolved nutrients. If phytoplankton graze bacteria and thereby are able to incorporate bacterial P in their own cell metabolism they do, indeed, have access to a P source which they do not have to share with the bacteria.

Table 8.1 summarizes the reaction of phytoplankton and bacterioplankton to nutrient additions and demonstrates how different components of the microplankton are restricted by different nutrients in Lake Örträsket. When analyses of phytoplankton response to nutrient enrichment are extended to

➔

Fig. 8.4. Effect of N and P enrichment on different functional groups of phytoplankton in Lake Örträsket. Treatments significantly higher than the control are denoted by an asterisk (one-tailed t-test $p < 0.1$). N+P treatment for potential mixotrophs on 3 August is 130 μg C l^{-1}. For experimental details see Fig. 8.2 and Jansson et al. (1996)

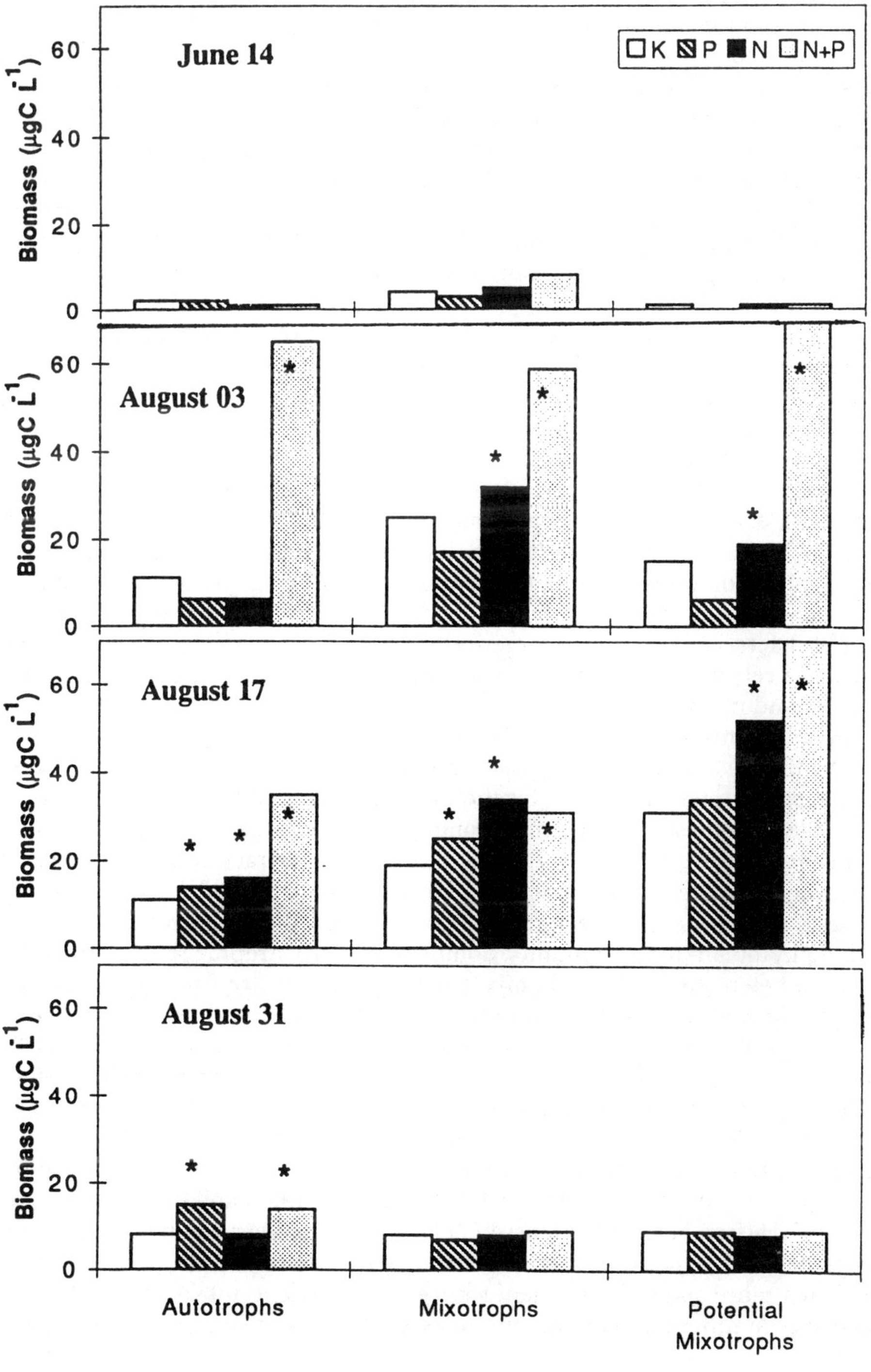
K
P
N
N+P
June 14
August 03
August 17
August 31
Biomass (μgC L^-1)
0
20
40
60
*
Autotrophs
Mixotrophs
Potential Mixotrophs

include the reaction of different species and functional groups it is obvious that the phytoplankton cannot be treated as an entity when discussing limiting nutrients (Fig. 8.4). Obligate autotrophic phytoplankton, which occupy a small share of the total phytoplankton, are mostly limited by P or N+P. Autotrophic phytoplankton are probably dependent on inorganic P for P uptake (Cembella et al. 1984; Jansson 1993). Since they share this source with the bacteria and cannot successfully compete for low concentrations, the results in Table 8.1 may offer part of the explanation why autotrophs are so scarce in humic lakes. The mixotrophic or potentially mixotrophic phytoplankton (in principle, chrysophyceans and cryptophytes, respectively, see Jansson et al. 1996) are clearly N-limited or N+P-limited. Since these groups are dominant among phytoplankton during most of the summer it is likely that they are responsible for the increases in phytoplankton biomass which follow N addition in Fig. 8.4 and Table 8.1.

Mixotrophic or potentially mixotrophic species, that are N-limited, probably obtain nutrients from grazing on bacteria, which induces a deficiency of N relative to P, since bacteria have higher P:N ratio than phytoplankton (Jansson et al. 1996). Mixotrophy may be a suitable strategy for phytoplankton in humic lakes to coexist with P-limited bacteria. As discussed above, phytoplankton probably have to use another P source than the bacteria and a radical solution to the problem of finding a suitable source is to consume P-rich bacteria. It may be energetically favourable for phytoplankton to let bacteria retrieve and concentrate nutrients instead of using energy for activation and maintenance of efficient uptake systems for low concentrations of inorganic nutrients (cf. Jansson 1993).

A mixture of heterotrophic (bacteria) and mixotrophic plankton may also increase the total energy mobilization in nutrient-limited humic lakes. A reasonable hypothesis is that allochthonous dissolved organic carbon serves as an energy source for bacteria and allows them to extract available nutrients from sources like, e.g., humic compounds, which are unavailable or less available to phytoplankton. Bacteria therefore become the major nutrient source for a phytoplankton community, dominated by mixotrophic species. In addition to heterotrophic growth, mixotrophic plankton also use light as an energy source and thereby add to the energy mobilization in the lake. Liberation of photosynthetically generated DOC by excretion, lysis and grazing will increase the potential for bacterial production and nutrient mobilization. This hypothesis offers a simple mechanistic model (Fig. 8.5) on how bacteria/phytoplankton coexistence in humic lakes is regulated in order to optimize the total planktonic productivity of the lake. The model is supported by documented conditions in humic lakes (see Jones 1992) and by experimental results (Hessen et al. 1994; Jansson et al. 1996). However, the model must be regarded as a hypothesis that needs to be tested. For example, it is necessary to learn more about the nutrient sources and nutrient uptake strategies and capacity in different plankton and to assess the autotrophic vs heterotrophic

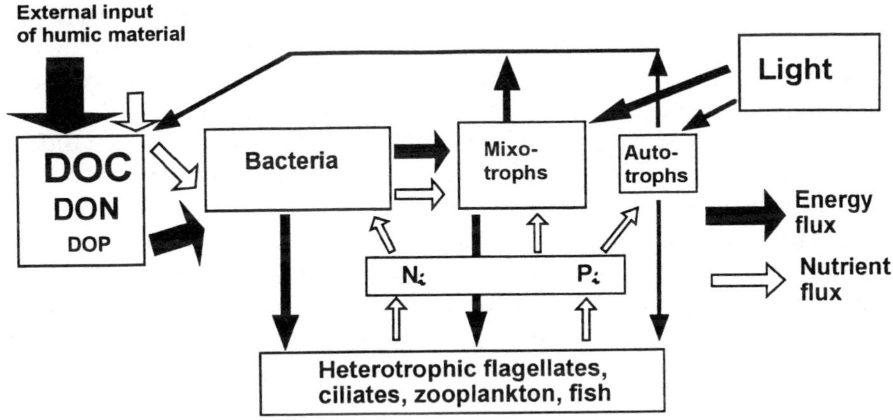

Fig. 8. 5. Major pathways of energy and nutrients in the light-saturated zone of a humic lake. Constructed from experimental data and field data from Lake Örträsket. *DOC* dissolved organic carbon; *DON* dissolved organic nitrogen; *DOP* dissolved organic phosphorus; N_i inorganic nitrogen; P_i inorganic phosphorus.

energy mobilization in mixotrophic species and the whole humic lake ecosystem.

If the model for nutrient retrieval and production of bacterioplankton and phytoplankton in Fig. 8.5 offers a correct interpretation of the outcome of the nutrient enrichment experiments in humic lakes (Figs 8.2, 8.4 and Table 8.1), it means that both P and N limit the total production in humic lakes. This simultaneous P and N-limitation is not the same as the balance between N and P limitation sometimes reported for clearwater oligotrophic lakes where the N and P supply ratios correspond to the demand in plankton. In these lakes, an increase of either N or P should have little immediate effect on lake productivity. In humic lakes, an increase in P can primarily stimulate bacterial energy mobilization, while an increase in N can stimulate photosynthetic energy mobilization by phytoplankton. P or N enrichment both result in an increase in total production.

8.6 Influence of Light Climate on Nutrient Uptake in Bacterioplankton and Phytoplankton

The nutrient flux between algae and bacteria schematically outlined in Fig. 8.5 is dependent on utilization of different energy sources: allochthonous dissolved organic carbon for bacteria and light for phytoplankton. Fig. 8.5 is, therefore, relevant for the light-saturated zone of humic lakes only. Since the light climate in humic lakes is restricted due to the adsorption of light by humic compounds (cf. Ramberg 1979; Heyman 1983), light as well as nutrients may limit primary production. True planktonic species are entirely dependent on the effective light climate, i.e. the mean light intensity in the epilimnion, for their photosynthesis. Phytoflagellates, which can move vertically, are less dependent on light climate since they can swim to the surface layer for photosynthesis. Therefore, even the flux of nutrients between different compartments of microplankton should, to a large extent, be a function of light availability.

The effective light climate is inversely proportional to humic content and water colour (Jones 1992). However, since heat adsorption by humic material decreases the depth of the mixed layer, the effects of coloured substances on the effective light climate may be that reduced light penetration can be counterbalanced by decreased depth of the mixed layer so that the mixed zone becomes the same as the trophogenic zone (Jones 1992; Jones, this Vol.). However, this is true only for small lakes where wind-induced mixing has little effect on the stratification. In larger humic lakes, the mixed layer is deeper than the light-saturated zone. The effective light climate will, at a given water colour, be determined by the depth of the mixed zone.

In lakes with a deep epilimnion it must be assumed that a large share of nutrients is allocated to bacteria because the mean light intensity is too poor to permit high primary production. A test of this hypothesis was made with data from Lake Örträsket. Lake Örträsket has a mixed layer, the depth of which varies between 12 and 15 m in most summers. The mean effective light climate during summer is 10 $\mu E\ m^{-2}\ s^{-1}$ (Bergström, unpublished data), which is about one order of magnitude lower than in shallow humic lakes with approximately the same colour (cf. Heyman 1983). A calculation of the vertical variation in nutrient demand in bacterioplankton and phytoplankton was carried out based on the following analyses and assumptions.

- Bacterial production in terms of carbon per litre was measured every second week (leucine incorporation) during the observation period on composite samples from the epilimnion.
- Primary production ($H^{14}CO_3$-incorporation) in terms of carbon per litre was measured every second week at each half metre in the 0 to 4 m surface layer (photosynthesis was never recorded below 3.5 m).
- The total bacterial and primary production was calculated for the whole epilimnion and observation period. The mean depth of the epilimnion during the observation period was 14 m.
- Bacterial production was assumed to be uniformly distributed within the epilimnion, and calculated for the 0 to 1 m, 0 to 2 m, 0 to 4 m and 0 to14 m layers respectively
- Primary production was calculated for the same layers.
- The total demand for P and N for bacterial and primary production was calculated from the theoretical C:N:P ratio (weight) for bacteria of 19:3:1 (Caron et al. 1990) and for phytoplankton of 41:8:1 (Redfield 1958).

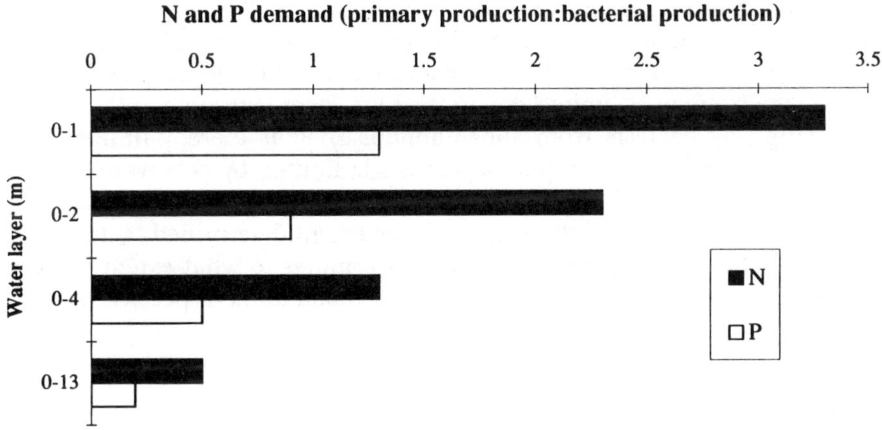

Fig. 8. 6. Relation between uptake of N and P in autotrophic phytoplankton and bacterioplankton in different water layers of the epilimion in Lake Örträsket (calculated for the total uptake in the epilimnion during June–September 1995)

Fig. 8. 6 shows how the availability of light to planktonic organisms is critical for the P and N flux in algal-bacterial communities. The calculation indicates that P uptake in primary producers was lower than in bacterioplankton in all water layers below 1 m. In the surface (0 to 1 m) layer the P demand for primary production was slightly higher than for bacterioplankton production. In the whole epilimnion (0 to 13 m layer) of Lake Örträsket, the uptake of P in primary producers is approximately 20% of that in bacteria. The corresponding figures for N are higher which, e., means that in the 0 to 1 m layer the demand for N by autotrophic phytoplankton is three times that of bacterioplankton. However, in the whole epilimnion, the bacteria N demand was twice the N demand of primary producers. Fig. 8. 6 demonstrates that the light climate and stratification are important lake characteristics influencing the utilization of limiting nutrients in different components in the microbial food web. Light climate and input of allochthonous DOC determine to what degree the limiting nutrients are used for heterotrophic or autotrophic energy mobilization in humic lakes.

8.7 Conclusions

Bacterioplankton in humic lakes are nutrient-limited because the supply of energy from allochthonous DOC is high enough to relieve the bacteria from the dependence of autochthonous DOC generated by phytoplankton as an energy source. Phosphorus is most likely the prime limiting nutrient for bacterial production. P-limited bacterioplankton in humic lakes most likely outcompete phytoplankton for bioavailable dissolved P. The phytoplankton community in humic lakes is often dominated by mixotrophic phytoflagellates. Mixotrophic phytoflagellates retrieve nutrients by grazing on bacteria and thereby become N-limited due to the high P-content in bacteria. Energy mobilization by bacteria from allochthonous DOC is thereby limited by P and energy mobilization by photosynthesis is limited by N. An increase in the P supply or N supply in humic lakes may, therefore, lead to higher production.The effective light climate, to a large extent determined by the water colour and the depth of the epilimnion, determines to what extent limiting nutrients are used mainly by bacteria and phytoplankton, respectively.

8.8 Summary

It is wellknown that P and N control the growth of phytoplankton. Other environmental factors than nutrients, e.g. light and temperature affect the growth of phytoplankton by direct influence on metabolic activity but also by controlling the nutrient uptake and utilization of plankton. Light generated energy thus drives the enhanced P-uptake in algae during P-starvation. Other nutrients (N, Si, Fe, or C) can control the growth and nutrient uptake of phytoplankton. N is considered to be the most critical limiting nutrient next to P. Empirical relationships between nutrient loading and trophic state demonstrate good correlations for N, although not as good as for P. Nutrient enrichment experiments often shows that there is a close balance between P and N limitation in oligotrophic lakes. That N and P can be interactive in determining growth conditions and species composition for algae is indicated by the highly variable N:P ratios among phytoplankton species. N can also be single limiting nutrient, during periods of the summer in eutrophic lakes when inorganic nitrogen sources are exhausted by uptake in plankton and denitrification in shallow sediments.

Humic substances affect nutrient uptake and production of bacterioplankton and phytoplankton in several ways. Humic material serves as an energy and carbon source for bacteria and thus stimulates bacterial production. Phytoplankton production is restricted by poor effective light climate in humic lakes as a result of light absorption by humic substances.The ratio between bacterial production/biomass and phytoplankton production/biomass is therefore higher in humic lakes than in other types of non-perturbed lakes.

The input of energy in the form of allochthonous DOC in humic lakes relieves the bacterioplankton from dependance of DOC released from phytoplankton and bacterial growth become P-limited instead of C-limited in humic lakes. Since bacteria outcompete phytoplankton for low concentrations of dissolved P, the availability of P to phytoplankton should be extremely low in humic lakes.

The phytoplankton community is often dominated by mixotrophic flagellates in humic lakes. Mixotrophs can retrieve nutrients by grazing on bacteria and the assimilation of ingested nutrients is efficient, especially for P. The dominance of mixotrophs in humic lakes may therefore be explained by their superior ability to retrieve nutrients by bacterial grazing in a situation where they otherwise must compete with both bacteria and other phytoplankton for dissolved nutrients.

The fact that bacteria have a higher P:N ratio than phytoplankton in combination with the efficient P retrieval of the mixotrophs can lead to that grazing induce a relative excess of P in mixotrophic phytoplankton. The mixotrophs thereby become N-limited. Therefore, the production of bacterioplankton in humic lakes is P-limited while the production of the dominant phytoplankton group is N-limited during large periods of the summer.

Stratification and effective light climate determine to what extent available nutrients are used for bacterial production or primary production. In epilimnia which are deeper than the photic zone relatively more nutrients are used for bacterial production than when the epilimnion is thin and equals the depth of the photic zone.

References

Arvola L (1984) Vertical distribution of primary production and phytoplankton in two small lakes with different humic concentration. Holarct Ecol 7:390–398

Azam F, Fenchel T, Field JG, Gray JS, Meyer-Reil LA, Thingstad F (1983) The ecological role of water-column microbes in the sea. Mar Ecol Prog Ser 10:257–263

Beijerinck MW (1890) Kulturversuche mit Zoochlorellen, Lichenogonidien und anderen Algen. Bot Ztg 48:725–739, 741–754, 757–768, 781–785

Blomqvist P, Heyman U, Grundström R (1981) The structure of the pelagic ecosystem in L. Siggeforasjön. Scr Limnol Ups Coll 18, Scr A22

Blomqvist P, Pettersson A, Hyenstrand P (1994) Ammonium-nitrogen: a key regulatory factor causing dominance of non-nitrogen fixing cyanobacteria in aquatic systems. Arch Hydrobiol 132:141–164

Boraas ME, Estep KW, Johnsson PW, Sieburth JM (1988) Phagotrophic phototrophs: the ecological significance of mixotrophy. J Protozol 35:249–252

Carlsson P, Segatto AZ, Granéli E (1993) Nitrogen bound to humic matter of terrestrial origin – a nitrogen pool for coastal phytoplankton? Mar Ecol Prog Ser 97:105–116

Caron DA, Porter KG, Sanders RW (1990) Carbon, nitrogen and phosphorus budgets for the mixotrophic phytoplagellate *Poterioochromonas malhamensis* (Chrysophyceae) during bacterial ingestion. Limnol Oceanogr 35:433–442

Cembella AD, Antia NJ, Harrison PJ (1984) The utilization of inorganic and organic phosphorus compounds as nutrients by eukariotic microalgae: a multidisciplinary perspective. Part 1 Crit Rev Microbiol 10:317–391

Currie DJ (1990) Large scale variability and interactions among phytoplankton, bacterioplankton and phosphorus. Limnol Oceanogr 35:1437–1455

Currie DJ, Kalff J (1984a) A comparison of the abilities of freshwater algae and bacteria to acquire and retain phosphorus. Limnol Oceanogr 29:298–310

Currie DJ, Kalff J (1984b) The relative importance of bacterioplankton and phytoplankton in phosphorus uptake in freshwater. Limnol Oceanogr 29:311–321

Currie DJ, Kalff J (1984c) Can bacteria outcompete phytoplankton for phosphorus? A chemostat test. Microb Ecol 10:205–216

DeHaan HR, Jones R, Salonen K (1990) Abiotic transformations of iron and phosphates in humic lake water, revealed by double isotope labelling and gel filtration. Limnol Oceanogr 35: 491–497

delGiorgio PA, Peters RH (1993) Balance between phytoplankton production and plankton respiration in lakes. Can J Fish Aquat Sci 50:282–289

Guildford SJ, Healey FP, Hecky RE (1987) Depression of primary production by humic matter and suspended sediments in limnocorral experiments in Southern Indian Lake, northern Manitoba. Can J Fish Aquat Sci 4:1408–1417

Hessen DO (1992) Dissolved organic carbon in a humic lake: effects on bacterial production and respiration. Hydrobiologia 229:115–123

Hessen DO, Nygaard K, Salonen K, Vähätalo A (1994) The effect of substrate stoichiometry on microbial activity and carbon degradation in humic lakes. Environm Int 20:67–76

Heyman U (1983) Relation between production and biomass of phytoplankton in four Swedish lakes of different trophic status and humic content. Hydrobiologia 101:71–88

Heyman U, Lundgren A (1988) Phytoplankton biomass and production in relation to phosphorus: some conclusions from field studies. Hydrobiologia 170:211–228

Jansson M (1988) Phosphate uptake and utilization by bacteria and algae. Hydrobiologia 170: 177–190

Jansson M (1993) Uptake, exchange and excretion of orthophosphate in phosphate starved *Scenedesmus quadricauda* and *Pseudomonas* K7. Limnol Oceanogr 38:1162–1178

Jansson M, Blomqvist P, Jonsson A, Bergström A-K (1996) Nutrient limitation of bacterioplankton, autotrophic and mixotrophic phytoplankton, and heterotrophic nanoflagellates in Lake Örträsket, a large humic lake in northern Sweden. Limnol Oceanogr 41:1552–1559

Johansson JÅ (1983) Seasonal development of bacterioplankton in two forest lakes in central Sweden. Hydrobiologia 101:71 -88

Jones RI (1990) Phosphorus transformations in the epilimnion of humic lakes: biological uptake of phosphate. Freshw Biol 23:323–337

Jones RI (1992) The influence of humic substances on lacustrine planktonic food chains. Hydrobiologia 229:73–91

Jones RI (1994) Mixotrophy in planktonic protists as a spectrum of nutritional strategies. Mar Microb Food Webs 8:87 -96

Kirchman DL (1994) The uptake of inorganic nutrients by heterotrophic bacteria. Microb Ecol 28:255–271.

Lindell M, Graneli W, Tranvik L (1995) Enhanced bacterial growth in response to photochemical transformation of dissolved organic matter. Limnol Oceanogr 40:195–199

Lindström K, Rodhe W (1978) Selenium as a micronutrient for the dinoflagellate *Peridinium cinctum* fa.*westii*. Mitt Int Verein Limnol 21:168–173

Moran MA, Hodson RE (1990) Bacterial production on humic and nonhumic components of dissolved organic carbon. Limnol Oceanogr 35:1744–1756

Ohle W (1935) Organische Kolloide in ihrer Wirkung auf dem Stoffhaushalt der gewässer. Naturwissenschaften 23:480–484

Porter KG (1988) Phagotrophic phytoflagellates in microbial food webs. Hydrobiologia 159: 89–97

Pringsheim EG (1963) Farblose Algen: Ein Beitrag zur Evolutionsforshung. Fischer.

Ramberg L (1979) Relationships between phytoplankton and light climate in two Swedish forest lakes. Int Rev Gesamten Hydrobiol 64:749–782

Redfield AC (1958) The biological control of chemical factors in the environment. Am Sci 46: 205–221

Rodhe W (1948) Environmental requirements of freshwater planktonic algae. Symb Bot Ups 10 (1), 149 pp

Rothaupt KO (1992) Stimulation of phosphorus-limited phytoplankton by bacterivorous flagellates in laboratory experiments. Limnol Oceanogr 37:750–759

Rothaupt KO, Güde H (1992) The influence of spatial and temporal concentration gradients on phosphate partitioning between different size fractions of plankton: further evidence and possible causes. Limnol Oceanogr 37:739–749

Sanders RW, Porter KG (1988) Phagotrophic phytoflagellates. Adv Microb Ecol 10:167–192

Sakamoto M (1966) Primary production of phytoplankton community in some Japanese lakes and its dependence on lake depth. Arch Hydrobiol 62:1–28

Salonen K, Jokinen S (1988) Flagellate grazing on bacteria in a small dystrophic lake. Hydrobiologia 161:203–209

Schindler DW (1977) Evolution of phosphorus limitation in lakes. Science 195:260–262

Shapiro J (1988) Introductory lecture at the international symposium "Phosphorus in Freshwater Ecosystems", Uppsala, Sweden in October 1985. Hydrobiologia 170:9–17

Skuja H (1948) Taxonomie des Phytoplanktons einiger Seen in Uppland. Schweden. Symb Bot Ups 9 (3):1–399.

Skuja H (1956) Taxonomische und biologische Studien über das Phytoplankton schwedische Binnengewässer. Nova Acta Reg Soc Sci Uppsal. 16 (3):1–404

Sterner RW (1989) N:P resupply by herbivores: zooplankton and the algal competitive arena. Am Nat 136:209–229

Stewart AJ, Wetzel RG (1982) Influence of dissolved humic materials on carbon assimilation and alkaline phosphatase activity in natural algal-bacterial assemblages. Freshwat. Biol. 12: 369–380

Stockner JG (1991) Autotrophic picoplankton in freshwater ecosystems: the view from the summit. Int Rev Gesamten Hydrobiol 76:483–492

Tilman DS (1977) Resource competition between planktonic algae: an experimental and theoretical study. Ecology 58:338–348

Tilman DS, Kilham S, Kilham P (1982) Phytoplankton community ecology: the role of limiting nutrients. Annu Rev Ecol Syst 13:349–372

Tipping E (1981) The adsorption of aquatic humic substances by iron oxides. Geochim Cosmochim Acta 45:191–199

Tranvik LJ (1988) Availability of dissolved organic carbon for planktonic bacteria in oligotrophic lakes of different humic content. Microb Ecol 16:311–322

Tranvik LJ (1989) Bacterioplankton growth, grazing mortality and quantitative relationship to primary production in a humic and a clearwater lake. J Plankton Res 11:985–1000

Tranvik LJ, Porter KG, Sieburth JMcN (1989) Occurrence of bacterivory in *Cryptomonas*, a common freshwater phytoplankter. Oecologia 78:473–476

Vadstein O, Jensen A, Olsen Y, Reinertsen H (1988) Growth and phosphorus status of limnetic phytoplankton and bacteria. Limnol Oceanogr 33:489–503

Vollenweider RA (1968) Scientific fundamentals of the eutrophication of lakes and flowing waters, with particular reference to nitrogen and phosphorus as factors in eutrophication. OECD, DAS/CSI/68.27, Paris, 274 pp

III Humus and Secondary Production

9 The Role of Microbial Extracellular Enzymes in the Transformation of Dissolved Organic Matter in Humic Waters*

Uwe Münster and Henk De Haan

9.1 Introduction

Non-living natural organic matter (NOM) comprises in most aquatic environments the largest organic matter (OM) fraction, and it harbours also the largest organic carbon pool (Allen 1976; Wetzel 1983; Steinberg and Münster 1985; Thurman 1985; Hedges 1992; Münster 1993). Furthermore, NOM has been recognized as the most important carbon and energy resource in the detritus food chain processes (Williams 1981; Hobbie 1992; Wetzel 1992). Non-predatory processing and transfer of organic carbon in parallel with the classical grazing food chain are significant fractions of NOM passing and channelled through the aquatic food chain (Azam and Cho 1987; Cho and Azam 1988; Williams 1990; Ducklow 1991). The NOM pools can be subdivided into dissolved (DOM) and particulate organic matter (POM). Both of these summarize the major part of the organic detritus pools (Wetzel 1983; Thurman 1985). Besides NOM utilization by phagotrophic microbes such as flagellates (Sherr 1988; Sanders 1991; Tranvik et al. 1993), osmotrophic microorganisms such as bacteria and fungi are the major sinks for NOM, and by utilizing primarily DOM resources (Benner et al. 1986, 1989; Azam and Cho 1987; Cho and Azam 1988) they are closing the link between the classical grazing food chain and the microbial loop (Pomeroy 1974; Sorokin 1977; Williams 1981; Azam et al. 1983; Wetzel 1992). The ratio of living to non-living OM in many aquatic environments has been estimated in the range of 100 to 10:1 and it empasizes the importance of NOM sources in aquatic ecosystem processes (Allen 1976; Wetzel 1984, 1990; Farrington 1992; Hobbie 1992; Münster and Albrecht 1994). The flow of OM through the aquatic food chain is mostly dominated at first by the photosynthetic production of OM, the utilization and transformation of NOM by microheterotrophs via the microbial food web, and finally its transference into the higher food chain, partially respired and/or its sedimentation to the bottom (Fig. 9.1).

* Dedicated to Prof. Drs. J. Overbeck, Max-Planck-Institute for Limnology, on the occasion of his 75th birthday

Ecological Studies, Vol. 133
Hessen, Tranvik (eds) Aquatic Humic Substances

Two sources of OM can be identified in aquatic ecosystems: autochthonous-produced OM and allochthonous-produced OM (Romankevich 1984; Wetzel 1984; Thurman 1985). In most aquatic environments, autochthonous-produced OM are believed to be the major OM source (Allen 1976; Gagosian and Lee 1981; Romankevich 1984; Hedges 1992). However, in some cases allochthonous-produced OM can be a significant OM pool in aquatic ecosystem processes (Gjessing 1976; Steinberg and Münster 1985). This is especially valid in freshwaters in large parts of the northern boreal areas (NBA) and in many tropical ecosystems (Wissmar et al. 1981; Schindler et al. 1992). In these environments, humic substances (HS) are the major carbon source in allochthonously derived OM in such aquatic ecosystems, as rivers and head water lakes (Gjessing 1976; Steinberg and Münster 1985; Salonen et al. 1992). Because of its chemical complexity (Schnitzer and Khan 1972; De Haan and De Boer 1979; Stevenson 1982; De Haan 1983; Harvey et al. 1983; Hayes et al. 1989) and its high abundances, HS are important environmental factors and have been a major concern in many freshwater studies in the past (Povoledo and Golterman 1975; Gjessing 1976; Aiken et al. 1985; Frimmel and Christman 1988; Hayes et al. 1989; Perdue and Gjessing 1990). The characterization of HS and their contribution to functioning aquatic ecosystems are still one of the most challenging and difficult tasks in water research.

Three types of trophic exploitation of NOM by microheterotrophs in aquatic environments can be defined (Fig. 9.2): (1). A rapid uptake and assimilation of biologically labile, dissolved organic matter (LDOM) and its conversion into biomass and/or its partial respiration to CO_2 (Azam and Cho 1987); (2). A slow uptake, assimilation and respiration of NOM by microheterotrophs (Alexander 1975; Stabel and Steinberg 1976; Stabel et al. 1979; Geller 1985b, 1986; Benner et al. 1986, 1989, 1996; McKnight et al. 1990, 1991; Moran and Hodson 1990) due to the polymeric character of refractory/recalcitrant dissolved organic matter (RDOM). Except for the phagotrophic NOM utilizations (Sherr 1988; Tranvik et al. 1993), this second utilization process requires mostly an extracellular cleavage step of the polymeric OM such as by microbial enzymes (De Haan 1974, 1976, 1983; Billén 1984; 1991; Chróst 1990, 1991; Hoppe 1991; Münster 1991; Wetzel 1991) or by photolytic reactions (Gjessing and Gjerdahl 1970; Kramer 1979; Geller 1985a; Salonen and Tulonen 1990; Mopper et al. 1991; Salonen and Vähätalo 1994; Lindell et al. 1995; Wetzel et al. 1995; Granéli et al. 1996); (3). In some cases NOM is not utilized as sole compounds but requires co-substrates (Horvath 1972; Alexander 1975; Schmidt and Alexander 1985; Schink et al. 1992; Swift 1992), and in many cases this is seen during the degradation of xenobiotics (Chaudhry 1994). All three types of organic substrate utilizations have been traced and identified as existing in aquatic environments (Gagosian and Lee 1981; Steinberg and Münster 1985; Hedges 1992; Lee and Wakeham 1992; Münster and Albrecht 1994 and cited references therein). Notably, the rate and intensities of the biodegradation of NOM depend largely on its "organic matter reactivity" (ORM) of the inherent NOM compounds (Alexander 1975; Hedges 1992; Swift 1992).

Former studies on OM in pelagial water bodies and in sediments have shown that at least two to three different OM fractions with distinct reactivities to microbial degradation can be defined (Westrich and Berner 1984; Billén and Servais 1989; Burdige 1991). Accordingly, due to the biodegradability of these OM fractions, Westrich and Berner (1984) have defined a first order rate model, their *"G-model"*, and Burdige (1991) has formulated his "mixed-model", and both models were successfully tested in marine sediments. For pelagic water bodies, the *"HSB-model"* was developed and tested by Billén and Servais (1989) with a similar approach for studies of DOM decomposition and utilization by bacterioplankton in freshwater and marine environments, which included notably the extracellular hydrolysis of biopolymers by bacteria. For reverine microbes, lentic and lotic environments including their catchments and ecotones Sinsabaugh et al. (1994a, b, 1997) have approved the importance of microbial enzymes in decomposition processes and have developed and applied a microbial enzyme based decomposition rate model (Microbial Allocation of Rescourses among Community Indicator Enzymes, *"MARCIE*-model") for a better understanding, quantification and prediction in detritus processing. However, these approaches were mostly lacking the part of the chemical characterization and identification of the labile and the refractory OM-fractions (LDOM and RDOM respectively) and the possible contribution of their structural-chemical properties to explain their differences in the OMR.

Formerly combined chemical and microbial approaches to NOM degradation studies in eutrophic and humic lakes revealed that the LDOM pool comprises mostly low molecular weight dissolved organic carbon (LMWDOC) compounds and represents only 1–20% of the DOC in a eutrophic mesohumic lake (Chróst et al. 1989, Münster and Chróst 1990) and only 1–5% in polyhumic waters (Münster 1991, 1993). For this fraction, good correlations were obtained between, e.g., [^{14}C]-glucose uptake and the amount of dissolved free glucose or the dissolved free amino acid (DFAA) concentration and the [^{14}C]-DFAA uptake (Münster and Chróst 1990; Münster 1993). However, the majority of the DOC was RDOM and comprised 80–90% of DOC as high molecular weight dissolved organic carbon (HMWDOC). This HMWDOC was a mixture of different biopolymers (such as proteins, polynucleotides, polysaccharides, lipids, ligninocellulosic material) and geopolymers [e.g. dissolved humic matter, (DHM); Münster 1985; Münster and Albrecht 1994]. Interestingly, part of this HMWDOC fraction could have been directly utilized by phagotrophic microheterotrophs such as flagellates (Sherr 1988; Tranvik et al. 1993; Kristoffersen et al. 1996). However, the main fraction of the HMWDOC fraction is not easily available to phagotrophic and osmotrophic assimilation and comprises a high amount of the recalcitrant or refractory dissolved organic matter (RDOM).

The trophic and ecophysiological roles of this RDOM in energy and food web processes are still poorly understood (Allen 1976; Steinberg and Münster 1985; Wetzel 1990, 1992; Hobbie 1992; Schindler et al. 1992). Physical and chemical methods characterized part of such RDOM in deep marine waters with turnover times of about 1000–6000 years (Druffel and Williams 1990; Bauer et al. 1992), which is probable beyond any of our biological experimental time scales, whereas some microbiological approaches found much lower turnover times (about 1–10 h) for bulk DOM from marine upper surface waters (Kirchman et al. 1991; Amon and Benner 1996; Wheeler et al. 1996). Although both approaches have collected and traced probable different NOM resources and they may not be directly comparable, these discrepancies and mismatch in the exact chemical and biological characterization of the DOM reflect our poor knowledge about the trophic role of especially refractory or recalcitrant organic carbon pools in aquatic environments (De Haan 1974, De Haan 1983; Allen 1976; Stabel and Steinberg 1977; Rich and Wetzel 1978; Stabel et al. 1979; Gagosian and Lee 1981; Rich 1984; Geller 1985a, b, 1986; McKnight et al. 1985, 1991; Steinberg and Münster 1985; Hedges 1992; Martin and Fitzwater 1992; Münster 1993; Münster and Albrecht 1994).

Two main routes can be drawn by which RDOM sources could be processed and finally utilized in aquatic environments (Fig. 9.3): (1). Via a photolytic cleavage mechanism, which releases LDOM compounds from RDOM resources such as simple organic acids (Mopper et al. 1991; De Haan 1993; Salonen and Vähätalo 1994; Lindell et al. 1995; Graneli et al. 1996; Wetzel et al. 1996); (2). Via an enzymatic cleavage pathway of biopolymers into oligo- and monomeric substances by bacteria and in the digestive tracts of micro- and macrograzers (Sherr 1988; Tranvik et al. 1993; Kristoffersen et al. 1996), but essentially finally filling up the low concentration level of the LDOM pools (Hoppe 1983; Billén 1984, 1991; Chróst 1989, 1990, 1991, 1994; Chróst et al. 1989; Münster 1991; Wetzel 1991). However, it is not yet clear which cleaving process is more important and more effective in the processing of NOM in aquatic environments. Probably they are even acting in a synergistic manner (Münster 1991, 1993). This study will contribute to a better understanding of NOM processing, especially in freshwaters from the NBA which carry largely 90–95% of their NOM as RDOM, and raise important questions and problems which still are waiting to be answered and solved. The authors have reviewed the literature back to about 20 years ago, especially such papers which could contribute to explaining and understanding the role of HS in microbial enzymatic NOM processing and its subsequent utilization in marine and freshwaters. As there were rather little data available from aquatic ecosystem studies on HS-enzyme interactions, the authors have also included in some aspects of this chapter results from terrestrial habitat studies, especially when HS-microbial enzyme interactions had to be evaluated in their role in nutrient cycling and biodegradations. Due to time limits, the authors could not completely review the latest literature on terrestrial ecosystems and may have missed some. However, in a recent review by Schinner and Sonnleitner (1996) and Morra (1997) on soil ecosystems, there appeared no fundamentally new aspects.

9.1.1 A Look Backwards

9.1.1.1 The Discovery of Biocatalysis

The idea that enzymes may take part as essential biomolecules/catalysts in life cycles was born during 1752–1783 when an Italian priest named Lazzaro Spallanzani and the famous French scientist René Antoine Ferchaut de Réaumur made their first experiments with metal-box encapsulated pieces of meat as prey for birds (*Milvus milvus* and *Buteo buteo*). Their idea lasted an additional 60–70 years until the Swedish scientist J. J. Berzélius (1836) formulated his view about the catalytic theory in plant metabolic processes. Berzélius can be called the father of the view of the biocatalytic function of enzymes. In 1822, L. Pasteur was the first to identify microbial processes during fermentation reactions and the importance of involved enzyme reactions. In 1930, about 80 enzymes were already known and described, but during 1984 more than 2500 enzymes were identified and classified, and this number is increasing rapidly and is certainly much higher today, which may show the importance of these biomolecules in life cycle reactions and biological processes. To prevent imprecise names, characterization and classifications of those enzymes, in 1961 an international enzyme commission (the EC) was established, and since then all enzymes are classified and characterized according to the recommendations and rules of the EC (Table 9.1).

9.1.1.2 The Aquatic Aspect

The present view that enzymes act as catalysts in aquatic life cycle processes and especially during the degradation of OM in aquatic environments is not as new as one might at first assume. Already at the beginning of this century

Table 9.1. Overview of enzyme classifications according to the EC rules (Enzyme Nomenclature 1975, 1984 and later revisions 1986–1990)

EC No.	EC Name	Example of Reactiontype	Type of reaction	Used synonymes
1	Oxidoreductase	$AH_2 + B \Leftrightarrow A + BH_2$	Redox reaction	Oxidase
2	Transferase	$AB + C \Leftrightarrow A + BC$	Transfer reaction	
3	Hydrolase	$AB + H_2O \Leftrightarrow AOH + BH$	Hydrolysis reaction	
4	Lyase	$AB \Leftrightarrow A + B$	Lysis-Synthesis reaction	Synthases
5	Isomerase	$ABC \Leftrightarrow BAC$	Isomerase reaction	
6	Ligase	$A + B + ATP \Leftrightarrow AB + ADP + P_i$	Phosphorylation reaction	Synthetases

Symbols: $\Leftrightarrow$ this arrow shall indicate the reaction in both directions at equilibrium conditions

Fermi (1906) noted the existence of proteolysis in stagnant water pools and Harvey (1925) pointed out the possible involvements of catalase and oxidase in marine waters, as was later emphasized also by Kreps (1934), who stated that extracellular enzymes from bacteria, plants and animals may exist in seawaters, participating in nitrate reduction and ammonia oxidation. Only 4 years later, Steiner (1938) deduced the existence of phosphatase in the phosphorus cycle in freshwaters. There were probably more of such studies in the past and the list may be incomplete, but these few examples may show and exemplify the early beginnings of enzyme studies in aquatic habitats.

Although experimentally and methodologically limited in their possibilities, those early aquatic scientists deduced and determined the role of microbial enzymes in aquatic ecosystem processes as rather self-reliant. It took an additional 30–50 years until Overbeck (1961) and Reichardt et al. (1967) identified free phosphatase in freshwaters and characterized its role in phosphorus cycling and uptake by freshwater algae. After this landmark in studies on free enzymes in aquatic ecosystems, a series of studies showed that enzymes may play important roles in nutrient cycling and biodegradations in aquatic ecosystems (partially reviewed by Chróst 1990 and Hoppe 1991). Especially since the introduction of better model substrates and more sensitive measurement techniques, a large number of studies have been undertaken to understand better the role of microbial enzymes in aquatic ecosytems. This knowledge was partially summarized in a specially dedicated workshop on *Microbial Enzymes in Aquatic Environments* during July 23–27, 1989 at Ringberg Castle, Bavaria, Germany, organized and conducted by the Max-Planck-Institute for Limnology, Department of Microbial Ecology, Plön, Germany, sponsored by the Deutsche Forschungsgemeinschaft and the Max Planck Society, Munich, Germany. This workshop was attended by leading experts in this field and they presented their results and discussed the role of microbial enzymes in different aquatic habitats. The outcome of this workshop was a book edited and published by Springer Verlag in Brock/Springer Series in Contemporary BioScience, entitled *Microbial Enzymes in Aquatic Environments* (ed. RJ Chróst 1991) and the reader is referred to this for further reading and more detailed information on this topic.

9.2 Definitions, Terms, Classifications, Measurements and Techniques

Scientifically enzymes are classified and named according to the recommendations of the EC rules (Enzyme Nomenclature 1975, 1984, 1986–1990, Table 9.1). However, although such rules have existed along time, in aquatic ecosystem studies very few reports have followed such rules (cf. Chróst 1990, 1991; Hoppe 1991 and cited references therein). This may be due to the complexity of aquatic ecosystem structure and functions and the coexistence of many different types of organisms contributing to the enzyme pool sizes and activities, which may have influenced researchers in the past to use sometimes synonymous and confusing terms, names and uncertainties regarding the definitions of the kind of enzymes measured, and why they were measured in that way and not under "standard" conditions dealing with aquatic microbial ecology questions. This misuse is most evident in the definitions where these enzymes were located or acting and who the main contributors were.

9.2.1 Definitions Related to Microbial Enzymes in Aquatic Environments

In a review, Chróst (1990), in his definitions and the characterization of microbial enzymes regarding their locations and place of actions, followed those definitions of Karnovsky (1986 and cited references therein) and defined them as "ectoenzymes" and "extracellular enzymes" (Table 9.2), depending on whether the enzymes are cell-surface-bound or dissolved free enzymes. In earlier studies, Hoppe (1983) and Chróst et al. (1989) called them "exoenzymes", which defined all enzymes outside the cell membranes as particle bound and dissolved free enzymes. Other researchers called them only "extracellular enzymes", which included all enzymes outside the cytoplasmatic cell membranes, but attached to the cell surface or in the periplasmatic space and/or truely dissolved free enzymes (Burns 1978, 1982, 1983; Aaronson 1981; Priest 1984, 1992; Sinsabaugh et al 1992a, b). The term

Table 9.2 Overview of the most commonly used terms and definitions of microbial enzymes in aquatic environments

Name	Definitions	Comments	References
Ecto- and extracellular enzymes	Microbial enzymes cleaving biopolymers outside the plasma membrane (ecto-enzymes) and as dissolved free enzymes (extracellular enzymes)	Private and operational definition, without official legitimations, e.g. by the E C, but partially used in aquatic microbial ecology publications	Chróst (1990, 1991)
Extracellular enzymes	Microbial enzymes cleaving biopolymers after they have passed the plasma membrane, but in contact with mother cells or as dissolved free enzymes in the growth medium	Operational definition and most commonly used definition in biochemistry, microbiology and soil science publications	Burns (1978, 1982, 1983), Aaronson (1981), Priest (1984, 1992), Sinsabaugh et al. (1992a, b)
Exoenzymes	Microbial enzymes cleaving biopolymers outside the plama membrane but in contact with mother cells, particles or as dissolved free enzymes	Private and operational definition, without official legitimations, e.g. by the E C, but temporally used in publications in marine microbiology	Hoppe (1983) Chróst et al. (1989)

"ectoenzyme" has been defined by Karnovsky (1986 and references therein) as those enzymes located in special cells in animal or human tissues, participating in signalling of nerve reactions and certain cells for key metabolic processes such as ATP and peptide functions. Although Karnovsky's definitions are rather convincing mostly for those human and animal cell lines and purposes, it can be highly questionable whether these definitions are appropriate for aquatic environments, where algae and bacteria are the main contributors to microbial extracellular enzyme (MEE) activities.

From these three terms, at least two conflict with other synonymous terms which are regulary used to characterize, e.g., enzyme cleaving patterns, such as exo-/endo cleaving patterns in biopolymer processing (see below), and ecto-/endo is used in biological tissues to differentiate between ecto- and endo positions of cells in plant, animal and human tissues and ectomycorrhiza communities. Of these two overlapping and partially confusing terms, the authors have favoured and finally applied in all of their studies the term *"Microbial Extracellular Enzymes"* (*MEE*), as it was and still is regulary used in biochemistry and microbiology and it was also proposed by Burns (1978, 1982, 1983), Aaronson (1981), Priest (1984, 1992), Hoppe (1991), Meyer-Reil (1991), and Wetzel (1991). This definition is more general/ operational and more practical in ecological contexts; it is hardly possible without any pre-manipulations of the microbial community structures to differentiate between cell-surface-bound enzymes on free living bacteria and particle-bound-bacteria and their enzymes, or between particle-attached enzymes and dissolved free enzymes in lake water or in intestines of filter feeders (grazers) which are released or excreted as particle, and colloidal-associated enzymes or as dissolved free enzymes. They all can be partially mixed up with those from their prey.

Even more complicated is the situation with enzymes from phagotrophic microorganisms and fungi, which can excrete vesicles with vacuoles enriched with enzymes (Aaronson 1981), or with the cellulosome complex in cellulose-degrading bacteria and fungi, which consist of a set of different types of enzymes, and those cellulosomes may even act independently from their host cells (Wood and McCrea 19979; Ljungdahl 1989; Wood and Gracia-Campayo 1990; Priest 1992). However, all of these enzyme sources may occur and exist at the same time in an aquatic environment. Most of these enzymes have passed the cytoplasmatic membrane and can be defined as truely "extracellular enzymes", partially in contact and in communication with their producers, but also partially completely separated from their place of biosynthesis and producers and acting independently in their environments.

Similar confusions in the definitions and the use of terms for microbial enzymes exist in soil sciences, where microbiologists use terms for microbial enzymes in addition to "extracellular-", "endo-", "exo-", "ecto-" and "immobilized" enzymes also such as "abiontic" and "biontic" enzymes to differentiate between living-cell-associated enzymes ("biontic") and non-living-cells-associated and free enzymes ("abiontic") (Schinner and Sonnleitner 1996).

Soil microbiologists seem to define all enzymes associated with living cells as "ectoenzymes" in the sense of Chróst's definition (Chróst 1990) and call them "biontic" enzymes, whereas all non-living-cell and particle-associated enzymes are called "abiontic" enzymes (Schinner and Sonnleitner 1996). Although soil and sediments differ significantly in their physical-chemical structures and microbial enzymes may be exposed certainly to quite different microenvironments compared with those in water columns, it is rather surprising to define highly structured and efficiently acting biomolecules/biocatalysts with a definite biological origin as "abiontic"; this may demonstrate the rather diffuse knowledge and view of the ecology and biochemistry of microbial enzymes in their natural habitats. Obviously, there is an urgent demand for a precise and stringent protocol and clear definitions for studies on microbial enzymes in aquatic environments, a task for a special expert group in this field.

9.2.2 Some Problems Related to Terms in the Sense of Enzyme Cleaving Patterns

Degradation of biopolymers such as starch, cellulose, lignin, proteins, polynucleotides, lipids and geopolymers is hardly facilitated by only one type of enzyme system; biopolymers are mostly degraded by a set of different enzymes working in a synergistic and cooperative way during the cleaving process (Wood and McCrae 1979; Burns 1983; Priest 1992; Winkelmann 1992). Principally two types of cleaving patterns can be observed during enzymatic attacks on biopolymers: an exo and an endo cleaving pattern (1). Exo cleaving reactions occur more frequently and at specific sites and they follow certain reaction types from the end of the polymeric matrix such as the realease of glucose and maltose or cellobiose in starch and cellulose degradations (Ljungdahl 1989) (2). Endo cleaving patterns occur more randomly and are less specific in their locations. They release oligomeric and smaller polymeric subfractions and pieces from the biopolymers (Wood and Garcia-Campayo 1990), which in turn are finally processed by exo cleaving enzymes (Ljungdahl 1989). These two terms were therefore defined as being important in detritus processing (Sinsabaugh and Linkins 1988, 1990; Sinsabaugh et al. 1992a, b; Sinsabaugh and Findlay 1995) and are proposed in this context as being reserved and strictly followed in enzyme studies on NOM in order to understand and characterize precisely biopolymer processing in aquatic environments.

9.2.3 Measurements and Techniques

Generally, studies and results on aquatic ecosystem structure and functions are highly biased by the selected and applied methods and assays (Wetzel and Likens 1991; Hobbie 1993, 1994; Paul 1993; Kemp 1994). This problem holds even more true for studies on microbial enzymes in aquatic environments (Hoppe 1991, 1993). Contrary to enzyme studies in biochemistry with stringent assay controls and procedures such as strict control of temperatures, ion compositions and pH conditions (Bergemeyer 1990, Dixon and Web 1979; Manafi et al. 1991; McFeters et al. 1995), such standard enzyme assays do not exist and are not always applied and followed in aquatic enzyme research. A large number of model substrates have been currently used in microbial enzyme activity measurements in freshwaters, marine environments and sediments (Table 9.3), with spectrophotometric, fluorometric, radiometric and viscosimetric detection methods and variable incubation conditions and protocols (reviewed by Billén 1984, 1991; Chróst 1990; Hoppe 1991).

Therefore, due to the acknowledged importance of NOM processing via MEE, Hoppe (1993), as an early pioneer in this field, has developed and proposed a general protocol to be followed during the application of all fluorogenic substrate analogues or model substrates for studies of MEE in aquatic habitats. These fluorogenic molecular substrate probes (FMPs) allow shorter incubation times due to their high sensitivity in measurement techniques with spectrofluorimeters and/or in analogy with microtitreplate readers (Schmidt-Biegel and Obst 1989; Haugland 1996, Manafi et al. 1991; Hoppe 1993; McFeters et al. 1995), a valuable advantage compared with chromogenic substrate analogs with spectrophotometer readings, which need 10–100 times longer incubation times, as has been usually applied in the past (Manafi et al. 1991). Also radiolabelled oligomers and polymers such as polysacharides and proteins can provide highly sensitive assay techniques as have been successfully applied by Hollibaugh and Azam (1983), Ammerman (1991), Billén (1984, 1991), Hernandez et al. (1996).

Although not specially dedicated to humic waters, some of these special assays have been developed for tracing phosphoester cleaving mechanisms in marine and freshwaters, which seem to be very elegant and promising tools also in applications for humic waters. In these studies, very sensitive radiotracer methods with ^{32}P-labelled γATP and ^{32}P-labelled glucose-6-phosphate have been used to differentiate between phosphomonoesterase (PME) and 5'-nucleotidase (5'-DNAse) activities from algae and bacteria in the same water samples (Table 9.3). These selected techniques and substrates were very powerful in discriminating between algae and bacteria MEE contribution and their preferences for certain organic phosphorus and/or carbon resources (cf. references in Table 9.3). A generally accepted protocol is still waiting to be defined for MEE studies in aquatic environments; this is certainly a severe deficiency because the published data from different studies and environments are hardly comparable. However, as a first step towards more standardized MEE measurements and assays, the protocol as outlined by Hoppe (1993) could be followed.

9.2.4 Recommended Official Classifications of MEE According to Type of Reaction During Catalysis

Due to the large variability of the biopolymer composition in NOM, a large number of enzyme systems, enzyme reaction patterns and regulation of catalysis patterns have evolved during evolution and in different life cycles. Therefore, the Enzyme Commision (EC) of the International Union of Biochemistry (IUB) has written down rules and has made recommendations as to how enzymes should be classified, named and characterized according to their reaction and cleaving patterns (Enzyme Nomenclature 1975, 1984, and Enzyme Nomenclature Supplement and Revision in Eur J Biochem 1986, 1989, 1990). Following these rules and recommendations, the identified enzymes are now classified into six major groups as listed in Table 9.1. Depending on the nature of the substrates and their co-substrates/enzymes during enzyme catalysis, the six enzyme classes are further subdivided into subclasses such as into about 17 oxidoreductases and about 11 hydrolases or at present even much more. However, as this topic is too large to be detailed in this chapter, the reader is referred to current handbooks and proceedings of enzyme nomenclature and methods (Methods in Enzymology Vol. 1–286, 1955–1996; or Boyer et al. 1959–1983; Bergemeyer 1990).

9.3 The Role and Significance of MEE in Aquatic Environments

The majority of non-living organic matter (NOM) in aquatic environments is of polymeric nature (Allen 1976; Wetzel 1983; Thurman 1985) and therefore not easily available to the main consumer and sinks of organic carbon, nitrogen and phosphorus such as bacteria and fungi (Ducklow et al. 1986; Ducklow 1991; Azam et al. 1994). Principally two mechanisms can be identified as responsible for biopolymer cleavage and processing as partially oulined in Fig. 9.3: a photolytic and an enzymatic cleavage. Both exist in aquatic ecosystems and have been studied in the past. Photolytic NOM degradation is an abiotic process, not under the control of the biota and not directly coupled and regulated due to its nutrient demand, community structures and food web interactions, but more dependent on lake water chemistry, lake physics, its location and some other external factors. This cleavage aspect is reviewed and discussed in another chapter of this book. Enzymatic cleavage of biopolymers is a biologically controlled process and directly linked to the nutrient bioavailability and dynamics, trophic interactions and food web structures (Figs. 9.1–9.3).

There has been mutual consensus among aquatic scientists that MEE are key factors in aquatic ecosystem processes, contributing to nutrient cycles, organic matter processing and biodegradations (reviewed by Billén 1991; Chróst 1990, 1991; Hoppe 1991; Azam et al. 1994). However, most of these

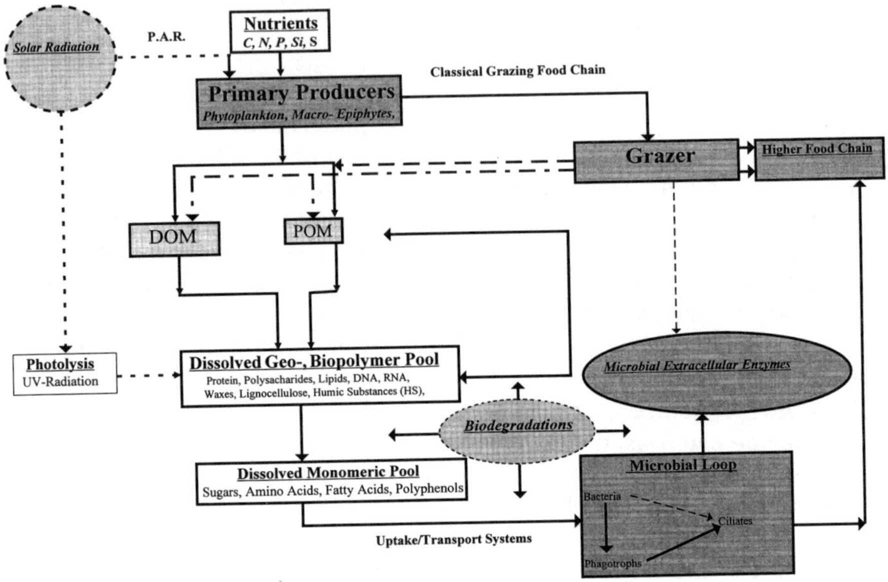

Fig. 9.1. Principal nutrient and organic matter flow in aquatic environments

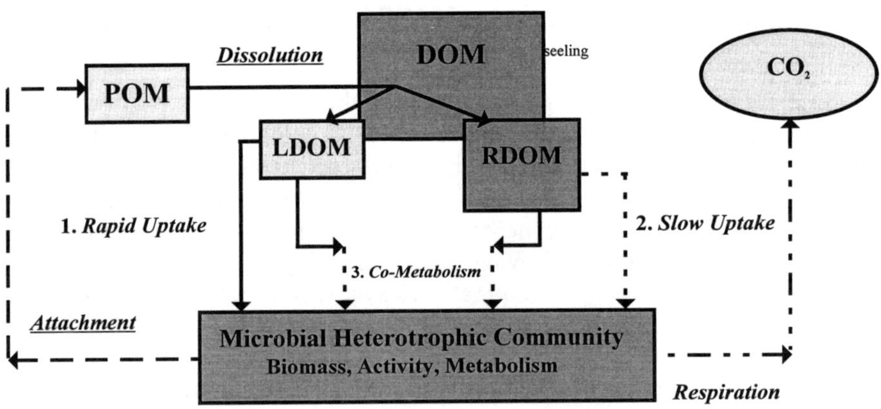

Fig. 9.2. Principal mechanisms of microbial detritus processing and utilization in aquatic environments

studies on MEE have been carried out in marine and/or temperate freshwaters with moderate or higher nutrient but low HS contents (Table 9.3). Except for terrestrial ecosystems, where MEE have been studied under different aspects of nutrient and HS contents (reviewed by Burns 1978, 1982, 1983; Schinner and Sonnleitner 1996; Morra 1997), rather few studies have considered the role of MEE in HS enriched aquatic environments (Münster 1991,

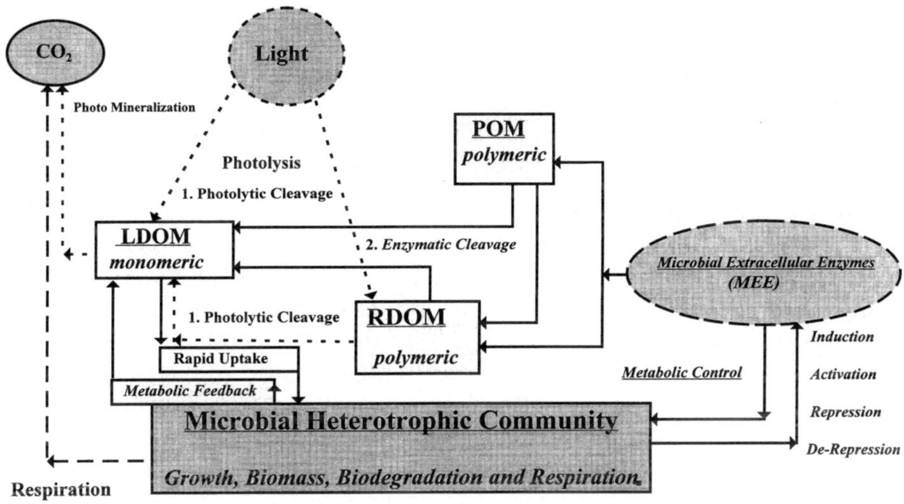

Fig. 9.3. Principal cleaving mechanisms of non-living organic matter (NOM) in aquatic environments. (1). Photolytic cleavage. (2). Enzymatic cleavage

1992a, b, 1994, 1997b, c; Wetzel 1991; Edling and Tranvik 1996). However, HS-rich waters cover large areas in the NBA and tropical ecosystems, which are important factors in global climate change aspects (Schindler and Bayley 1993, Schindler et al. 1990, 1992, 1996; Dixon 1992; Milner & Oswood 1996; Post et al. 1992). Freshwaters in the NBA have significant lower nutrient contents (Salonen et al. 1992; Münster 1997a,c) and different food web structures (cf. Salonen et al. 1992 and another chapters in this book). MEE in humic rich waters are probably modified in their activities and distribution (Wetzel 1991; Münster 1991, 1997b, c). Such high-HS-carrying freshwaters are very abundant and important habitats in the NBA such as in Russia, Scandinavia (Gjessing 1976), Canada (Schindler and Bayley 1993; Schindler et al. 1990, 1992) and some northern parts of America e.g. in Alaska (Milner & Oswood 1996) and partially in tropical freshwater ecosystems, such as in the Amzon (Wissmar et al. 1981; Sioli 1984).

The main focus of this chapter therefore will be to review and discuss the role of MEE in HS-rich freshwaters. This review is based on about 20 years of literature back search and on some selected lakes from NBA in Scandinavia such as in Norway, Sweden and Finland because, from these lakes the authors could collect sufficient data to discuss the role of MEE in the transformation of DOM in aquatic environments. Therefore. this review will be partially based on literature data and on the authors own studies on the composition and utilization of DOM by microheterotrophs in typical humic and small forest lakes with different nutrient levels and environmental impacts.

9.3.1 Role of MEE in Humic Waters

In many detritus-rich freshwaters two fundamental NOM cleavage pathways by MEE can be defined and appear to be the most relevant ones (Sinsabaugh and Linkins 1988, 1990; Münster 1991, 1997b, c; Sinsabaugh et al. 1992a, b, 1994a, b; Jackson et al. 1995; Sinsabaugh and Findlay 1995):

(1) A hydrolytic cleavage pathway via hydrolytic enzymes, e.g. acid and/or alkaline phosphatase activities (APA), aminopeptidase (AMPA), lipase/esterase and glycosidase (GlycAse) activities, which cleave, e.g. phosphoester bonds, peptide bonds, lipidester and glycosidic bonds; (2). An oxidative cleavage pathway, e.g. by phenol oxidases and phenol peroxidases which can cleave especially aliphatic carbon bonds and aromatic ring structures and their inherent carbon bonds (De Haan 1974, 1976, 1983; Sinsabaugh et al. 1992a+b; Münster et al. 1997a, b) such as in recalcitrant geopolymers as generally found in lignocellulosic compounds and probably also in humic substances (HS). However, these oxidative enzyme systems can also participate in the opposite directions, e.g. in polymerization reactions (Hayaishi 1962, Saunders et al. 1964), a fact which has been seldom studied in aquatic environments. Although not so much intensively studied as in marine and in temperate lakes and rivers, both, hydrolytic and oxidative cleavage patterns are probably keeping an equal central position in the microbial biopolymer processing of NOM in humus-rich waters. However, their measurements require carefully designed assays and instrumentations, as will be outlined below.

9.3.1.1 MEE Measurements in Humic Waters

Humic waters contain different organic solutes and concentrations compared with clear water, which may affect the *in situ* enzyme activities and reaction patterns. Therefore, the development and final design of the enzyme protocols are crucial especially in humus rich waters.

9.3.1.1.1 Hydrolytic Enzyme Systems

Measurements of hydrolytic enzymes (EC 3.1, EC 3.2 and EC 3.4) under *in situ* conditions have become important tools in aquatic microbial ecology to understand better the microbial utilization of carbon, nitrogen and phosphorus sources and the abundances and classifications of the involved MEE of those cleaving processes. However, in humic waters MEE activity measurements may be biased by the amount and composition of HS (Münster 1996b; Münster et al. 1989) and need therefore careful assay preparations. Due to the generally low MEE activities in HS-rich waters, chromogenic substrates probably need to long incubations, e.g. 1–6 days, even in eutrophic waters according to Boon (1989, 1990), and may be therefore more biased compared to fluorogenic MEE substrates, which require only 10–60-min incubations (Hoppe 1993). However, measurements of fluorogenic substrates in humic-rich waters suffer under the adsorption, quenching and pH effects of

HS present in the water (Münster et al. 1989; Bittl and Babenzien 1996; Edling & Tranvik 1996). This effect was found to vary with lake water depths and also on a seasonal scale, which was mostly due to the change in pH and HS contents (Münster et al. 1989; Edling & Tranvik 1996). Careful control of pH and quenching effects by HS is needed during MEE measurements with fluorogenic substrates (Münster et al. 1989; Hoppe 1993; Edling & Tranvik 1996). This can be achieved by preparing calibration assays for each sample measurement under well controlled pH conditions, calibration and fluorimeter settings (Chróst & Krambeck 1986; Münster et al. 1989; Hoppe 1993; Edling & Tranvik 1996). Special problems may arise with samples from anaerobic environments and fluorogenic substrate assay applications, as has been observed by Hoppe et al. (1988).

Although not specially developed in humus-rich waters, for true field experiments, where the incubation is made at the lake and enzyme activitiy cannot be measured immediately after incubations, Chróst and Velimirov (1991) have developed assays to measure MEE activities in a mesohumic lake, marine samples and brackish water after incubations combined with deep freezing techniques. They found nearly 95–100% fluorescence recovery of fluorogenic substrate after incubation combined with 1–10-day storage at -20°C, in case samples that cannot be measured immediately after incubations (Chróst and Velimirov 1991). Another important area of MEE measurements is related to attached microbes on solid surfaces, in biofilms and sediments (Meyer-Reil 1986, 1991; Goulder 1990; Hoppe 1991, 1993; Jones and Lock 1989; Marxen and Witzel 1991; Marxen and Fiebig 1993; Marxen and Schmidt 1993 and cited references therein). These MEE measurements require special assay developments and incubation techniques and they have to be tested new for each environment. One of the main disadvantages of fluorogenic substrates is their partially low water solubilities and their instability due to light effects and self-decompositions. About 1–10% of methylumbelliferyl substrates decay and decompositions have been observed within 5–14 days of storage in sterile glass bottles at -20°C measured with high performance liquid chromatography (HPLC) techniques combined with fluorescence detections (Münster, unpubl. results). This aspect is especially crucial when kinetic measurements are applied, maximal reaction velocity (V_{max}) and Michaelis-Menten constant (K_m) values, are calculated and saturation levels are estimated.

Therefore, the application of fluorogenic substrate assays for MEE measurements needs a very careful and strict control of substrate quality, incubation conditions and fluorescence readings/calibrations. A more strict, standardized assay preparation procedure for MEE measurements in aquatic environments is therefore highly desirable to get a better experimental basis for more reliable and comparable data from different habitats. Such standard assays and their outcomes will also have more and stronger impacts on quantifying microbial processes and carbon flux models (Sinsabaugh et al. 1994a, b).

9.3.1.1.2 Oxidizing Enzyme Systems

It has been formerly shown by De Haan (1974, 1976, 1983) that benzoate oxidizing bacteria can contribute via oxygenase enzyme systems in the biodegradation of RDOM pools in humic lake water. Therefore, in addition to hydrolytic MEE also oxidizing enzymes like phenoloxidase (EC 1.10.3.2 and 1.14.18.1) and phenol peroxidase systems (EC 1.11.1.7) may play an even larger role in NOM processing in HS-rich waters, due to the predominance of RDOM compound rich in HS. However, these oxidizing enzyme systems can promote both depolymerization and polymerization reactions (Hayaishi 1962). Although many studies have been carried out in soil enzymes with these two type of enzyme systems (Burns 1983; Schinner and Sonnleitner 1996 and references therein), rather little is known about their contribution in aquatic environments. This is because, in addition to the phenol oxidase/phenol peroxidase enzyme system as studied by Sinsabaugh et al. (1992a, b) during degradation of RDOM sources in lenthic and lotic environments, enzymatic degradation of RDOM compounds such as they exist in lignocellulosic material includes a set of four to five other oxidative enzyme systems such as Mn-dependent peroxidase (EC 1.11.1.13), ligninperoxidase (EC 1.11.1.14), glucose oxidase (EC 1.1.3.4), glyoxal oxidase (EC 1.1.3.13), laccase (EC 1.10.3.2) and some other still hardly known enzyme systems (reviewed by Zeikus 1981; Fiechter 1993; Orth et al. 1993, Reddy 1993).

Sinsabaugh and coworkers (Sinsabaugh and Linkins 1988, 1990; Sinsabaugh et al. (1992a, b; Sinsabaugh and Findlay 1995) have applied two different assays to trace oxidative enzymatic cleaving activities by using L-3,4-dihydroxyphenylalanine (L-DOPA) as tracer substrate, incubated with H_2O_2 additions for phenol peroxidase and without H_2O_2 additions for phenol oxidase activitiy measurements (Sinsabaugh et al. 1992a, b). This assay followed a similar procedure as applied in soil microbiology (Schinner and Sonnleitner 1996). It is a rather unspecific assay for oxidative enzyme activities, as Sinsabaugh and co-workers have tried to address in their publications (Sinsabaugh et al. 1992a, b). However, the assays are simple to prepare and may be also cheap, because only L-DOPA, H_2O_2 as co-substrate and acetate buffers are needed and the results can be measured after 1–12 hours incubation in a spectrophotometer at 460 nm as usually available in laboratories. According to these assays, phenol oxidase varied between 1.2 and 255 μmol/g AFDM/h (AFDM = ash free dry mass, cf. Table 9.3) and phenol peroxidase varied between 6 and 500 μmol/g AFDM/h in lotic communities in river ecosystems (Sinsabaugh and Linkins 1988, 1990; Sinsabaugh et al. 1992a, b; Sinsabaugh and Findlay 1995).

These data emphasize the important contribution of oxidative MEE and their synergistic activities in NOM processing. However, the authors did not specify precisely in their protocols and instructions what kind of reference compound or system they have used to calibrate this enzymatic reaction under *in situ* conditions; this aspect could be important in order to specify which carbon bonds were cleaved and which carbon sources were finally utilized by the indigenous microbial community and used in carbon flux calculations (Sinsabaugh et al. 1994a, b). They also did not discuss the possibility that their substrate could have been potentially used as carbon sources for microheterotrophs before the enzyme system started to work. However, this is certainly a general problem with oxidizing enzyme substrate assays, because there are always potential reverse effects, such as catalytic polymerizations with the same substrate-enzyme systems (Hayaishi 1962). Probably ^{14}C-radiolabelled substrates combined with detailed and sensitive analytical equipment and techniques can bring more light to these complex systems of enzymatic oxidative NOM processing, as Benner et al. (1986, 1989) and Schoenberg et al. (1990) have applied in their studies on NOM processing in marshes and acid environments.

Another assay for oxidative enzymatic activities in humic lake waters was developed by Münster et al. (1997a), by using veratrylalcohol (3,4-dimethoxybenzoic acid, VeraOH) as naturally occurring tracer substrate for ligninperoxidase (LiP) activity measurements. This was formerly developed by Tien and Kirk (1984, 1988) in lignin degradation studies by white rot fungi and since then it has very often been used in ligninperoxidase activitiy (EC 1.11.1.14) measurements (reviewed by Schoemaker and Leisola 1990). This assay can be easily used in LiP studies with white rot fungi and spectrophotometric techniques (Tien and Kirk 1984). However, a more precise and sophisticated assay is possible and recommended with high performance liquid chromatography (HPLC), as reported by Schoemaker and Leisola (1990). Such assay has been further developed, modified and applied for LiP measurements in humic lake waters by Münster et al. (1997a). According to Münster et al. (1997a, b) it is possible with this assay to trace *in situ* peroxidase activities with veratrylalcohol (VeraOH) and HPLC techniques (Fig. 9.4).

They compared the validity of this assay in cultures with *Phanerochaete chrysosporium as* model and reference organism producing LiP and an isolated bacteria strain from a mesohumic lake. This assay was successful in tracing lignin peroxidase (LiP) activities in the liquid growth medium of both cultures, because a similar decrease in VeraOH was found in *Phanerochaete chrysosporium* and one lake isolate (Fig. 9.4a, b), but with a lower activitiy in the lake water bacterium growth medium. Münster et al. (1997a) have also tested the chemical stability of VeraOH under different environmental conditions such as pH, H_2O_2 contents and light impacts, and concluded that VeraOH may be a surrogate tracer substrate for peroxidase measurements in humic lakes and followed the seasonal vatiation of LiP in a small polyhumic forest lake (Fig. 9.4c). However, the disadvantage of this approach was the relatively long incubation time (4–7 days under *in situ* lake water conditions) and the potential degradation and utilization of VeraOH as carbon substrate during this incubation by bacteria (Münster et al. 1997b), which is equally valid for the amino acid L-DOPA as substrate for phenol oxidase and phenol peroxidase, as Sinsabaugh et al (1992a, b) have used. To bypass this problem many oxidative enzyme studies with cultured micoorganisms used different azo, triphenyl methane, heterocyclic and polymeric

Table 9.3. Overview of MEE activities in different aquatic environments (part 1/2)

No	Location		Range of phosphoester hydrolyzing enzyme activity (PEHE nmol $l^{-1}h^{-1}$)				Range of carbohydrate hydrolyzing enzyme activity (CHE nmol $l^{-1}h^{-1}$)										
	Marine	Freshwater	acPME	alPME	PDE	5'-DNAse	αGluc Ase	βGluc Ase	αGal Ase	βGal Ase	αMan Ase	βMan Ase	βXyl Ase	CBH	Cellu lAse	βNAGl Ase	Tab. ref. no
1		Mesotr. lake		212–393		10–86											1
2		Oligo-mesotr. lake				1–86											2
3	Eutr. estuary					0.06–0.3											3
4	Mesotr. fjord			6–687		5–97											4
5	Mesotr. coast					0.03–0.6											5
6		Eutr. lake		10–49				1.2–9.3									6
7		Hypertr. lake		546–762													7
8		Eutr. lake															8
9		Eutr. lake						69–710									9
10		Eutr. river		750–5208			4.2–146	4.2–142								110–233	10
11		Eutr. river		25-450	3.8–51	2–59											11
12		Eutr. reservoir					0.7–5	2–8								1.3–10	12
13		Eutr. reservoir														0.34–63	13
14	Mesotr-fjord			3–390			1-36	1–24								0.8–15	14
15	Mesotr. coast	Eutr. pond						0.8–120									15
16		Acid poly-humic lake	12–103				0.1-23	4.4–61.3									16
17		Eutr. lake		1–90	0.1–0.8			5.1–11.3									17

18		Eutr. lake						5–82									18
19		Acid humic lake	1.5–175					4.4–68									19
20		Acid poly-humic lake	945–1706														30
21		Acid poly-humic lake	51–162				0.1–90	4.4–68	0.1–117	0.1–65	0.1–102	0.1–81					32
22		Acid poly-humic lake	10–150				0.1–45	1–54	0.15–67	0.1–46	0.1–23	0.1–21					33
23		Eutr. reservoir					0.4–5	1.1–7.4								1.4–10.1	34
24		Eutr. reservoir		6.7–727													34
25		Eutr. lake						178–645									25
26	Mesotr. coast						0.05–0.91	0.31–2.1									26
27a		River/creek sed.		77–159													27
28a		River/creek sed.						35–77									28
29a		River/creek sed.		4.6–88				5.1.6.8									29
30		Mesotr.-eutr. lake		19.2–44.4													30
31		Mesotr.-eutr. lakes															31
32	Mesotr.-eutr. coastal water																32
33	Pacific water			1.14–7.8													33
34		Mesotr.-eutr.-lake		10–400													34

No	Location		Range of phosphoester hydrolyzing enzyme activity (PEHE nmol $l^{-1}h^{-1}$)				Range of carbohydrate hydrolyzing enzyme activity (CHE nmol $l^{-1}h^{-1}$)										
	Marine	Freshwater	acPME	alPME	PDE	5'-DNAse	αGluc Ase	βGluc Ase	αGal Ase	βGal Ase	αMan Ase	βMan Ase	βXyl Ase	CBH	Cellu lAse	βNAGl Ase	Tab. ref. no
35		Mesotr. lake		20–330													35
36		Eutr. lake		50.4–426													36
37b		Periphyton		12–43				1.4–5.9									37
38		Acid oligotr.-lake	36–1554														38
39		Neutral-acid-lake	6–984														39
40		Oligotr. lake		6–624													40
41		Acid-alkaline lake	10.2–408	1.8–264													41
42		Mesotr. lake		78–288		18–60											42
43		Oligotr. lake		300–480													43
44		Mesotr.-eutr. lake		60													44
45		Eutr. lake		2400–10700													45
46		Mesotr.-eutr. lakes		0.42–28													46
47		Eutr. lake	5.4–25	6.3–35													47
48		Mesotr. lake		22–204													48
49		Mesotr. lake	3600–10300	1800-5800													49
50		Acid humic lake	8.4–1475														50

51c, d		River/lotic epilython	0.4–57.5				2.6-65.5	0.5–36.5					1.5–13.4	0.3–8	147–3065	0.9–37	51
52c, d		River/lotic-community						22–42						2–10	5–47		52
53c, d		Watersheds/ catchments						1–25					1–6	1-3	1000–14000		53
54c, d		River/lotic-community	1–180	1–230				1–180							10–1550		54
55e		Estuary/ sediments						7.4–119.2					2.2–22.5	0.2–18.9	160–8200	4.9–91.2	55
56e		Coastal wetland						1–35						0.5–7	1-7		56
57		Rivers		12.8–140				3.2–72.5		0.9–19.1			12.8–140				57
58f		River biofilm						50–450					5–200			50–8000	58
59		Acid-alkaline humic lakes						0.1–4.2									59
60		Acid humic lakes															60

Table 9.3. Overview about MEE Activities in different Aquatic Environments (part 2)

No	Location		Range of esterase and lipase activity (nmol l^{-1} h^{-1})		Range of exo- and endopeptidase activity (nmol l^{-1} h^{-1})		Range of oxidase and peroxidase activity (nmol l^{-1} h^{-1})		Type of selected/applied substrates				Detection methods			Tab. ref. no
	Marine	freshwater	Ester Ase	Lip Ase	Leu Ampa	Prote Ase	Phenol Ase	Peroxid Ase	Chromogenic	Fluorogenic	Radioactive	Others	Photom.	Fluor.	LSC	
1		Mesotr. lake								MUF-PO_4	[γ-^{32}P]-ATP			x	x	1
2		Oligo-mesotr. lake									[γ-^{32}P]-ATP				x	2
3	Eutr. estuary										[γ-^{32}P]-ATP				x	3
4	Mesotr. fjord									MUF-PO_4	[γ-^{32}P]-ATP			x	x	4
5	Mesotr. coast.										[γ-^{32}P]-ATP				x	5
6		Eutr. lake			50–542					MUF-PO4, MUF-βGluc, Leu-AMC				x		6
7		Hypertr. lake								3-O-methyl-fluorescein-PO_4				x		7
8		Eutr. lake		53–754					pNP-palmitate				x			8
9		Eutr. lake			86–1132				pNP-βGluc, Leu-pNA				x			9
10		Eutr. river		42–500	145–47250	33–129			pNP and pNA substrates				x			10
11		Eutr. river							pNP substrates				x			11
12		Eutr. reservoir								MUF-βGluc, MUF-βNAGLA				x		12
13		Eutr. reservoir								MUF-βNAGLA				x		13
14	Mesotr.-fjord				3–91					MUF- substrates, Leu-AMC				x		14
15	Mesotr. coast.	Eutr. pond								MUF-βGluc.				x		15
16		Acid poly-humic lake			6.5–26					MUF substrates, Leu-AMC				x		16

17		Eutr. lake			112–201					MUF substrates, Leu-AMC				x		17
18		Eutr. lake								MUF-βGluc.				x		18
19		Acid humic lake			6.5–123					MUF substrates, Leu-AMC				x		19
20		Acid poly-humic lake			3.7–25					MUF-PO_4, Leu-AMC				x		20
21		Acid poly-humic lake			6.5–26					MUF-substrates, Leu-AMC				x		21
22		Acid poly-humic lake			25–62					MUF substrates, Leu-AMC				x		22
23		Eutr. reservoir								MUF substrates				x		23
24		Eutr. reservoir														24
25		Eutr. lakes								MUF-PO_4				x		25
26	Mesotr. coast				378–954					MUF-βGluc, Leu-AMC				x		26
27a		River/creek sed.			5.6–63.1					MUF substrates, Leu-AMC				x		27
28a		River/creek sed.								MUF-PO_4				x		28
29a		River/creek sed.								MUF-βGluc				x		29
30		Meso-eutr. lake									[^{32}P]-glucose-6-PO_4				x	30
31	Mesotr-eutr. coastal water				176–5041					LLβNA				x		31
32		Mesotr. lake			10–1200	1–880				LLβNA	[^{14}C]-protein			x	x	32
33	Pacific water									3-O-methyl-fluorescein-PO_4				x		33

No	Location		Range of esterase and lipase activity (nmol l^{-1} h^{-1})		Range of exo- and endopeptidase activity (nmol l^{-1} h^{-1})		Range of oxidase and peroxidase activity (nmol l^{-1} h^{-1})		Type of selected/applied substrates				Detection methods			Tab. ref. no
	Marine	freshwater	Ester Ase	Lip Ase	Leu Ampa	Prote Ase	Phenol Ase	Peroxid Ase	Chromogenic	Fluorogenic	Radioactive	Others	Photom.	Fluor.	LSC	
34		Meso-eutr.-lake								pNP-PO_4			x			34
35		Mesotr. lake								MUF-PO_4				x		35
36		Eutr.-lake								MUF-PO_4				x		36
37b		Periphyton								pNP-substrates, Leu-pNP			x			37
38		Acid oligotr. lake			65–150					MUF-PO_4			x			38
39		Neutral-acid-lake								MUF-PO_4				x		39
40		Oligotr. lake								3-O-methyl-fluorescein-PO_4				x		40
41		Acid-alkaline lake								MUF-PO_4				x		41
42		Mesotr. lake								3-O-methyl-fluorescein-PO_4	[α-^{32}P]-ATP			x	x	42
43		Oligotr. lake								3-O-methyl-fluorescein-PO_4				x		43
44		Mesotr.-eutr. lake								MUF-PO_4				x		44
45		Eutr. lake								pNP-PO_4			x			45
46		Mesotr.-eutr. lake							pNP-PO_4				x			46
47		Eutr. lake							pNP-PO_4				x			47
48		Mesotr. lake								3-O-methyl-fluorescein-PO_4				x		48

49		mesotr.-lake								MUF-PO_4				x		49
50		Acid humic lake							pNP-PO_4				x			50
51 c,d		River/lotic epilython					1.2–235.5	23.2–480.8	pNP-substrates, L-DOPA			CMC viscos.	x			51
52 c,d		River/lotic community					2–9	6–36	pNP-PO_4, L-DOPA			CMC viscos.	x			52
53 c,d		Watersheds/ catchments					1–25		pNP substrates, L-DOPA	βNA substrates		CMC viscos.	x	x		53
54 c,d		River/lotic community					1–255	1–500	pNP substrates, L-DOPA			CMC viscos.	x			54
55e		Estuary/ sediments					7–426	7–614	pNP substrates, L-DOPA				x			55
56e		Coastal-wetland					5–255	10–500	pNP substrates, L-DOPA				x			56
57		Rivers			99–986					MUF substrates, Leu-AMC				x		57
58f		River-biofilm			5–100					MUF substrates, Leu-/Pep-AMC				x		58
59		Acid humic-lake								MUF-βGluc.				x		59
60		Acid humic lake								MUF substrates, Pep-AMC				x		60

Abbreviations

Codes for differently used dimensions of enzyme activities

No. 27a, 28a, 29a: a stands for the dimension nmol/ml sediment/h

No. 37b: b stands for the dimension nmol cm^{-2} h^{-1}; cm^{-2} stands for the area colonized by periphyton

No. 51c,d; 52c,d; 53c,d; 54c,d: c stands for the dimension µmol/g AFDM/h; and d stands for units/g AFDM/h; AFDM stands for ash free dry mass

No. 55e, 56e: e stands for the dimension mol g^{-1} OM h^{-1}; OM stands for organic matter

No. 58f: in this line f stands for the dimension nmol cm^{-2} h^{-1}; cm^{-2} stands for area colonized by periphyton ("Aufwuchs") in the biofilm

The values in the columns stand for the reported ranges of enzyme activities in nmol l^{-1} h^{-1} measured in the water and habitats, except those specified with codes a–f

Codes for measured enzymes and their substrates:

acPME: acid phosphomonoesterase; alPME: alkaline phosphomonoesterase; PDE: phosphodiestrase; 5'-DNA: 5'-nucleotidase

αGlucAse: α-glucosidase; βGlucAse: β-glucosidase; αGalAse: α-galactosidase; βGalAse: β-galactosidase; αManAse: α-mannosidase

βManAse: β-mannosidase; βXylAse: β-xylosidase; CBH: cellobiohydrolase; CellulAse: cellulase; βNAGLAse: N-acetyl-β-glucosamidase; EsterAse: esterase; LipAse: lipase

LeuAmpa: leucine aminopeptidase; ProteAse: endo-peptidase/protease; PhenolAse: phenol oxidase; PeroxidAse: phenol peroxidase

Abbreviation for type of selected/applied substrates:

MUF-PO_4: 4-methyl-umbelliferyl-phosphate; MUF-βGluc: 4-methyl-umbelliferyl-β-D-glucopyranoside; MUF-βNAGLA: 4-methyl-umbelliferyl-N-acetyl-β-D-glucosaminpyranoside

MUF-substrates: different 4-methyl-umbelliferyl substrates; Leu-AMC: L-leucyl-7-amidomethylcoumarine; Pep-AMC: peptidyl-7-amidomethylcoumarine;

L-DOPA: L-3,4-dihydroxy-phenylalanine; CMC: carboxymethylcellulose, used as substrate for cellulase activity mesurements with a viscosimeter;

βNA-substrates: β-N-acetyl substrates; [γ-^{32}P]-ATP: at gamma-position ^{32}P-labelled adenosine triphosphate; [α-^{32}P]-ATP: at alpha-position ^{32}P-labelled adenosine triphosphate;

[^{32}P]-glucose-6-PO4: ^{32}P-labelled glucose-6-phosphates; [^{14}C]-protein: [^{14}C]-labelled bovine serum albumine (BSA), hämoglobine and others;

pNP-PO_4: para-nitrophenyl-phosphate; pNP-substrates: different para-nitrophenyl-substrates, such as from acetic acid, butyric acid, stearic acid, sugars and others;

LLβNA: L-leucyl-β-naphthylamide; 3-O-methyl-fluorescein-PO_4: 3-O-methyl-fluorescein-phosphate; Leu-pNA: L-leucine para-nitranilid

Abbreviations for the detection methods:

Photom.: photometrically measured enzyme activities; Fluor.: fluorimetrically measured enzyme activities;

viscos.: viscometric cellulase activity measurements with carboxymethylcellulose (CMC);

LSC: liquid scintillation counting technique for mesurements with [^{14}C]- and [^{32}P]-labelled substrates

Abbreviations for locations

Hypertr., Eutr., Mosotr., Oligotr.: hypertrophic, eutrophic, mesotrophic and oligotrophic waters respectively; sed.: sediment

Reference List for Table 9.3:

1. Siuda W, Güde H (1994) A comparative study on 5'-nucleotidase (5'-nase) and alkaline phosphatase (APA) activities in two lakes. Arch Hydrobiol 131:211–229
2. Cotner JB Jr, Wetzel RG (1991) 5'-nucleotidase activity in a eutrophic lake and an oligotrophic lake. Appl Environ Microbiol 57:1306–1312

3. Tamminen T (1989) Dissolved organic phosphorus regeneration by bacterioplankton: 5'-nucleotidase activity and subsequent phosphate uptake in a mesocosm enrichment experiment. Mar Ecol Prog Ser 58:89–100
4. Thingstad TF, Skoldal FEF, Bohne RA (1993) Phosphorus cycling and algal-bacterial competition in Sandsfjord, western Norway. Mar Ecol Prog Ser 99:239–259
5. Ammerman JW, Azam F (1985) Bacterial 5'-nucleotidase in aquatic ecosystems: a novel mechanisms of phosphorus regeneration. Science 227:1338–1340
6. Chróst RJ, Münster U, Rai H, Albrecht D, Witzel KP, Overbeck J (1989) Photosynthetic production and exoenzymatic degradation of organic matter in euphotic zone of an eutrophic lake. J Plankton Res 11:223–242.
7. Newman S, Aldridge FJ, Philips EJ, Reddy KR (1994) Assessment of phosphorus availability for natural phytoplankton populations from a hypertrophic lake. Arch Hydrobiol 130: 409–427
8. Gajewski AJ, Chróst RJ, Siuda W (1993) Bacterial lipolytic activity in an eutrophic lake. Arch Hydrobiol 128:107–126
9. Gajewski AJ, Chróst RJ (1995) Production and enzymatic decomposition of organic matter by microplankton in a eutrophic lake. J Plankton Res 17:709–728
10. Boon PI (1990) Organic matter degradation and nutrient regeneration in Australian freshwaters: II. Spatial and temporal variation, and relation with environmental conditions. Arch Hydrobiol 117:405–436
11. Boon PI (1993) Organic matter degradation and nutrient regeneration in Australian freshwaters: III. Size fractionation of phosphatase activity. Arch Hydrobiol 126:339–360
12. Vrba J, Nedoma J, Šimek K, Seda J (1992) Microbial decomposition of polymer organic matter related to plankton development in a reservoir: activity of α-, β-glucosidase, and β-N-acetylglucosaminidase and uptake of N-acetylglucosamine. Arch Hydrobiol 126:193–211
13. Nedoma J, Vrba J, Hejzlar J, Šimek K, Straškrabová V (1994) N-acetylglucosamine dynamics in freshwater environments: concentration of amino sugars, extracellular enzyme activities, and microbial uptake. Limnol Oceanogr 39:1088–1100
14. Hoppe HG, (1983) Significance of exoenzymatic activities in the ecology of brackish water: measurements by means of methylumbelliferyl-substrates. Mar Ecol Progr Ser 11:299–308
15. Somville M (1984) Measurement and study of substrate specificy of exoglucosidase activity in eutrophic water. Appl Environ Microbiol 48:1181–1185
16. Münster U, Einiö P, Nurminen J (1989) Evaluation of the measurements of extracellular enzyme activities in a polyhumic lake by means of studies with 4-methylumbelliferyl-substrates. Arch Hydrobiol 115:321–337
17. Chróst RJ, Krambeck HJ (1986) Fluorescence correction for measurements of enzyme activity in natural waters using methylumbelliferyl-substrates. Arch Hydrobiol 106:79–90.
18. Chróst RJ (1989) Characterization and significance of β-glucosidase activity in lake water. Limnol Oceanogr 34:660–672
19. Münster U (1992a) Microbial extracellular enzyme activities and biopolymer processing in two acid polyhumic lakes. Arch Hydrobiol Ergebn Limnol 37:21–32
20. Münster U (1992b) Microbial extracellular enzyme activities in humex lake Skjervatjern. Environ Int 18:637–647
21. Münster U, Einiö P, Nurminen J, Overbeck J (1992a) Extracellular enzymes in a polyhumic lake: important regulators in detritus processing. Hydrobiologia 229:225–238
22. Münster U, Nurminen J, Einiö P, Overbeck J (1992b) Extracellular enzymes in a small polyhumic lake: origin, distribution and activities. Hdrobiologia 243/244:47–59
23. Vrba J (1992) Seasonal extracellular enzyme activities in decomposition of polymeric organic matter in a reservoir. Arch Hydrobiol Ergebn Limnol 37:33–42
24. Vrba J (1993) Enhanced activity of alkaline phosphatases-phytoplankton response to epilimnetic phosphorus depletion. Wat Sci Tech 28:15–24.
25. Chróst RJ, Velimirov B (1991) Measurement of enzyme kinetics in water samples: effect of freezing and soluble stabilizer. Mar Ecol Progr Ser 70:93–100

26. Karner M, Rassoulzadegan F (1995) Extracellular enzyme activity:indications for high short-term variability in a coastal marine ecosystem. Microb Ecol 30:143–156
27. Marxen J, Schmidt HH (1993) Extracellular phosphatase activity in sediments of the Breitenbach, a central european mountain stream. Hydrobiologia 253:207–216
28. Marxen J, Witzel KP (1991) Significance of extracellular enzymes for organic matter degradation and nutrient regeneration in small streams. In: Chrost RJ (ed) Microbial enzymes in aquatic environments. Brock/Springer series in contemporary biosciences. Springer, Berlin Heidelberg New York, pp 270–285
29. Marxen J, Fiebig DM (1993) Use of perfused cores for evaluating extracellular enzyme activity in stream-bed sediments. FEMS Microbial Ecol 13:1.12
30. Hernandez I, Hwang SJ, Heath RT (1996) Measurement of phosphomonoesterase activity with radiolabelled glucose-6-phosphate. Role in the phosphorus requirement of phytoplankton and bacterioplankton in a temperate mesotrophic lake. Arch Hydrobiol 137: 265–280
31. Jacobsen TR, Rai H (1991) Aminopeptidase activity in lakes of differing eutrophication. In: Chróst RJ (ed) Microbial enzymes in aquatic environments. Brock/Springer series in contemporary bioscience. Springer, Berlin Heidelberg New York, pp 155–164
32. Billén G (1991) Protein degradation in aquatic environments. In: Chróst RJ (ed) Microbial enzymes in aquatic environments. Brock/Springer series in contemporary bioscience. Springer, Berlin Heidelberg New York, pp 123–143
33. Perry MJ (1972) Alkaline phosphatase activity in subtropical central north Pacific waters using a sensitive fluorometric method. Mar Biol 15:113–119
34. Gage M, Gorham E (1985) Alkaline phosphatase activity and cellular phosphorus as an index of the phosphorus status of phytoplankton in Minnesota lakes. Freshw Biol 15:227–233
35. Elser JJ, Elser M, Carpenter SR (1986) Size fractionation of algal chlorophyll, carbon fixation and phosphatase activity: relationships with species-specific size distribution and zooplankton community structure. J Plankton Res 8:365–383
36. Chróst RJ, Overbeck J (1987) Kinetics of alkaline phosphatase activity and phosphorus availability for phytoplankton and bacterioplankton in lake Plußsee (North German Eutrophic Lakes). Microb Ecol 13:229–248
37. Scholz O, Boon PI (1993) Alkaline phosphatase, aminopeptidase and β-D-glucosidase activities associated with billabong periphyton. Arch Hydrobiol 126:429–443
38. Jansson M, Olsson H, Broberg O (1981) Characterization of acid phosphatases in the acidified lake Gårdsjön, Sweden. Arch Hydrobiol 92:377–395
39. Olsson H (1988) Phosphatases in lakes - characterization, activity and ecological implications. Ph. D. Thesis, Uppsala University, Uppsala
40. Wetzel RG (1981) Longterm dissolved and particulate alkaline phosphatase activity in a hardwater lake in relation to lake stability and phosphorus enrichments. Verh Int Verein Limnol 21:369–381
41. Tabata M, Tachibana W, Suzuki S (1988) pH dependence of phosphatase activity in freshwater lakes. Japan Journ Limnol 49:93–98
42. Bentzen E, Taylor WD (1991) Estimating organic P utilization by freshwater plannkton using [^{32}P]ATP. J. Plankton Res 13:1223–1238
43. Stewart AJ, Wetzel RG (1982) Influence of dissolved humic materials on carbon assimilation and alkaline phosphatase activity in natural algal-bacterial assembledges. Freshw Biol 12: 369–380
44. Cotner JB Jr, Wetzel RG (1992) Uptake of dissolved inorganic and organic phosphorus compounds by phytoplankton and bacterioplankton. Limnol Oceanogr 37:232–243
45. Boavida MJ, Heath RT (1988) Is alkaline phosphatase always important in phosphate regeneration. Arch Hydrobiol 111:507–518
46. Jones JG (1972) Studies on freshwater micro-organisms: phosphatase activity in lakes of different degrees of eutrophication. J Ecology 60:777–791
47. Halemejko G, Chróst RJ (1984) The role of phosphatases in phosphorus mineralization during decomposition of lake phytoplankton blooms. Arch Hydrobiol 101:489–502

48. Pick FR (1987) Interpretations of alkaline phosphatase activity in lake Ontario. Can J Fish Aquat Sci 44:2087–2094
49. Berman T, Wynne D, Kaplan B (1990) Phosphatase revisited: analysis of particle-associated enzyme activities in aquatic systems. Hydrobiologia 207:287–294
50. Cotner JB Jr, Wetzel RG (1988) Potential phosphate release from phosphomonoesters by acid phosphatase in a bog lake. Arch Hydrobiol 111:329–338
51. Sinsabaugh RL, Linkins AE (1988) Exoenzyme activity associated with lotic epilithon. Freshw Biol 20:249–261
52. Sinsabaugh RL, Linkins AE (1990) Enzymatic and chemical analysis of particulate organic matter from a boreal river. Freshw Biol 23:301–309
53. Sinsabaugh RL, Antibus RK, Linkins AR, McClaugherty CA, Rayburn L, Repert D, Weiland T (1992) Wood decompostions over a first-order watershed: mass loss as a function of ligno-cellulase activity. Soil Biol Biochem 24:743–749
54. Sinsabaugh RL, Weiland T, Linkins AE (1992) Enzymic and molecular analyis of microbial communities associated with lotic particulate organic matter. Freshw Biol 28:393–404
55. Sinsabaugh RL, Findlay S (1995) Microbial production, enzyme activity, and carbon turn-over in surface sediments of the Hudson river estuary. Microb Ecol 30:127–141
56. Jackson CR, Foreman CM, Sinsabaugh RL (1995) Microbial enzyme activities as indicators of organic matter processing rates in a lake Erie coastal wetland. Freshw Biol 34:329–342
57. Chappell KR, Goulder R (1995) A between-river comparison of extracellular-enzyme activity. Microb Ecol 29:1–17
58. Jones SE, Lock MA (1989) Hydrolytic extracellular enzyme activity in heterotrophic biofilms from two contrasting streams. Freshw Biol 22:289–296
59. Edling H, Tranvik L (1996) Effects of pH on β-glucosidase activity and availability of DOC to bacteria. Arch Hydrobiol Ergebn Limnol 48:123–132
60. Münster U (1991) Extracellular enzymes in eutrophic and acid polyhumic lakes. In: Chróst RJ (ed) Microbial enzymes in aquatic environments. Brock/Springer series in contemporary bioscience. Springer, Berlin Heidelberg New York, pp96–122

dyes to trace oxidase and peroxidase activities in growth media (Glenn and Gold 1983; Gold et al. 1988; Ollikka et al. 1993). The authors tested also these dyes as tracer substrate for oxidizing enzyme activities with indigenous microbial communities in humic waters and were not much convinced of the reliability of those results (Münster unpublished results); because those tracer dyes also suffer from the long incubation time (5–10 days) and their rather unspecific indication of what kind of oxidative enzyme activities and cleaving pattern they are indicating. Further, in NOM degradations under *in situ* conditions many different enzymes are acting synergistically, and, e.g. LiP is certainly only one of the whole set of different oxidative enzyme systems contributing to NOM biodegradations. Therefore, most of those oxidative enzyme assays suffer under low sensitivity and their limited specificy. Therefore, due to the importance of oxidative enzyme cleaving of RDOM, the authors were still searching for alternative peroxidase substrates and assays to trace specific enzymatic oxidative and peroxidative activities in humic waters under *in situ* conditions.

A quite different oxidative enzyme assay was developed by Palenik and Morel (1990a, b) for tracing the utilization of organic nitrogen sources by phytoplankton via a plasmalemma redox system, e.g. an L-amino acid oxidase (probably from EC 1.4 or EC 1.6 or EC 1.16.3 classes). This is a cell sur-

A

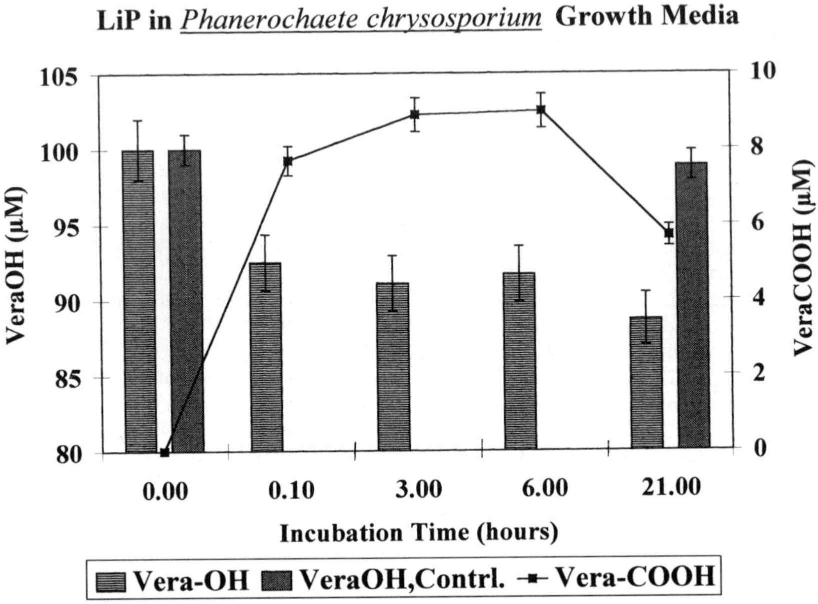

B

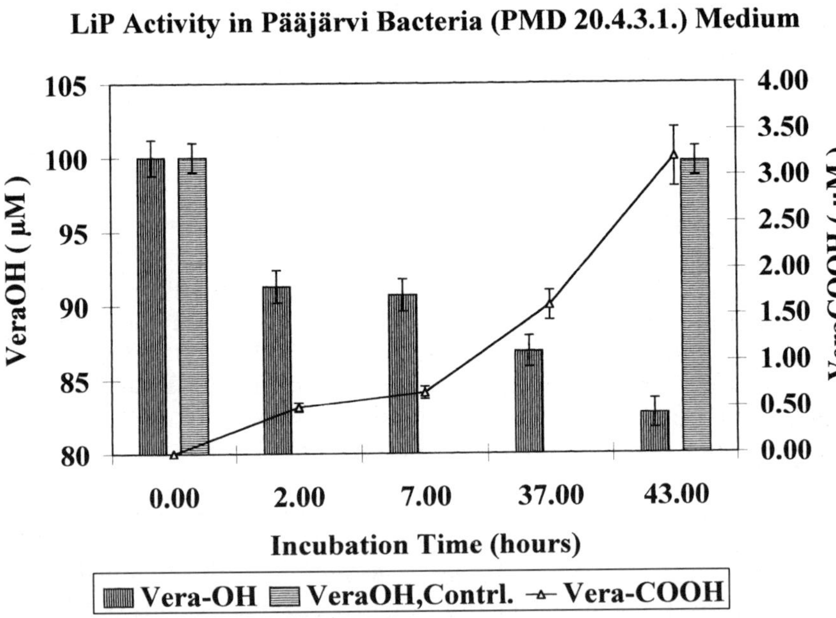

Fig. 9.4. Tracing of peroxidase activities from boreal freshwater microbes based on photometric and fluorimetric assays. (**A**). Ligninperoxidase (LiP) in the growth medium of reference organism *Phanerochaete chrysosporium.* (B). LiP in growth medium of an isolated bacteria strain from a Finnish mesohumic lake (Lake Pääjärvi, southern Finland) (PMD 20.4.3.1. is the code number of the isolated bacteria strain from Pääjärvi and indicates the date of isolation and the medium for isolation). LiP was measured directly in the growth medium from (**A**) and (**B**) at room temperature with veratrylalcohol (VerOH) as LiP tracer substrate according to Tien and Kirk (1984, 1988). High performance liquid chromatography (HPLC) with UV-detection at 260 nm was used to trace the fate of VeraOH and its oxidation product veratryl acid (VeraCOOH) in the

C

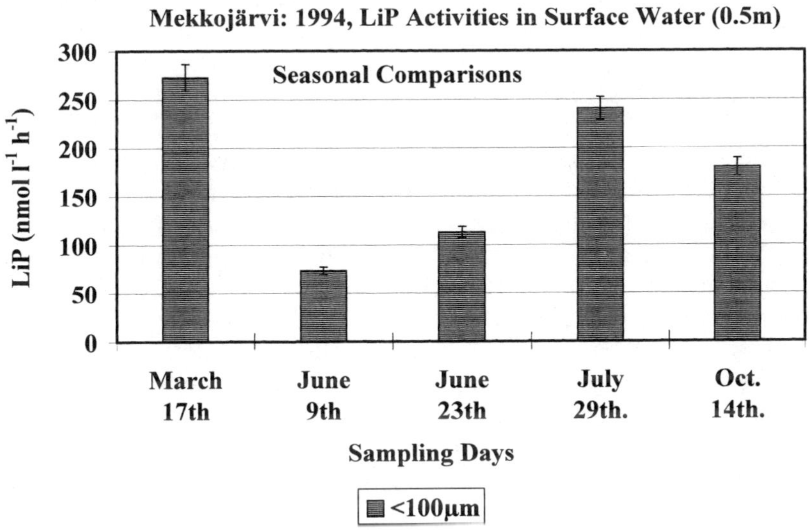

D

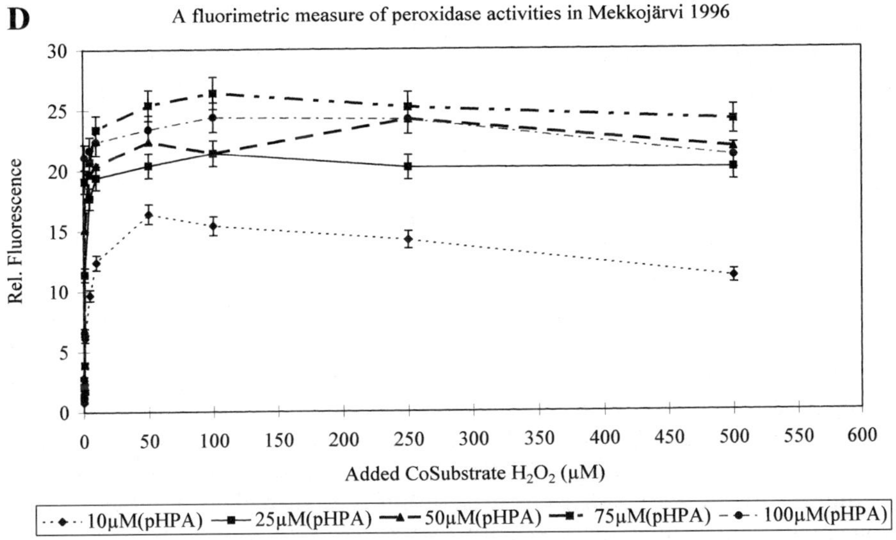

growth medium according to Münster et al. (1997a). (**C**). LiP activities in a polyhumic forest lake. Concentrations of VeraOH and VeraCOOH were measured over time series in 100µm mesh size prefiltered lake water and calculated as LiP activities to monitore and to trace the potential possibility of *in situ* LiP activities by indigenous microbial communities in polyhumic lake Mekkojärvi, southern Finland (modified from Münster et al. 1997b). (**D**). A fluorimetric measure of peroxidase activity in an indigenous microbial community from polyhumic Lake Mekkojärvi with a modified technique from Palenik and Morel (1990a,b). Different concentrations of para-hydroxypropionic acid (pHPA) were used in combination with H_2O_2 to trace the presence and activity of an unspecific peroxidase in 100 µm pore size prefiltered lake water at room temperature. The plots show the relative increase in fluorescence at different pHPA and H_2O_2 concentrations after 95 min. incubation time, indicating the presence of microbial peroxidase activites in the water sample

face bound enzyme which oxidizes amino acids and organic amines into ammonium and organic acids, from which only ammonium is assimilated but not the remaining organic reaction products such as ketoacids in the case of amino acids. This assay is based on the production of H_2O_2 by algae (Stevens et al. 1973; Jones et al. 1987; Palenik and Morel 1988; Palenik et al. 1987) and the conversion of non-fluorescent p-hydroxypropionic acid (pHPA) to a fluorescent dimer of pHPA. The amino acid oxidase activity was monitored fluorimetrically in the presence of H_2O_2 as studied by Palenik and Morel (1990a, b) in algal cultures and marine waters, where nitrogen can be a growth limiting factor for phytoplankton (Palenik et al. 1987, 1988/89). This approach was later modified and improved by Pantoja et al. (1993) by synthesizing specific FMPs based on luciferin yellow (LY) derivatives of lysine, cadaverine and other amino compounds as potential marker/reporter substrates for the cell surface oxidation of amino compounds by phytoplankton (Pantoja et al. 1993; Pantoja and Lee 1994).

The authors considered these assay techniques as an alternative approach and very promising tool to trace the role of peroxidase activities in humic waters and their contribution to RDOM degradations (Fig. 9.5). They have used the principles of the former assay from Palenik and Morel (1990a, b) but with small modifications to trace peroxidase activitiy in humic waters with fluorimetric techniques. Although the fluorimetric method of this assay was expected to be much more sensitive compared with that of the LiP HPLC assay, the incubation time to receive enough activities was not much shorter compared with the LiP assay and lasted also 3–6 days. This may imply that oxidative enzymatic degradation of NOM polymers in these humic waters by oxidase/peroxidase activities was rather slow compared with hydrolytic enzymes such as phosphomonoesterase, glycosidase and aminopeptidase activities. Five different pHPA concentrations were amended with different H_2O_2 concentrations (1, 5, 10, 25 50, 100, 250 and 500 μM) to follow the production of the fluorescent pHPA-dimer due to peroxidase activity as shown in Fig. 9.4d. From this experiment we could find a clear fluorescence response with increasing H_2O_2 addtions and increasing pHPA cocentrations following a kind of kinetic plot. Obviously 10–100 μM H_2O_2 additions and 50–100 μM pHPA were sufficient to saturate the peroxidase activity in this experiment, which may again suggest that peroxidase activity is an existing biocatalyis in those humic waters. However, it was not yet clear whether

➔

Fig. 9.5. In between variation of acid phosphomonoesterase (acPME) in boreal freshwaters. (A). Seasonal variation and comparison of acPME in small Finnish forest-lakes: Mekkojärvi (MKJ), Horkkajärvi (HORKKJ), Alinen Mustajärvi (ALMU) and Syrjänalunen (SYRJAL) at 0-1 m depths during 1988. The lakes have decreasing humic substances contents as listed here in the legend text (modified from Münster 1994). (B). Seasonal variation of acPME in the polyhumic lake Mekkojärvi with maximum, minimum and average values at 0-1m depth with weekly sampling during 1994. (C). Spring bloom aspect of acPME in lake Mekkojärvi with daily sampling in 1988 (modfied from Münster 1997c). All lakes are located in the Evo-forest research area in southern Finland with no direct human impacts.

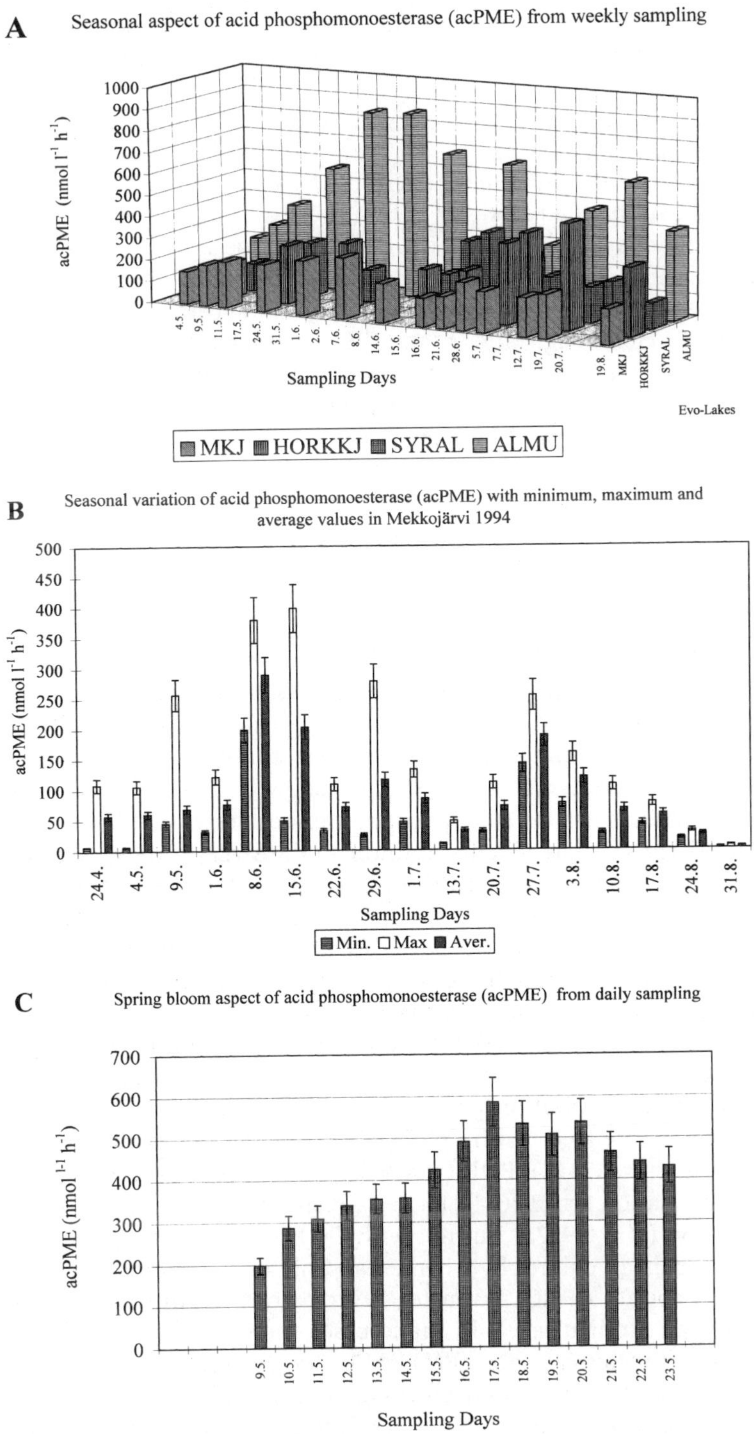
A
Seasonal aspect of acid phosphomonoesterase (acPME) from weekly sampling
acPME (nmol l-1 h-1)
1000
900
800
700
600
500
400
300
200
100
0
4.5. 9.5. 11.5. 17.5. 24.5. 31.5. 1.6. 2.6. 7.6. 8.6. 14.6. 15.6. 16.6. 21.6. 28.6. 5.7. 7.7. 12.7. 19.7. 20.7. 19.8.
MKJ HORKKJ SYRAL ALMU
Sampling Days
Evo-Lakes
MKJ HORKKJ SYRAL ALMU
B
Seasonal variation of acid phosphomonoesterase (acPME) with minimum, maximum and average values in Mekkojärvi 1994
acPME (nmol l-1 h-1)
500
450
400
350
300
250
200
150
100
50
0
24.4. 4.5. 9.5. 1.6. 8.6. 15.6. 22.6. 29.6. 1.7. 13.7. 20.7. 27.7. 3.8. 10.8. 17.8. 24.8. 31.8.
Sampling Days
Min. Max Aver.
C
Spring bloom aspect of acid phosphomonoesterase (acPME) from daily sampling
acPME (nmol l-1 h-1)
700
600
500
400
300
200
100
0
9.5. 10.5. 11.5. 12.5. 13.5. 14.5. 15.5. 16.5. 17.5. 18.5. 19.5. 20.5. 21.5. 22.5. 23.5.
Sampling Days

photometric or fluorimetric assays provide more sensitive protocols. The present data basis is probably still to small to make general conclusions.

9.3.1.2 Distribution and Variation of MEE Activities in Marine and Freshwater Environments

To evaluate the role of MEE in aquatic environments such as in marine and freshwaters it is recommended to study at first the seasonal and vertical distribution and variations (Billén 1991; Chróst 1991; Hoppe 1991, 1993; Wetzel 1991). In most of those studies it has been shown that MEE activities vary over large time scales owing to seasonal, vertical and catchment aspects (Wetzel 1981, 1992; Hollibaugh and Azam 1983; Sinsabaugh et al 1992a, b, Münster 1997c). Highest MEE were found in the trophogenic and the tropholytic water layers. However, also large particles and aggregates such as "marine snow" can be colonized by bacteria and can contribute with their MEE activities significantly to "marine snow" solubilizations, nutrient dissolutions and carbon processing (Karner and Herndl 1992; Smith et al. 1992; Azam et el 1994 and references therein). Similar studies are rather scarce in freshwaters, where most studies have been carried out with planktonic organisms and some in sediments.

9.3.1.2.1 Seasonal and Daily MEE Aspects in Humic Waters

In humic waters seasonal aspects of MEE activities have been studied by Münster et al (1992a, b) in small forest lakes located in southern Finland with different environmental impacts such as nutrient contents, acidity, forest managements and HS contents. Those studied showed that in meso- and polyhumic lakes MEE such as acid phosphomonoesterase (acPME), α,β-glycosidase (α,βGlycAse) and leucine-aminopeptidase (LeuAMPA) showed significant seasonal variations and depth profiles (Münster 1992a, b, 1997c). Surprisingly highest AcPME was observed in spring, soon after ice break, although dissolved inorganic phosphorus (DIP) and total phosphorus (P_{tot}) were probably not limiting growth factors for phyto- and bacterioplankton (Fig. 9.6a, b). These distribution patterns have been confirmed in recent studies in the same lake with weekly sampling intervals (Münster 1997c). Again, highest acPME was measured about 1–2 weeks after ice melted from the lake surfaces. The distribution pattern of acPME over the seasonal and vertical variation correlated with phosphorus fractions such as P_{tot} ($r^2 = 0.7426$), Porg ($r^2 = 0.7412$), POP ($r^2 = 0.6552$), microbial activities such as community respiration (r^2=0.6621), bacteria respiration (r^2=0.8398), netto primary production ($r^2 = 0.9572$) but not with DOP, DIP and microbial biomass parameter such as chlorophyll contents (Chl) and bacteria numbers (acriflavine direct counts-AFDC) (Münster 1997c). However, this correlation was not valid for all studied lakes. In many cases there was no correlation between phosphorus resources and acPME (Münster 1997c).

During short term studies on nutrient dynamics on daily sampling basis in a mesohumic lake it was noted that acPME followed the development of a

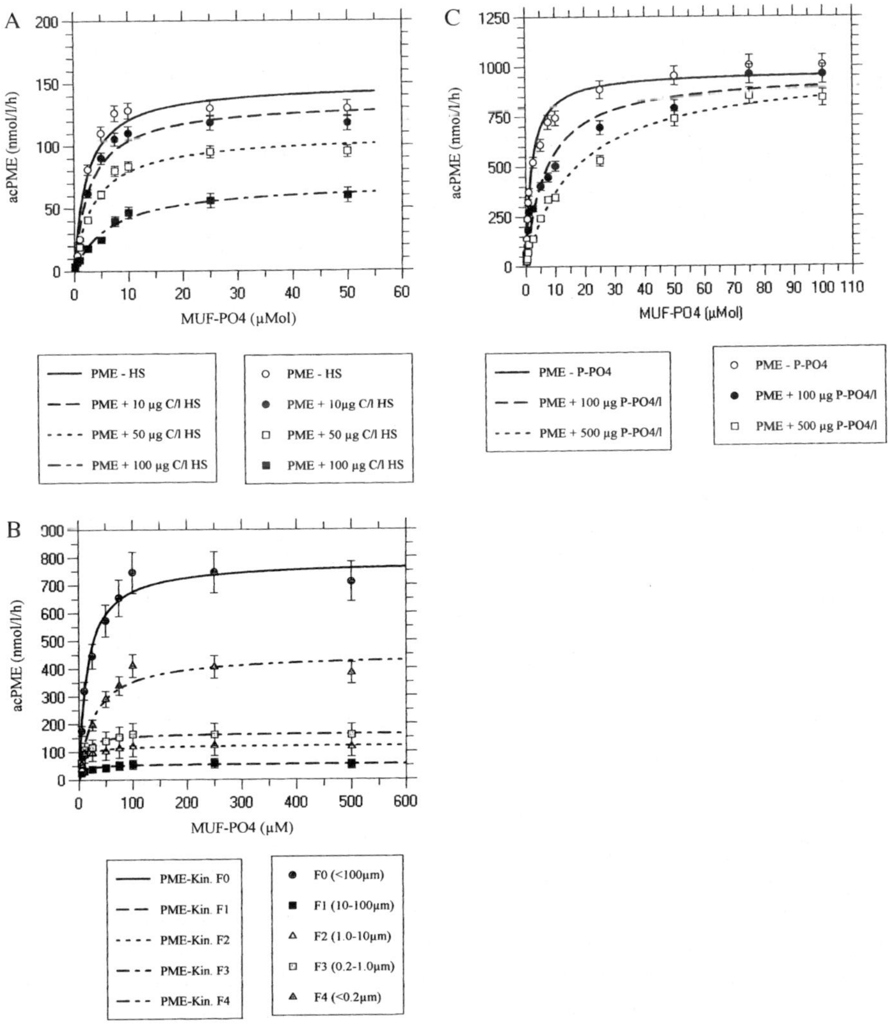

Fig. 9.6. Enzyme kinetics of acid phosphomonoesterase (acPME) in boreal freshwaters. (A). Inhibition effects of humic substance (HS) additions on acPME kinetics in humic substances and acidification related experimental lake HUMEX-Lake Skjervatjern, Norway (Gjessing 1992). Purified HS from the same water sample was added as specified in the figure to 100 µm mesh size prefiltered original water sample and acPME kinetic was measured fluorimetrically according to Münster et al. (1989). (B). Inhibition effects of elevated phosphate concentrations on acPME. Lower, ambient phosphate additions are not plotted in the figure (cf. Table 9.4b). (C). Contrasting acPME kinetics in different particle size class associated and dissolved free acPME in HUMEX-lake Skjervatjern, Norway. Enzyme kinetic prameter such as V_{max}, K_m (cf Table 9.4a, b, c) were measured with Grafit software package from Erithacus (UK) on an IBM compatible personal computer (all graphs were modified from Münster 1997c)

spring bloom (Fig. 9.6c) with a mximum of acPME in may just after ice break, which confirmed former data. This sampling interval was intensified during diurnal studies on vertical migrations of planktonic organisms in a mesohumic forest lake in June, July and August 1996, where we found lowest acPME and LeuAMPE activities between 03:00–05:00 hours in the morning and not at the darkes period at 24:00–02:00 hours (Münster 1997c). This may suggest that MEE followed the light-dark cycles and the microbial migration and/or activity patterns in this lake.

Among the most often studied MEE in those humic waters, the following sequence and ranking can be stated according to the activities of the studied MEEs: acPME > esterase > lipase > exo-peptidase > phenol-/peroxidase > β-xylosidase > endo-peptidase > a,β-glucosidase > N-acety-β-glucosaminodase > acid phosphodiesterase > 5'-nucleotidase. This ranking was constant over the entire seasonal scale but varied between studied locations and type of lakes (Münster 1994, 1997b, c).

9.3.1.2.2 HS Effects on MEE: Inhibition, Complexation, Immobilization, Translocations and Reactivations

HS have been shown to have strong impacts on aquatic food web structures (Hessen 1985, 1992) and then modifiy microbial activities in freshwaters (Stewart & Wetzel 1982; McKnight et al. 1991). HS can also provide higher bacteria-carrying capacities compared with clear waters (Tranvik 1988, 1989, 1990, 1992). However, it is still not clear what the impact of HS is on MEE. Although similar studies have been carried out in soil ecosystems, little has been done in marine and freshwater environments. According to Wetzel (1991, 1992), HS can have significant impacts on the activities of MEEs in the littoral zone of freshwaters and their watersheds and interfaces. This aspect is especially important in context with global climate change effects, where especially the land-water interface may receive the highest impacts from temperature and light changes and can modifiy its microbial activities (Benner et al. 1989; Schindler et al. 1990, 1992, 1996) due to increased leaching effects from catchments and drainage areas (Rasmussen et al. 1989). Therefore, Wetzel (1991, 1992) raised some important questions about such events and discussed the effects of HS on MEE activities. Especially dissolved free MEE may have the strongest interactions with HS, due to their high solubility in the water and their susceptibility to inhibition, complexation, immobilization, precipitation and translocation by HS (Wetzel 1992). Indeed, dissolved free phosphatase have been detected in significant amounts in marine and freshwaters (Münster 1997c). About 30 to 90% of total phosphatese can exist as dissolved free enzymes (Overbeck & Babenzien 1963; Wetzel 1981, 1991; Stewart and Wetzel 1982; Chróst and Overbeck 1987; Münster 1997c). In small boreal forest lakes with high HS contents Münster (1991, 1992a, b) and Münster et al. (1992a, b) reported that 70–90% of acPME was non-particle bound, i.e. it existed as dissolved free acPME.

Table 9.4. Enzyme kinetics from acid phosphomonoesterase in the acidified basin-A from HUMEX-lake Skjervatjern, Norway (modified from Münster 1997c).

(A) Effects of Humic Substances (HS) Additions on Acid Phosphomonoesterase (acPME) kinetics in 1992 in the acidified basin-A

Range of Humic Substance Additions	PME Codes	V_{max} (MUF-PO_4) (nmol l^{-1} h^{-1})	K_m (MUF-PO_4) (μM)
0 μg C_{HS} l^{-1}	acPME - HS	149.1	2.5
10 μg C_{HS} l^{-1}	acPME + 10 μg C_{HS} l^{-1}	134.9	3.0
50 μg C_{HS} l^{-1}	acPME + 50 μg C_{HS} l^{-1}	108.0	3.7
100 μg C_{HS} l^{-1}	acPME + 100 μg C_{HS} l^{-1}	69.8	65.0

Abbreviations:

C_{HS}: organic carbon content of added purified humic substances (HS) from the same lake and water sample (Münster 1997c)

V_{max}: maximum reaction velocity at substrate saturation level from kinetic experiments

K_m: half saturation constant or Michaelis-Menten constant from kinetic experiments

acPME: acid phosphomonoesterase

MUF-PO_4: 4-methyl-umbelliferyl-phosphate

(B) Effects of ambient and P-PO_4 additions on acPME kinetics from Lake Skjervatjern 1992 in basin-A

Range of P-PO_4 Additions	Plotting Codes	V_{max} (MUF-PO_4) (nmol/l/h)	K_m (MUF-PO_4) (μM)
0 μg P-PO_4 l^{-1}	PME + 0 μg P-PO_4 l^{-1}	1044.7	2.2
10 μg P-PO_4 l^{-1}	PME + 10 μg P-PO_4 l^{-1}	1043.9	2.1
25 μg P-PO_4 l^{-1}	PME + 25 μg P-PO_4 l^{-1}	1045.1	2.3
50 μg P-PO_4 l^{-1}	PME + 50 μg P-PO_4 l^{-1}	1038.5	2.0
75 μg P-PO_4 l^{-1}	PME + 75 μg P-PO_4 l^{-1}	1029.8	2.8
100 μg P-PO_4 l^{-1}	PME + 100 μg P-PO_4 l^{-1}	977.7	6.9
500 μg P-PO_4 l^{-1}	PME + 500 μg P-PO_4 l^{-1}	878.8	26.3

(C) acPME kinetics from lake Skjervatjern during 1992 from basin-A in different size class fractions

Size Class Fractions	Size Class Codes	V_{max} (MUF-PO_4) (nmol l^{-1} h^{-1})	K_m (MUF-PO_4) (μM)
< 100 μm (Total ComunityPME)	F0	783.7 ± 27.9	16.0 ± 2.6
100–10 μm (meso-Zooplankton)	F1	54.9 ± 2.2	8.1 ± 1.8
10-1.0 μm (Phytoplankton)	F2	122.5 ± 5.1	6.8 ± 1.7
1.0-0.2 μm (Bacteriaplankton)	F3	165.3 ± 5.1	7.7 ± 1.3
< 0.2 μm (Dissolved Free PME)	F4	448.9 ± 28.3	27.6 ± 6.7

The view that HS may interact with proteins and notably with enzymes has its roots in plant physiology, where polyphenolic compounds inhibited enzyme activities in cell protoplasma after cell disintegrations (Loomis & Bataillle 1966; Andersen and Sowers 1968) and from soil microbiology (Ladd & Butler 1975; Burns 1978, 1982, 1983; Skujins 1967). The HS-enzyme interactions are mainly based on HS-metal-protein interactions, dipole-dipole and van der Waals forces via hydrogen bridge-bonding between the NH_2-and/or carboxylic groups at the protein/enzyme surface and the phenolic OH groups of the polyphenols originally from plant vacuoles (Andersen and Sowers 1968; Haslam 1974; McManus et al. 1985; Münster 1985; Steinberg and Münster 1985; Lähdesmäki and Piispanen 1988, 1992; Spencer et al. 1988).

A similar mechanism has been discussed by Wetzel (1991) for HS-enzyme interactions in freshwaters. It is assumed that dissolved free enzymes can be covered by polyphenolic compounds from HS at the protein surface and notably modify their catalytic potential and properties, but they also may act as a shield against inhibitors and proteolysis. This assumption has been verified in part by Wetzel and coworkers experimentally and they showed that HS have significant impacts on the kinetic properties of alkaline phosphatase (Stewart and Wetzel 1982; Wetzel 1991). A similar effect of HS has been observed by Münster (1994, 1997c) during studies on acPME from humic forest lakes where he showed that acPME had different kinetics when HS was added to the lake water as shown in Tab. 9.4, Fig. 9.6a. The opposite effect was observed, after the addition of polyvinyl pyrrolidone (PVP), a significant immobilizer of HS (Andersen and Sowers 1968), to lake water samples from polyhumic small forest lakes, which induced a higher acPME compared to the control without PVP additons (Münster unpublished results). Both experiments supported the assumption that HS may have modified the MEE activities by changing their catalytic properties, e.g. their catalytic centre. Similar observations have been experimentally verified with soil enzymes and the effects of HS on their catalytic properties (reviewed by Ladd and Butler 1975; Burns 1978, 1982, 1983; Schinner and Sonnleitner 1996). This implies that HS must be considered as important reactants with MEE in many freshwaters of the NBA and can modify the catalytic properties of the MEE or shield them against inactivation and biodegradation.

9.3.1.2.3 Kinetic Aspects of MEE Studies in Natural Habitats

In addition to studies on temporal and vertical variations and distribution of MEE in aquatic habitats, more detailed information about the role, properties and function of MEE can be gained from studies on enzyme kinetics, when varying amounts of substrate (S) are added under constant reaction conditions and the catalytic reaction velocity (v) is measured. Theoretically and according to textbooks in biochemistry and enzymology (Dixon and Web 1979; Bergemeyer 1990; Zubay 1993), such kinetic approaches to characterize enzyme catalysis should be carried out only with clearly defined reaction conditions and isolated and purified enzymes. Ideally the enzyme catalysis will follow the classical Michaelis-Menten (MM) kinetics, which can be mathematically formulated in the well-known formula:

$$V = \frac{V_{max} \times S}{K_m \times S}$$

From this equation two important parameters can be calculated: the maximal velocity (V_{max}) of enzyme catalysis at substrate saturation level and the Michaelis-Menten constant (K_m) at half-substrate saturation level. Both parameters are very useful to characterize an enzyme in its catalytic properties and its interactions with the applied substrate and its inhibitors.

Ecologically orientated studies on MEE should measure enzyme activities under *in situ* conditions, i.e. without any manipulations and disturbances. Therefore, MEE kinetics in natural habitats under *in situ* conditions are the main goals in such kinetic approaches. However, there are only very examples of how such approaches have been successful, because the classical MM kinetic was not always working under such *in situ* conditions and mathematical corrections and modifications in the above formula were necessary to calculate V_{max} and K_m more reliable (Billén 1991), similar to the developement of substrate uptake by bacteria in natural habitats (Overbeck 1975; Button 1985, 1986 and cited refrences therein). Although enzyme kinetics can be calculated and evaluated at the beginning of such studies with rather simple graphic tools and a pocket calculator, only powerful computer software can provide a reliable basis for enzyme data calculations; examples now available include Enzfitter from Biosoft Elsevier, Grafit from Erithacus, or within statistical programs such as Systat or the computing program Mathcad from Mathsoft. These software packages provide convenient tools to generate enzyme kinetic transformations such as those from Lineweaver-Burke, Eadie-Hofstee, Hanes or even our own specially designed transformation plots. The main outcome of such kinetic approaches is, however, to trace what kind of reaction type followed the enzyme catalysis and what the main factors modifying the catalysis were.

This aspect is especially important for MEE activities in humic waters, where HS are expected to act as modifiers in MEE catalysis (Wetzel 1991, Münster 1997b, c). This was approved by using the Grafit software package (Erithacus UK) to determine the effects of HS on V_{max} and K_m of acPME from a mesohumic lake as shown in Fig. 9.6a. and Table 9.4a. From these plots it became evident that HS do change the kinetic parameters such as K_m and V_{max} underlining the assumption that HS act as modifier in biocatalysis in humic waters. Another example of valuable information obtained from kinetic data is shown in Fig. 9.6b and Table 9.4b, where acPME activities kinetic approaches provided valuable kinetic data about inhibition effects from ambient DIP concentrations on phosphatase activities. Acid phosphomonoesterase was i.e. not inhibited by ambient DIP concentrations in small Finnish forest lakes in the range of 1–50 µg $P\text{-}PO_4$ l^{-1}, which is the natural DIP level in most boreal surface waters (Münster 1997c), but above 50–75 µM

P-PO_4 acPMW was significantly inhibited. This is an important aspect related to *in situ* enzyme activity regulations such as induction/expressions, repression, derepression and environmental impacts. Another important aspect related to MEE activities is connected to questions such as: do have particle associated and dissolved free MEE different enzyme-substrate interactions, kinetics, or are they rather identical or similar? As one example in our studies on acPME we have measured kinetics from different particle size associated acPME and dissolve acPME in a mesohumic lake water as shown in Fig. 9.6c and Table 9.4c. These data show that particle associated acPME had not onyl lower V_{max} values but more intrestingly lower K_m values campared with that from dissolved free acPME (Table 9.4c). This suggest that the particle associated acPME had stronger binding and higer affinity to the model-substrate MUF-PO_4 compared with the dissolved free acPME, an important information to explain and understand the ecophysiological role of dissolved free MEE in aquatic environments.

9.3.1.2.4 Metabolic Regulations of MEE: Induction, Repression, Derepression and Environmental Impacts

Another central question in experiments with MEE in aquatic environments is how MEE are regulated in their activities at the cellular, the organism and the environmental level. In temperate lakes and in marine environments with few HS impacts the three best known and studied MEE are the phosphatase (EC 3.1.3.1), β-glucosidase-(EC 3.2.1.21) and leucine aminopeptidase (EC 3.4.1.1) systems as reviewed by Billén (1983, 1991), Chróst (1990, 1991, 1994) and Hoppe (1991). According to these reports, MEE are regulated at the cellular level on the basis of induction, expression, repression and de-repression mechanisms (Chróst 1990, 1991, 1994; Azam et al. 1994). This view is based on many experimental data from laboratory and on some *in situ* MEE measurements and implies that microbial cells such as bacteria and algae sensitize very effectively the bioavailability of required key nutrients such as phosphorus, nitrogen and carbon and regulate their MEE activities directly across the cell surface at the genetic level (cf. Chróst 1990, 1991; Azam et al. 1994 and cited references therein).

As the scientific knowledge about MEE regulations under *in situ* conditions in aquatic environments is based on rather few *in situ* experimental data sets, the view of the regulation and function of MEEs in nature with nutrient pure aquatic habitats is largely deduced from the present knowledge about enzyme expressions and regulations gained from bacterial and algal cultures, such as the regulation of phosphorus metabolism in model organisms, i.e. when *E. coli* is grown under DIP limitations below 0.16μM, its alkaline phosphatase is induced and the Pho-regulon expressed (Torriani-Gorini et al. 1994). According to present knowledge of the structure and function of the Pho-regulon of *E. coli*, a set of proteins/enzymes and co-factors are activated which sensitize and regulate the bioavailability of external P resources and its transport and flux accross the cell wall/membranes (Torriani-Gorini et al. 1994). Phosphoester hydrolysing enzymes such as PME play key roles in the external, extracellular regulation of P bioavailabilities. A similar mechanism and regulation can be assumed for aquatic microbes.

However, in aquatic environments, both bacteria and algae are competitors for the same key nutrient resources and they have to tune their enzymatic-metabolic machinery to gain access to those available but limited nutrient resources. This has been recently shown with an elegant ^{32}P-radio tracer assay by Hernandez et al. (1996), in which they observed that bacteria and algae may take different dissolved organic phosphorus (DOP) pools as key nutrients due to their different cellular C:N:Pratios and affinities to the available DOP compounds. According to this view, which was at first developed and confirmed by Ammerman and Azam (1985), bacteria may use P resources from different external P pools such as from dissolved DNA, by synthesizing cell-surface bound 5'-DNAse with high affinity to those DNA-phosphoester bonds (Ammerman 1991), whereas algae exploit other resources. All these results confirm that the key regulations of MEE activities are located mainly at the cell surfaces, where MEE are under strict control of the metabolic/genetic machinery (Chróst 1990, 1991, 1994; Azam et al. 1994).

However, in humic environments there is experimental evidence that phosphoester cleaving enzymes such as PME can exist in high quantity in non-cell-surface or non-particle-bound status (Wetzel 1981, 1991, 1992; Münster 1992a, b, 1994, Münster 1997b, c). This PME fraction is not under such strict genetic control as cell-surface-bound MEE, and the ecophysiological role of this dissolved PME fraction can only be speculated. Due to the predominance of HS in land-water interfaces and littoral zones, Wetzel (1990, 1991, 1992) discussed the role of immobilized and dissolved MEE as important regulators in cleaving reactions in freshwaters with long storage capacity in biocatalysis for phosphorus, nitrogen and carbon biopolymers. Although not yet completely verified experimentally, this view is congruent with the role of free enzymes in soil ecosystems (Skujins 1967; Ladd & Butler 1975; Burns 1978, 1982, 1983; Lähdesmäki and Piispanen 1988, 1992; Schinner and Sonnleitner 1996; Morra 1997) where so-called free abiontic enzymes, although not under the strict genetic/metabolic control of their producers, displayed higher resistance against inactivation by environmental factors such as pH, heavy metals, proteolysis, temperature, complexation and light effects. Similar properties have been observed for PME in humic waters with 50–60% activities after storage at room temperature in the darkness (Münster 1997c). It can be assumed that dissolved free and immobilized MEE in humic waters may react faster to short substrate pulses and variations compared with cell-surface-bound MEE, and they may have a key function in the initiation of biocatalysis and can act as inducers for the cell-surface-bound MEE by providing signal/inducer compounds for the biosynthesis of MEE in the cells as it is well known for the role of cellobiose in cellulose biocatalysis (Wood and McCrae 1979; Ljungdahl 1989). However, there is still a rather superficial knowledge of the ecophysiological role of HS (Visser 1985) and of free MEE in humic environments and their biocatalytic role and impacts in NOM biodegradations.

9.3.2 Main Contributors of MEE in Humic Waters

In many publications on MEEs in aquatic environmets it is stated that bacteria and algae are the main contributors to MEE activities (reviewed by Chróst 1990, 1991; Billén, 1991; Hoppe 1991, 1993). However, in most of these assumptions, conclusions were made on the basis of pre-filtrations and fractionations of the microbial communities and their contribution to the total, community MEE activities. Additional support for this view was gained from statistical and correlation data analysis, where MEE activities correlated rather well with bacteria production (Chróst 1989; Chróst et al. 1989), algal biomass, cell constituents and other activities (cf. Billén 1991; Chróst 1991, 1994; Hoppe 1991). However, aquatic food webs are highly complex and many contributors and other pathways by which the MEE pools could be filled up also from higher organisms may exist. In addition, the filtering technique is rather crude to address the question of what are the main contributors to MEE activities. It may be practicable in this case that most MEEs are particle-associated and only a small fraction (1–20%) may exist as dissolved free MEE fraction as has been regularly found in most marine environments (reviewed by Hoppe 1991).

However, a problem emerges when a larger fraction of the MEE is in the dissolved free MEE fraction, as has often been found for PME in some freshwaters and humic lakes (Wetzel 1981, 1991; 1992, Münster 1992a, b, 1994, 1997c; Münster et al. 1992a+b). As one often found example the acPME distribution in different particle size fractions is shown in Fig. 9.7. From this picture it becomes evident that 50–60% of community acPME (passing 100 μm mesh size plancton net) is existing as dissolved free acPME, passing 0.2 μm pore size Nuclepore membrane filter, when we assume that all living organisms and particles are retained on such filter. However, in some cases Münster (1992a, b, 1994) also reported a larger PME fraction in the 1–2 μm pore size class, which was dominated by flagellates. This observation is in agreement with data from Nagata and Kirchman (1992), who found significant amounts of released macromolecules which contained acid phosphatase from flagellates in marine waters. According to Sherr (1988) and Sherr and Sherr (1994), phagotrophs may be one of the most important contributors to nutrient regenerations, such as phosphorus and nitrogen by grazing on NOM or bacteria, a major sink of phosphorus in many freshwaters, and after digestion they release free phosphorus compounds and phosphatase into the waters (Jürgens and Güde 1990). In addition, Vrba et al. (1992) and Nedoma et al. (1994) identified flagellates as important contributors to cell wall hydrolyzing enzymes such as N-acetyl-β-glucosamidase. A similar effect can be expected by zooplankton feeding activities (Bochdansky et al. 1995). Phytoplankton in many humic freshwaters are dominated by flagellates (Ilmavirta

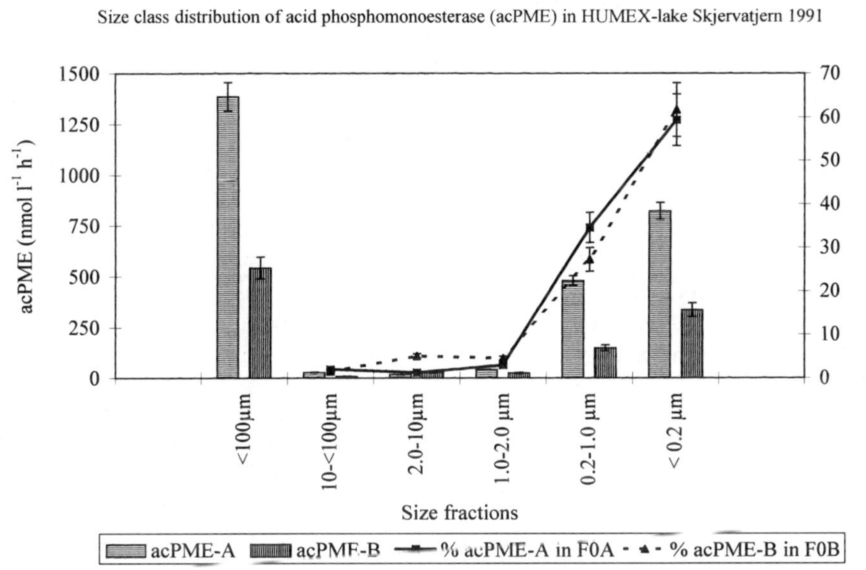

Fig. 9.7. Size class distribution of acPME in humic substances related experimental lake, HUMEX-Lake Skjervatjern, Norway (Gjessing 1992), from the acidified basin-A and the control basin-B at 0-1 m depth during summer 1991. Columns show acPME activity in the < 100 µm size fraction (total community activity passing through a 100 µm plankton net) and the acPME activities in five sub-fractions (F1-F5) according to their retention on the corresponding Nuclepore filters in nmol l^{-1} h^{-1}. Lines show the percentual contribution of each fraction in relation to the total (<100 µm, F0) community activity (modified from Münster 1992b)

1988) and they may be potential resources not only for phosphatase but also for protease and glycosidase.

In addition to micrograzers such as phagotrophs, other planktivorous organisms such as rotifers, ciliates and even larger zooplankton can contribute to dissolved free MEE pool sizes due to their grazing activities and digestion of bacterioplankton and algae. However, from such faeces and exudates it is rather speculative to address the origin of the released and remaining MEE in the water. A very specific identification method has to be made based on, e.g., immunoreactions, on protein amino acid sequences analysis and/or on the genetic basis in order to identify reliable the origins of MEE. Phyto- and bacteriophages can contribute to microplankton mortality in aquatic environments and may thus indirectly also influence the MEE composition and activities due to cell leakages (cf. Azam et al. 1994). With respect to biodegradation of RDOM sources in humic waters, especially aquatic fungi may be another powerful contributor to the MEE activities (Bärlocher 1981). However, this group of microheterotrophs has been seldom studied in pelagic humic water bodies, a missing link probably in nutrient regeneration and carbon fluxes studies.

9.3.3 MEE in Aquatic Food Webs

The picture of aquatic food web structure and function has dramatically changed during the last 10–20 years of aquatic research, mainly due to the early view of Pomeroy (1974) of the role of microbial processes in marine waters and the establishment, formulation and refinement of the concept of the microbial loop by Azam et al. (1983, 1994). Within this concept, MEE have been later recognized as important regulators of nutrient and carbon fluxes (Billén 1983, 1991; Azam et al. 1994; Hoppe 1991). This rather clear view of marine ecosystems may be also valid for and transferable to freshwater ecosystems (Chróst 1990, 1991, 1994). However, freshwaters are probably also more heterogeneous and variable in their structure and function compared with the large water bodies in oceans. Especially freshwaters from the NBA and partially also in tropical rain forest ecosystems, external factors can modify this general view of the carbon flow and food web interactions.

The predominance of HS, e.g. in many freshwaters of the NBA, has been shown to modify the current view of carbon flow and food web structures (cf other chapters in this book). The generally observed and accepted close metabolic coupling between autotrophic-heterotrophic microorganisms in marine and many temperate freshwater environments, seem to be less effective and quantitatively of minor importance in humic waters (Salonen 1981, Salonen et al. 1983, 1992; Hessen 1985, 1992; Tranvik 1988, 1989, 1992 and cited references therein). The greater amount of RDOM sources from allochthonous DOM and POM with highly complex chemical structures (Christman and Gjessing 1883; De Haan 1983; Harvey et al. 1983; Hayes et al. 1989) will certainly encourage more studies of different and even new biodegradation pathways and may involve other structured MEE pools with different efficiencies and mechanisms compared with marine waters and temperate freshwaters (Münster 1997b). Due to the different nutrient level and pool sizes in humic waters (Salonen et al. 1992; Münster 1991, 1997c), a higher heterotrophic activity via grazer impacts can modify MEE composition and activities, and other efficiencies may be expected. However, at present few fundamental studies and results are available to evaluate the principal differences between the role of MEE in clear water and humic waters (Münster 1991, 1997b, c).

9.3.4 MEE in Biogeochemical Cycles

Essentially four key elements can be defined as of primary importance in global biogeochemical aspects: carbon, nitrogen, sulfur and phosphorus. Among those four elements, only phosphorus moves mainly via waterways globally and locally. The other three elements are highly mobile, exist also as gases and differ in this aspect from phosphorus. A large amount of those four elements is located in organic matter (OM) with primarily polymeric structures (Harvey et al. 1983; Romankevich 1984; Steinberg and Münster

1985, Thurman 1985; Hedges 1992; Lee and Wakeham 1992). Microheterotrophs such as flagellates, bacteria and fungi are the main agents in mobilizing those four key elements. Due to the predominance of macromolecular structures in OM on earth, extracellular cleavage is compulsory to mobilize these key nutrients. In addition to photolytic cleavage of biopolymers in aquatic environments, flagellates, bacteria and fungi and their extracellular biocatalyst are probably the driving forces in the transformation and degradation of OM globally and locally (Azam et al. 1994). Due to their large biomass and high bioactivity, especially bacteria in marine waters are quantitatively certainly of greater importance compared with those in freshwaters. However, freshwaters of the NBA and most large rivers and freshwater drainage areas are also important sources of and sinks for carbon and nutrients. Therefore, there is a demand to improve the knowledge of the biological mechanisms participating in these cycles. Studies on the distribution, variation and regulation of MEE in marine waters and freshwaters can contribute to improving the basic knowledge of the mobility and pathways by which these key nutrients are mobilized and transfomed in aquatic habitats.

According to the present knowledge, phosphatase, protease and glycosidase enzyme systems are currently the best-known and studied extracellular enzyme systems participating in and contributing to aquatic nutrient cycling and food web processes (Billén 1984, 1991, Chróst 1990, 1991, 1994; Hoppe 1991; Wetzel 1991; Azam et al. 1994, Sinsabaugh et al. 1997). Organic sulphur hydrolysing enzymes such as arylsulphatase are less studied in open waters and more applied in sediments and soil ecosystems (Burns 1983; Meyer-Reil 1986 1991; Schinner and Sonnleitner 1996). Based on these four enzyme systems, probably only 10–20% of the DOC processing can at present be explained and quantified. Although photolytic degradation and cleavage of DOC can be significant (Kramer 1979; Geller 1985a; Mopper et al. 1991; De Haan 1993; Lindell et al. 1995; Granéli et al. 1996; Wetzel et al. 1995), these mechanisms are rather limited to the upper surface layers (Salonen and Vähätalo 1994). In deep and large water bodies such as in oceans and large and deep lakes, the main DOC pool is not penetrated by light and therefore less exposed to photochemical cleavage reactions. In shallow water bodies as frequently found in freshwaters of the NBA this mechanisms can be higher but it is only restricted to the upper 10–50 cm depths (Gjessing and Gjerdahl 1970; De Haan 1983; Salonen and Vähätalo 1994; Wetzel et al. 1995). The largest polymeric organic matter pool requires, however, biocatalytic reaction mechanisms such as MEE.

Rather little is known about these enzyme systems that are participating in the cleavage and processing of RDOM sources but with slow turnover times. According to Sinsabaugh et al. (1992a, b), oxidative enzymes such as phenoloxidase and phenol peroxidase may be strong contributors to such RDOM processing. Sinsabaugh and coworkers have studied oxidative enzyme activities in lenthic and lotic freshwater environments and explained OM mass losses in a riparian and reverine watershed area and in boreal river ecosystem based on the measurements of the oxidative enzyme and lignocellulosic enzyme systems (Sinsabaugh et al. 1994a, b, 1997). They correlated the oxidative and cellulolytic enzyme activities with microbial biomass and activities which was highly dependent upon seasonal and local variables. They also correlated enzyme activities with oxygen consumption and carbohydrate contents from particulate organic matter (POM) fractions in a boreal river (Sinsabaugh and Linkins 1990).

In another experiment, Sinsabaugh et al. (1992a, b) studied the enzyme contributions to litter decomposition in riparian and lotic sites in forested watersheds, and, based on cumulative principal component analysis, they showed that mass losses of OM were 83% due to enzymatic lignocellulose degradation in the watershed area (Sinsabaugh et al. 1997). Using similar oxidative-hydrolytic enzyme processing of RDOM approach combined with amino acid and aromatic carbon utilization in a boreal headwater lake in southern Finland, Münster et al. (1997b) calculated about 40–50% of microbial carbon processing and utilization in a small boreal polyhumic lake was supported by oxidative-hydrolytic enzyme activities, which was much higher than proteolytic and glycolytic activities calculations alone. This may indicate that in humic waters a significant carbon fraction may be also processed via oxidative enzyme processes and may enhance microbial recalcitrant aliphatic and aromatic carbon sources utilizations. However, the variability of these carbon flux pathways can be high and the methods still suffered from long incubation times, and the results may be therefore partially biased by the selection of the methods and assays. However, the incorporation of oxidative enzymatic processing of RDOM pools and the utilization of released aliphatic and aromatic carbon sources by microhetrotrophs seem to be significant (Sinsabaugh et al. 1992a, b, 1997, Münster et al. 1997b) and probably represents a further step towards quantifying more carbon flow processes in boreal freshwaters. Sinsabaugh and coworkers have demonstrated how it could work developing quantitative models including MEE activities for riparian lentic and lotic communities (Sinsabaugh et al. 1994a, b, 1995, 1997). Probably, at the moment this represents the most advanced status in the understanding of MEE contributions to biogeochemical processes and element cycling in lentic and lotic freshwater ecosystems, which could be a model approach for pelagic communities.

9.4 Problems, Open Questions and Future Aspects

Due to the data from the literature and the authors own experiences with MEE in humic waters, we will now raise some questions which we believe are essential to understanding and evaluating better the role of MEE in humic waters. There is stronger evidence to support the common view that microbial activities, nutrient fluxes and food web interactions are differently structured in humic waters compared with clear and eutrophic water bodies (cf. other chapter of this book). MEE are acknowledged as playing a central role in this network (Fig. 9.1–7). However, to evaluate, quantify and model their role at the cellular, community and ecosystem level we need a reliable scientific basis on which the data are generated. Therefore, the measurements of MEE in aquatic habitats need a sound experimental protocol, which can be used in different environments and give results comparable for different communties and habitats. This goal could be achieved by an international working/expert group to evaluate and standardize the selection of suitable substrates, incubation conditions, kinetic approaches and calculations and modelling. Such protocol developments should be strongly related to other important protocols and parameters, such as autotrophic and heterotrophic (e.g. algae, bacteria, fungi, micrograzer) production, respiration, mortality and growth in representative habitats and ecosystems. MEE-HS interactions seem to be more a fact than an exception or vision in most humic waters, as already generally acknowledged in soil microbiology. But we know rather little about their mechanisms, intensity, variability and impacts on MEE activities, kinetics and efficiency in humic waters.

Two main questions can be raised related to these topics: (1). How do aquatic HS affect microbial cell surface and membrane structures and their associated catalytic enzyme reactions, and what are the differences to that of dissolved free catalytic enzyme reactions? (2). Are there species- or community-specific aspects in these MEE-HS interactions, e.g. are certain types of enzymes more sensitive or do they display higher susceptibility to MEE-HS interactions compared with others, and is this habitat- or community-dependent or even MEE-specific, e.g. coded in the amino acid sequences, which includes even a genetic and species aspect.

It is a fact that in many humic waters a certain amount of MEE exist in non-particle-bound status and they can keep their catalytic activities over longer time scales, days or even weeks. We know, however, rather little about the nature of these biocatalysts and their role in aquatic food web processes and biogeochemical cycles. Therefore, we must ask: What is the physical-chemical basis for this astonishingly long viability or survival, and what is the biochemical basis for their catalytic activities? With modern instrumentation and sensitive enzyme substrates it is possible to measure the most common 10–12 different MEE within short time intervals and the MEE can be then ranked according to their activities. However, we do not know very much about synergistic aspects in this MEE community and could ask: Are these MEE ranking patterns constant or highly variable over time and habitats, and how are they dependent on the microbial community structure? An answer to such a question may imply or provide some view of how much of the MEE structure has a counterpart in the NOM pool compositions, or how much the MEE community activity reflects or mirrors the NOM structure and dynamics and the indigenous microbial community.

In most of the reviewed papers the MEE activities were related to bacteria and algae and a minor part to micro-/macrograzers. Most of the conclusions were based on filtration techniques, i.e. sizing the community with membrane filters, or in only a few cases on centrifugation techniques. This approach may work in clear water bodies with indigenous, easily sizeable communities. However, in most humic waters this filtration technique is unsatisfactory and may even produce artefacts because most of the phytoplankton consists of flagellates with auto-/mixo- and heterotrophic nutrition; due to this complex community with different size ranges and high sensitivity to any kind of mechanical manipulations, it is very difficult to attain reliable and representative size fractions, which can be used to answer the question: What are the main contributors from the microbial community to MEE activities and compositions? The interdependence of especially heterotrophic flagellates feeding as well on bacteria as on NOM may overlay and efface a clear and representative MEE pattern in time and space. To address the fundamental question about the origin and contributors to MEE activities in humic waters, we need more specific, reproducible and reliable techniques and approaches, as, e.g., provided now with molecular biology and genetic techniques.

With respect to increasing threat from global climate change and environmental aspects, there are four elements which should be the main targets in MEE studies: P, N, C and S. For all four elements enzyme systems exist which are participating in their mobilization, cycling and processing. However, to evaluate their individual role in these processes we need reliable, and standardized protocols which produce data that are repeatable in and comparable between the same habitats but also between different habitats from marine waters and freshwaters. Only this will provide the basis for a reliable interpretation of MEE activities on local and global scales and allow the development of MEE submodels and incorporate them into global carbon and nutrient flux scenarios.

9.5 Summary

The trophic exploitation of non-living organic matter (NOM) and its incorporated key nutrients in aquatic environments requires in many cases the presence and participation of extracellular biocatalysis. Microbial extracellular enzymes (MEE) are considered as the most effective biomolecules to catalyse biotransformations which make key nutrients such as phosphorus, nitrogen, sulphur and carbon in aquatic environments more easily available for rapid uptake by microbes. There is also increasing evidence that MEE are essential constituents for solubilization of particulate organic matter (POM) and for converting recalcitrant dissolved organic matter (RDOM) into labile dissolved organic matter (LDOM). Due to their central role in biotransformations, MEE have been recently also incorporated into carbon and nutrient flux models.

However, there is still a rather high variability in the applied protocol designs and the selection of the appropriate substrates to trace MEE under *in situ* conditions. In addition to the problems related to the lack of standardized measurement conditions, there exists also a rather diffuse selection of terms for identification of MEE, its main contributors and main modifier and inhibitors, especially in humic waters. Although the main MEE activities in marine and freshwater environments are thought to belong to bacterial communities and assemblages, the role of phagotrophic activities and other micrograzers in MEE compositions and activities is still little understood.

Measurements of MEE in aquatic environments have become more easier due to higher sensitivities with fluorogenic substrates and better instrumentations. However, due to the large variability of selected methods and protocols, the results are hardly comparable between different aquatic habitats and ecosystems. A stricter selection of substrates and reliable and standardized conditions of measurements and protocolls seem to be important steps in attaining more comparable data about MEE in aquatic environments.

Studies on the activities of MEE in aquatic environments have significantly increased during the last 10 years and have improved our understanding of nutrient cycling and carbon fluxes. However, we still know rather little about the regulators of MEE activities under *in situ* conditions, such as the effects of inorganic and organic nutrient compositions and the relative contribution of nutrient ratios such as the Redfield ratios of P:N:C to the expression, inhibition and de-repression of MEE. In humic waters, humic substances seem to play a key role in MEE activities, their inactivation, translocation and reactivation. There is also little knowledge of the control of MEE on genetic levels under *in situ* conditions, because these may function differently or may be modified compared with those from cultured microbes. Humic substances could be considered as important partners to those modifiers

ACKNOWLEDGEMENTS. This contribution was made possible by the support (UM) from the Max Planck Society, the Finnish Academy, the University of Helsinki, Maj and Tor Nessling Foundation, the Commission of the European Community (CEC) and Lammi Biological Station. Also we wish to express our thanks to D.-O. Hessen and L. Tranvik for inviting us to contribute to this special, edited book volume and finally the editorial support during the copy-editing procedure from Springer Verlag.

References

Aaronson S (ed) (1981) Chemical communication at the microbial level vol 1 CRC Press, Boca Raton

Aiken GR, McKnight DM, Wershaw RL, MacCarthy R (1985) Humic substances in soil, sediment and water. Geochemistry, isolation, and characterization. J. Wiley & Sons, New York

Alexander M (1975) Environmental and microbiological problems arising from recalcitrant molecules. Microb Ecol 2:17–27

Allen HL (1976) Dissolved organic matter in lakewater: characteristics of molecular weight size-fractions and ecological implications. Oikos 27:64–70

Ammerman JW (1991) Role of ecto-phosphohydrolases in phosphorus regeneration in estuarine and coastal ecosystems. In: Chróst RJ (ed) Microbial enzymes in aquatic environments. Brock/Springer Series in contemporary bioscience. Springer Berlin Heidelberg New York, pp 165–186

Ammerman JW, Azam F (1985) Bacterial 5'-nucleotidase in aquatic ecosystems: a novel mechanism of phosphorus regeneration. Science 227:1338–1340

Amon RMW, Benner R (1996) Bacterial utilization of different size classes of dissolved organic matter. Limnol Oceanogr 41:41–51

Andersen RA, Sowers JA (1968) Optimum conditions for bonding of plant phenols to insoluble polyvinylpyrrolidone. Phytochemistry 7:293–301

Azam F, Cho BC (1987) Bacterial utilization of organic matter in the sea. In: Fletcher M, Gray TRG, Jones JG (eds) Ecology of microbial communities, Cambridge University Press, Cambridge, pp 261–281

Azam F, Fenchel T, Field JG, Gray JS, Meyer-Reil LA, Thingstad F (1983) The ecological role of water column microbes in the sea. Mar Ecol Prog Ser 10:257–263

Azam F,Smith DC, Steward GF, Hagström Å (1994) Bacteria-organiv matter coupling and its significance for oceanic carbon cycling. Microb Ecol 28:167–179

Bärlocher, F (ed) (1981) The ecology of aquatic hyphomycetes. Ecological studies vol. 94. Springer, Berlin Heidelberg New York

Bauer JE, Williams PM, Druffel ERM (1992) ^{14}C activity of dissolved organic carbon fractions in the north-central Pacific and Sargasso Sea. Nature 357:667–670

Benner R, Moran MA, Hodson RE (1986) Biogeochemical cycling of lignocellulosic carbon in marine and freshwater ecosystems: relative contributions of procaryotes and eucaryotes. Limnol Oceanogr 31:89–100

Benner R, Lewis DL Hodson RE (1989) Biogeochemical cycling of organic matter in acid environments: are microbial degradative processes adapted to low pH? In: Rao SS (ed) Acid stress and aquatic microbial interactions, CRC Press, Boca Raton, pp 33–44

Benner R, Opsahl S, Chin-Leo G, Richey JE, Forsberg BR (1996) Bacterial carbon metabolism in the Amazon river system. Limnol Oceanogr 40:1262–1270

Bergemeyer HU (ed) (1990) Methoden der enzymatischen analyse, vols 1,2. Verlag Chemie, Weinheim

Billén G (1984) Heterotrophic utilization and regeneration of nitrogen. In: Hobbie JE, Williams PJ leB (eds) Hetrotrophic activity in the sea. NATO SAD. Plenum Press, New York, pp 313–355

Billén G (1991) Protein degradation in aquatic environments. In: Chróst RJ (ed) Microbial enzymes in aquatic environments, Brock/Springer Series in contemporary bioscience. Springer, Berlin Heidelberg New York, pp 123–143

Billén G, Servais P (1989) Modelisation des proces de degradation bactérienne de la matiére organique en milieu aquatique. In: Bianchi MC (ed) Micro-organismes dans les ecosystemes oceaniques. Masson, Paris, pp 219–245

Bittl T, Babenzien HD (1996) Microbial activity in an artificial divided acidic lake. Arch Hydrobiol Adv Limnol 48:113–121

Bochdansky AB, Puskaric S, Herndl GJ (1995) Influence of zooplankton grazing on free dissolved enzymes in the sea. Mar Ecol Progr Ser 121:53–63

Boon P (1989) Organic matter breakdown and nutrient regeneration in Australian freshwaters. I. Methods for exoenzyme assays in turbid aquatic environments. Arch Hydrobiol 115: 339–359

Boon P (1990) Organic matter breakdown and nutrient regeneration in Australian freshwaters. II: Spatial and temporal variation, and relation with environmental conditions. Arch Hydrobiol 117:405–436

Boyer PD, Lardy H, Mayrbäck K (eds) (1959–1983) The enzymes, vols. 1–16. Academic Press, New York San Francisco London

Burdige DJ (1991) The kinetics of organic matter mineralization in anoxic marine sediments. J Mar Res 49:727–761

Burns RG (ed) (1978) Soil enzymes. Academic Press, New York

Burns RG (1982) Enzyme activity in soil: location and possible role in microbial ecology. Soil Biol Biochem 14:423–427

Burns RG (1983) Extracellular enzyme-substrate interactions in soil. In: Slater JH, Whittenbury R, Wimpenny JWT (eds) Microbes in their natural environment. Cambridge University Press, Crambridge, pp 249–298

Button DK (1985) Kinetics of nutrient-limetid transport and microbial grpwth. Microb Rev 49:270–297

Button DK (1986) Affinity of organisms for substrate. Limnol Oceanogr 31:453–456

Chaudhry GR (ed) (1994) Biological degradation and bioremediation of toxic chemicals. Chapman and Hall, London

Cho BC, Azam F (1988) Major role of bacteria in biogeochemical fluxes in the ocean's interior. Nature 332:441-443

Christman RF, Gjessing ET (eds) (1983) Aquatic and terrestrial humic materials. Ann Arbor Science, Ann Arbor

Chróst RJ (1989) Characterization and significance of β-glucosidase activity in lake water. Limnol Oceanogr 34:660–672

Chróst RJ (1990) Ectoenzymes in aquatic environments: origin, activity and ecological significance. In: Overbeck J, Chróst RJ (eds) Advanced biochemical and molecular approaches to Aquatic microbial ecology. Brock/ Springer series in contemporary bioscience. Springer, Berlin Heidelberg New York pp 47–78

Chróst RJ (1991) Environmental control of the synthesis and activity of aquatic microbial ectoenzymes. In: Chróst RJ (ed.) Microbial enzymes in aquatic environments. Brock/Springer series in contemporary bioscience. Springer, Berlin Heidelberg New York, pp 29–59

Chróst RJ (1994) Microbial enzymatic degradation and utilization of organic matter. In: Overbeck J, Chróst RJ (eds) Microbial ecology of Lake Plußsee. Ecological studies. 105. Springer, Berlin Heidelberg New York, pp 118–174

Chróst RJ, Krambeck HJ (1986) Fluorescence correction for measurements of enzyme activity in natural waters using methylumbelliferyl-substrates. Arch Hydrobiol 106:79–90

Chróst RJ, Overbeck J (1987) Kinetics of alkaline phosphatase activity and phosphorus availability for phytoplankton and bacterioplankton in lake Plußsee (a north German eutrophic lake). Microb Ecol 13:229–248

Chróst RJ, Velimirov B (1991) Measurement of enzyme kinetics in water samples: effect of freezing and soluble stabilizer. Mar Ecol Progr Ser 70:93–100

Chróst RJ, Münster U, Rai H, Albrecht D, Witzel KP, Overbeck J (1989) Photosynthetic production and exoenzymatic degradation of organic matter in euphotic zone of an eutrophic lake. J Plankton Res 11:223–242

Colowick SP, Kaplan NO (founding eds, and serial eds) (1955–1996) Methods in Enzymology Vols.1–286. Academic Press, New York

De Haan H (1974) Effect of a fulvic acid fraction on the growth of a *Pseudomonas sp.* from Tjeukemeer (the Netherlands). Freshw Biol 4:301–310

De Haan H (1976) Evidence for the induction of catechol-1,2-oxygenase by fulvic acid. Plant Soil 45:129–136

De Haan H (1983) Use uf ultraviolet spectroscopy, gel filtration, pyrolysis/mass spectrometry and numbers of benzoate-metabolizing bacteria in the study of humification and degradation of aquatic organic matter. In: Christman RF, Gjessing ET (eds) Aquatic and terrestrial humic material. Ann Arbor Sci. Ann Arbor, pp 165–182

De Haan H (1993) Solar UV-light penetration and photodegredation of humic substances in peaty lake water. Limnol Oceanogr 38:1072–1076

De Haan H, De Boer T (1979) Seasonal variations of fulvic acids, amino acids, and sugars in Tjeukemeer (the Netherlands). Freshw Biol 4:301–310

Dixon M, Web EC (eds) (1979) Enzymes. Longman, London

Dixon RK (1992) Global carbon cycle and climatic change. ACS Symp Ser 483:375–378

Ducklow HW (1991) The passage of carbon through microbial foodwebs: results from flow network models. Mar Microb Food Webs 5:129–144

Ducklow HW, Purdie DA, Williams PJ, leB, Davis JM (1986) Bacterioplankton: a sink for carbon in a coastal marine plankton community. Science 332:865–867

Druffel ERM, Williams PM (1990) Identification of a depp marine source of particulate organic carbon using bomb ^{14}C. Nature 347:172–174

Edling, H, Tranvik L (1996) Effects of pH on β-glucosidase activity and availability of DOC to bacteria. Arch Hydrobiol Adv Limnol 48:123–132

Enzyme Nomenclature (1975) Recommendations (1972) of the International Union of Pure and Applied Chemistry and the International Union of Biochemistry. Elsevier, Amsterdam

Enzyme Nomenclature (1984) Recommendations of the Nomenclature Committee of the International Union of Biochemistry and the Nomenclature and Classification of Enzyme-Catalyzed Reactions. Academic Press, New York

Enzyme Nomenclature Supplementation and Revisions (1986–1990). In: Eur J Biochem 157: 1–26 (1986); 179:489–533 (1989); 187:263–281 (1990)

Farrington JW (1992) Marine organic geochemistry: review and challenges for the future. Mar Chem 39: 242pp

Fermi C (1906) The presence of enzymes in soil, water and dust. Zentralbl Bakteriol Parasitenk 26:330–334

Fiechter A (1993) Function and synthesis of enzymes involved in lignin degradation. J Biotechnol 30:49–55

Frimmel FH, Christman RF (eds) (1988) Humic substances and their role in the environment. Wiley, New York

Gagosian RB, Lee C (1981) Processes controlling the distribution of biogenic organic compounds in seawater. In: Duursma EK, Dawson R (eds) Marine organic chemistry. Evolution, composition, interactions and chemistry of organic matter in seawater. Elsevier, Amsterdam, pp 91–123

Geller A (1985a) Light-induced conversion of refractory, high molecular weight lake water constituents. Schweiz Z Hydrol 47:21–26

Geller A (1985b) Degradation and formation of refractory DOM by bacteria during simultaneous growth on labile substrates and persistent lake water constituents. Schweiz Z Hydrol 47: 27–34

Geller A (1986) Comparsion of mechanisms enhencing biodegradability of refractory lake waters constituents. Limnol Oceanogr 31:755–764

Gjessing ET (ed) (1976) Physical and chemical characteristics of aquatic humics. Ann Arbor Science, Ann Arbor

Gjessing ET (1992) The HUMEX project: experimental acidification of a catchment and its humic lake. Environ Int 18:535–543

Gjessing ET, Gjerdahl TG (1970) Influence of ultraviolet radiation on aquatic humus. Vatten 26:144–145

Glenn JK, Gold MH (1983) Decolorization of several polymeric dyes by the lignin-degrading basidiomycete *Phanerochaete chrysosporium*. Appl Environ Microbiol 45:1741–1747

Gold MH, Glenn JK, Alic M (1988) Use of polymeric dyes in lignin biodegradation assays. Methods Enzymol 161:74–78

Goulder R (1990) Extracellular enzyme activity associated with epiphytic microbiota on submerged stems of the red *Phragmites australis*. FEMS Microb Ecol 73:323–331

Graneli W, Lindell MJ, Tranvik LJ (1996) Photooxidative CO_2 production in lakes of different humic content. Limnol Oceanogr 41:698–706

Harvey GR, Boran DA, Chosal LA, Tokar JM (1983) The structure of marine fulvic and humic acids. Mar Chem 12:119–132

Harvey HW (1925) Oxidation in sea water. J Mar Biol Assoc UK 13:953–969

Haslam E (1974) Polyphenol-protein interactions. Biochem J 139:285–288

Haugland RP (ed) (1996). Molecular probes. Handbook of fluorescent probes and research chemicals. Molecular Probes, Eugene, 6th ed, 680 pp

Hayaishi O (ed) (1962) Oxygenases. Academic Press, New York

Hayes MHB, MacCarthy P, Malcolm RL, Swift RS (1989) Humic substances II: In search of structure. Wiley, Chichester

Hedges JI (1992) Global biogeochemical cycles: progress and problems. Mar Chem 39:67–93

Hernandez I, Hwang SJ, Heath RT (1996) Measurement of phosphomonoesterase activity with radiolabelled glucose-6-phosphate. Role in the phosphorus requirement of phytoplankton and bacterioplankton in a temperate mesotrophic lake. Arch Hydrobiol 137:265–280

Hessen DO (1985) The relation between bacterial carbon and dissolved humic compounds in oligotrophic lakes. FEMS Microbial Ecol 31:215–223

Hessen DO (1992) Dissolved organic carbon in a small humic lake: effects on bacterial production and respiration. Hydrobiologia 229:115–123

Hobbie JE (1992) Microbial control of dissolved organic carbon in lakes: research for the future. Hydrobiologia. 229:115–123

Hobbie JE (1993) A perspective on the ecology of aquatic microbes. In: Ford TE (ed.) Aquatic microbiology: an ecological approach. Blackwell, Oxford, pp 1–14

Hobbie JE (1994). The state of the microbe: a summary of a symposium honoring Lawrence Pomeroy. Microb Ecol 28:113–116

Hollibaugh JT, Azam F (1983) Microbial degradation of dissolved proteins in seawater. Limnol Oceanogr 28:1104–1116

Hoppe HG (1983) Significance of exoenzymatic activities in the ecology of brackish water: measurements by means of methylumbelliferyl-substrates. Mar Ecol Prog Ser 11:299–308

Hoppe HG (1991) Microbial extracellular enzyme activity: a new key parameter in aquatic ecology. In: Chróst RJ (ed) Microbial enzymes in aquatic environments, Brock/Springer series in contemporary bioscience. Springer Berlin Heidelber New York, pp 60–83

Hoppe HG (1993) Use of fluorogenic model substrates for extracellular enzyme activity (EEA) measurements of bacteria. In: Kemp PF, Sherr BF, Sherr EB, Cole JJ (eds) Handbook of methods in aquatic microbial ecology. Lewis, London, pp 509–512

Hoppe HG, Kim SJ, Gocke K (1988) Microbial decomposition in aquatic environments: combined processes of extracellular enzyme activity and substrate uptake. Appl Environ Microbiol 54:784–790

Horvath RS (1972) Microbial co-metabolism and the degradation of organic compounds in nature. Bact Rev 36:146–155

Ilmavirta V (1988) Phytoflagellates and their ecology in Finnish brown-water lakes. Hydrobiologia 161:255–270

Jackson CR, Foreman CM, Sinsabaugh RL (1995) Microbial enzyme activities as indicators of organic matter processing rates in a lake Erie coastal wetland. Freshw Biol 34:329–342

Jones GJ, Palenik BP, Morel FMM (1987) Trace metal reduction by phytoplankton: the role of plasmalemma redox enzymes. J Phycol 23:237–244

Jones SE, Lock MA (1989) Hydrolytic extracellular enzyme activity in heterotrophic biofilms from two contrasting streams. Freshw Biol 22:289–296

Jürgens K, Güde H (1990) Incorporation and release of phosphorus by planktonic bacteria and phagotrophic flagellates. Mar Ecol Progr Ser 59:271–284

Karner M, Herndl G (1992) Extracellular enzymatic activity and secondary production in free-living and marine snow associated bacteria. Mar Biol 113:341–347

Karnovsky ML (1986) Ectoenzymes: their modulation and similarity to certain enzymes of intracellular membranes. In: Karnovsky GW, Reddington M, Zimmermann H (eds) Cellular biology of ectoenzymes. Springer, Berlin Heidelberg New York, pp 3–13

Kemp PF (1994) A philosophy of methods development: the assimilation of new methods and information into aquatic microbial ecology. Microb Ecol 28:159–162

Kirchman DL, Suzuki Y, Garsdide C, Ducklow HW (1991) High turnover rates of dissolved organic carbon during a spring phytoplankton bloom. Nature 352:612–614

Kramer CJM (1979) Degradation by sunlight of dissolved fluorescing organic substances in the upper layers of the eastern Atlantic Ocean. Neth J Sea Res 13:325–329

Kreps E (1934) Organic catalysts or enzymes in sea water. In: James Johnstone Memorial Volume, University of Liverpool Press, Liverpool, pp 193–202

Kristoffersen K, Bernard C, Ekebom J (1996) A comparison of the ability of different heterotrophic nanoflagellates to incorporate dissolved macromolecules. Arch Hydrobiol Adv Limnol 48:73–84

Ladd JN, Butler JHA (1975) Humus-enzyme systems and synthetic, organic polymer-enzyme analogs. Soil Biochem 4:143–194

Lähdesmäki P, Piispanen R (1988) Degradation products and the hydrolytic enzyme activities in the soil humification process. Soil Biol Biochem 20:287–292

Lähdesmäki P, Piispanen R (1992) Soil enzymology: role of protective colloid systems in the preservation of exoenzyme activities in soil. Soil Biol. Biochem. 24:1173–1177

Lee C, Wakeham SG (1992) Organic matter in the water column: future research challenges. Mar Chem 39:95–118

Lindell MJ, Granéli W, Tranvik LJ (1995) Enhanced bacterial growth in response to photochemical transformation of dissolved organic matter. Limnol Oceanogr 40:195–199

Ljungdahl LG (1989) Mechanisms of cellulose hydrolysis by enzymes from anaerobic and aerobic bacteria. In: Coughlan MP (ed) Enzyme systems for lignocellulose degradation. Elsevier, London, pp 5–16

Loomis WD, Bataille J (1966) Plant phenolic compounds and the isolation of plant enzymes. Phytochemistry 5:423–438

Manafi M, Kneifel W, Bascomb S (1991) Fluorogenic and chromogenic substrates used in bacterial diagnostics. Microbial Rev 55:335–348

Martin JH, Fitzwater SE (1992) Dissolved organic carbon in the Atlantic, Southern and Pacific Oceans. Nature 356:699–700

Marxen J, Fiebig DM (1993) Use of perfused cores for evaluating extracellular enzyme activity in stream-bed sediments. FEMS Microbial Ecol 13:1.12

Marxen J, Schmidt HH (1993) Extracellular phosphatase activity in sediments of the Breitenbach, a central european mountain stream. Hydrobiologia 253:207–216

Marxen J, Witzel KP (1991) Significance of extracellular enzymes for organic matter degradation and nutrient regeneration in small streams. In: Chrost RJ (ed) Microbial enzymes in aquatic environments. Brock/Springer series in contemporary bioscience. Springer, Berlin, Heidelberg New York, pp 270–285

McFeters GO, Yu FP, Pyle BH, Stewart PhS (1995) Physiological assessment of bacteria using fluorochromes. J Microbiol Methods 21:1–13

McKnight DM, Behmel P, Francko DA, Gjessing ET, Münster U, Petersen Jr RC, Skulberg OM, Steinber CEW, Tipping E, Visser SA, Werner PW, Wetzel RG (1990) How do organic acids interact with solutes, surfaces, and organisms. In: Perdue EM, Gjessing ET (eds) Organic acids in aquatic ecosystems. Wiley, Chichester New York Brisbane Toronto Singapore, pp 223–243

McKnight DM, Aiken GR, Smith RL (1991) Aquatic fulvic acids in microbially based ecosystem: results from two desert lakes in Antarctica. Limnol Oceanogr 36:998–1006

McManus JP, Davis KG, Beart JE, Gaffey SH, Lilley TH, Haslam E (1985) Polyphenol interactions. Part 1. Introduction. Some observations on the reversible complexation of polyphenols with proteins and polysaccharides. J Chem Perkin Trans 2:1429–1438

Meyer-Reil LA (1986) Measurements of hydrolytic activity and incorporation of dissolved organic substrates by microorganisms in marine sediments. Mar Ecol Progr Ser 31:143–149

Meyer-Reil LA (1991) Ecological aspects of enzymatic activity in marine sediments. In: Chróst RJ (ed) Microbial enzymes in aquatic environments, Brock/Springer series in contemporary bioscience. Springer, Berlin Heidelber New York, pp 84–95

Milner AM, Oswood MW (eds) (1996) Freshwaters of Alaska, Ecological synthesis. Springer, Berlin Heidelberg New Yoek

Mopper K, Xianliang Z, Kieber RJ, Kieber DJ, Sikorski RJ, Jones RG (1991) Photochemical degradation of dissolved organic carbon and its impact on the oceanic carbon cycle. Nature 353: 60–62

Moran MA, Hodson RE (1990) Bacterial production on humic and nonhumic components of dissolved organic carbon. Limnol Oceanogr 35:1744–1756

Morra MJ (1997) Assessment of extracellular enzymatic activity in soil. In: Hurst CJ, Knudsen GR, McInnerney MJ, Stetzenbach LD & Walter MV (eds.) Manual of environmental microbiology. ASM Press, Washington, DC, pp 459–465

Münster U (1985) Investigations about structure, distribution and dynamics of different organic substrates in the DOM of Lake Plußsee. Arch Hydrobiol Suppl 70:429–480

Münster U (1991) Extracellular Enzyme Activity in Eutrophic and Polyhumic Lakes. In: Chróst RJ (ed) Microbial enzymes in aquatic environments, Brock/Springer series in contemporary bioscience. Springer, Berlin Heidelberg New York, pp 96–122

Münster U (1992a) Microbial extracellular enzyme activities and biopolymer processing in two acid polyhumic lakes. Arch Hydrobiol Ergebn Limnol 37:21–32

Münster U (1992b) Microbial extracellular enzyme activities in HUMEX lake Skjervatjern. Environ. Int. 18:637–647

Münster U (1993) Concentrations and fluxes of organic substrates in the aquatic environment. Antonie van Leeuwenhoek, J Gen Mol Microbiol 63:243–274

Münster U (1994) Studies on phosphatase activities in humic lakes. Environ. Int. 20:49–59

Münster U (1997a) Bioavailability of Nutrients. In: Eloranta P (ed) Limnology of Humic Lakes. Kluwer Academic Publisher, Dordrecht Boston London, in press

Münster U (1997b) Role of Microbes in Decomposition Processes. In: Eloranta P (ed.) Limnology of Humic Lakes. Kluwer Academic Publisher, Dordrecht Boston London, in press

Münster U (1997c) Phosphorus and phosphatase activities in boreal freshwaters. In: Whitton BA, Hernandez I (eds) Ecological aspects of phosphatase activities. Kluwer Academic Publisher, Dordrecht Boston London, in press

Münster U, Albrecht D (1994) Dissolved organic matter-analysis of composition and function by molecular-biochemical approach. In: Overbeck J, Chrost RJ (eds) Microbial ecology of Lake Plußsee. Springer, Berlin erg, New York, pp 24–62

Münster U, Chróst RJ (1990) Dissolved organic matter (DOM) in aquatic environments: origin, distribution, composition and microbial utilization. In: Overbeck J, Chróst RJ (eds) Advanced biochemical and molecular approaches to aquatic microbial ecology, Brock/Springer Series in contemporary bioscience. Springer, Berlin Heidelberg New York, pp 8–46

Münster U, Einiö P, Nurminen J (1989) Evaluation of the measurements of extracellular enzyme activities in a polyhumic lake by means of studies with 4-methylumbelliferyl-substrates. Arch Hydrobiol 115:321–337

Münster U, Einiö P, Nurminen J, Overbeck J (1992a) Extracellular enzymes in a polyhumic lake: important regulators in detritus processing. Hydrobiologia 229:225–238

Münster U, Nurminen J, Einiö P, Overbeck J (1992b) Extracellular enzymes in a small polyhumic lake: origin, distribution and activities. Hdrobiologia 243/244:47–59

Münster U, Heikkinen E, Salonen, K, De Haan H (1997a) Tracing peroxidase activity in a polyhumic lake. Acta hydrochim et hydrobiol in press

Münster U, Heikkinen E, Likolammi M, Järvinen M, Salonen K, De Haan H (1997b) Utilisation of polymeric and manomeric aromatic and amino acid carbon in a humic boreal forest lake. Arch Hydrobiol, Advances in Limnology, in press

Nagata T & Kirchman DL (1992) Release of macromolecular organic complexes by heterotrophic marine flagellates. Mar Ecol Prog Ser 83:233–240

Nedoma J, Vrba J, Hejzlar J, Šimek K, Straškrabová V (1994) N-acetylglucosamine dynamics in freshwater environments: concnetration of amino sugars, extracellular enzyme activities, and microbial uptake. Limnol Oceanogr 39:1088–1100

Ollikka P, Alhonmäki K, Leppänen VM, Glumhoff T, Raijola T, Suominen (1993) Decolorization of azo, triphenyl methane, heterocyclic, and polymeric dyes by lignin peroxidase isoenzymes from *Phanerochaete chrysosporium*. Appl Environ Microbiol 59:4010–4016

Orth AB, Royse DJ, Tien DM (1993) Ubiquity of lignin degrading peroxidases among various wood-degrading fungi. Appl Environ Microbiol 59:4017–4023

Overbeck J (1961) Die Phosphatasen von *Scenedesmus quadricauda* und ihre ökologische Bedeutung. Verh Int Verein Limnol 14:226–231

Overbeck J (1975) Distribution pattern of uptake kinetic responses in a stratified eutrophic lake. Verh Int Verein Limnol 19:2600–2615

Overbeck J, Babenzien HD (1963) Nachweis von freien Phosphatasen, Amylase und Saccharase im Wasser. Naturwissenschaften 50:571–572

Palenik B, Morel FMM (1988) Dark production of H_2O_2 in the Sargasso Sea. Limnol Oceanogr 33:1606–1611

Palenik B, Morel FMM (1990a) Amino acid utilization by marine phytoplankton. Limnol Oceanogr 35:260–269

Palenik B, Morel FMM (1990b) Comparison of cell-surface L-amino acid oxidase from several marine phytoplankton. Mar Ecol Progr Ser 59:195–201

Palenik B, Zafiriou OC, Morel FMM (1987) Hydrogen peroxide production by a marine phytoplankter. Limnol Oceanogr 32:1365–1369

Palenik B, Kieber DJ, Morel FMM (1988/89) Dissolved organic nitrogen use by phytoplankton: the role of cell-surface enzymes. Biol Oceanogr 6:347–354

Pantoja S, Lee C (1994) Cell-surface oxidation of amino acids in seawater. Limnol Oceanogr 39:1718–1726

Pantoja S, Lee C, Marecek JF, Palenik BP (1993) Synthesis and use of fluorescent molecular probes for measuring cell-surface enzymatic oxidation of amino acids and amines in seawater. Anal Biochem 211:210–218

Paul JH (1993) The advances and limitations of methodology. In: Ford TE (ed) Aquatic microbiology Balckwell, Oxford, p. 15–46

Perdue EM, Gjessing ET (eds) (1990) Organic Acids in Aquatic Ecosystems, Wiley, New.York

Pomeroy LR (1974) The ocean's food web, a changing paradigm. Bioscience 24:499–504

Post WM, Chavez F, Mulholland PJ, Pastor J, Peng TH, Prentice K, Webb T (1992) Climatic feedbacks in the global carbon cycle. ACS Symp Ser 483:392–412

Povoledo D, Golterman HL (eds) (1975) Humic substances, their structure and function in the biosphere. Proceedings of the international meeting on humic substances, Nieuwersluis, 1972. Pudoc, Wageningen

Priest FG (1984) Extracellular enzymes. Aspects Microbiol 9:1–79

Priest FG (1992) Synthesis and excretion of extracellular enzymes in bacteria. In: Winkelmann G (ed.) Microbial degradation of natural products, VCH, Weinheim, pp 1–26

Rasmussen JB, Godbout L, Schallenberg M (1989) The humic content of lake water and its relationship to watershed and lake morphometry. Limnol Oceanogr 34:1336–1343

Reichardt W, Overbeck J, Steubing L (1967) Free dissolved enzymes in lake water. Nature 216: 1345–1347

Reddy CA (1993) An overview of the recent advances on the physiology and molecular biology of lignin peroxidase of *Phanerochaete chrysosporium*. J Biotechnol 30:91–107

Rich PH (1984) Trophic-detrital interactions: vestiges of ecosystem evolution. Am Nat 123: 20–29

Rich PH, Wetzel RG (1978) Detritus in the lake ecosystem. Am Nat 112:57–71

Romankevich EA (ed) (1984) Geochemistry of organic matter in the ocean. Springer, Berlin Heidelberg New York

Salonen K (1981) The ecosystem of the oligotrophic lake Pääjärvi. 2. Bacterioplankton. Verh Int Verein Limnol 21:448–553

Salonen K, Tulonen T (1990) Photochemical and biological transformation of dissolved humic substances. Verh Int Verein Limnol 24:294

Salonen K, Vähätalo A (1994) Photochemical mineralisation of dissolved organic matter in lake Skjervatjern. Environ. Int. 20:307–312

Salonen K, Kolonen K, Arvola L (1983) Respiration of plankton in two small, polyhumic lakes. Hydrobiologia 101:65–70

Salonen K, Kairesalo T, Jones RI (eds) (1992) Dissolved organic matter in lacustrine ecosystems: energy source and system regulator. Hydrobiologia 229: 291pp

Sanders RW (1991) Trophic strategies among heterotrophic flagellates. Syst Assoc Spec Vol 45: 21–38

Saunders BC, Holmes-Siedle AG, Stark BP (eds) (1964) Peroxidase. Butterworths, London

Schindler DW, Bayley SE (1993) The biosphere as an increasing sink for atmospheric carbon: estimates from increased nitrogen deposition. Global Biogeochem Cycles 7:717–733

Schindler DW, Beaty KG, Fee EJ, Cruishank DR, DeBruyn ER, Findlay DL, Linsey GA, Shearer JA, Stainton MP, Turner MA (1990) Effects of climate warming on lakes of the central boreal forest. Science 250:967–970

Schindler DW, Mayley SE, Curtis PJ, Parker BR, Stainton MP, Kelly CA (1992) Natural and man-caused factors affecting the abundance and cycling of dissolved organic substances in precambrian shield lakes. Hydrobiologia 229:1–21

Schindler DW, Curtis PJ, Parker BR, Stainton MP (1996) Consequences of climate warming and lake acidification for UV-B penetration in North American boreal lakes. Nature 379:705–708

Schink B, Janssen PH, Frings J (1992) Microbial degradation of natural and of new synthetic polymers. FEMS Microbial Rev 103:311–316

Schinner F, Sonnleitner R (eds) (1996) Bodenökologie: Mikrobiologie und Bodenenzymatik, Grundlagen, Klima, Vegetation und Bodentyp. Springer, Berlin Heidelberg New York

Schmidt SK, Alexander M (1985) Effects of dissolved organic carbon and second substrates on the biodegradation of organic compounds at low concentrations. Appl Environ Microbiol 49: 822–827

Schmidt-Biegel, A, Obst U (1989) Rationelle fluorimetrische Bestimmungen von Enzymaktivitäten in vivo und der Biomasse (DNA) auf Mikrotiterplatten. Z Wasser Abwasser Forsch 22: 165–167

Schnitzer M, Khan SU (eds) (1972) Humic substances in the environment. Dekker, New York

Schoemaker HE, Leisola MSA (1990) Degradation of lignin by *Phanerochaete chrysosporium*. J Biotechnol 13:101–109

Schoenberg SA, Benner R, Armstrong A, Sobecky P, Hodson RE (1990) Effects of acid stress on aerobic decomposition of algal and aquatic macrophyte detritus: direct comparison in a radiocarbon assay. Appl Environ Microbiol 56:237–244

Sherr EB (1988) Direct use of high molecular weight polysaccharide by heterotrophic flagellates. Nature 335:348–351

Sherr EB, Sherr BF (1994) Bacterivory and herbivory: key roles of phagotrophic protists in pelagic food webs. Microb Ecol 28:223–235

Sinsabaugh RL, Linkins AE (1988). Exoenzyme activity associated with lotic epilithon. Freshw Biol 20:249–261

Sinsabaugh RL, Linkins AE (1990) Enzymatic and chemical analysis of particulate organic matter from a boreal river. Freshw Biol 23:301–309

Sinsabaugh RL, Findlay S (1995) Microbial production, enzyme activity, and carbon turnover in surface sediments of the Hudson River estuary. Microb Ecol 30:127–141

Sinsabaugh RL, Antibus RK, Linkins AR, McClaugherty CA, Rayburn L, Repert D, Weiland T (1992a) Wood decompositions over a first-order watershed: mass loss as a function of lignocellulase activity. Soil Biol Biochem 24:743–749

Sinsabaugh RL, Weiland T, Linkins AE (1992b) Enzymic and molecular analyis of microbial communities associated with lotic particulate organic matter. Freshw Biol 28:393–404

Sinsabaugh RL, Moorhead DL, Linkins AE (1994a) The enzymic basis of plant litter decomposition: emergence of an ecological process. Appl Soil Ecol 1:97–111

Sinsabaugh RL, Osgood MP, Findlay S (1994b) Enzymatic models for estimating decomposition rates of particulate detritus. J North Am Benthol Soc 13:160–169

Sinsabaugh RL, Findlay S, Franchini P, Fischer D (1997) Enzymatic analysis of riverine bacterioplankton production. Limnol Oceanogr 42:29–38

Sioli H (ed) (1984) The Amazon, Junk, Boston

Skujins JJ (1967) Enzymes in Soil. In: McLaren AD (ed) Soil biochemistry. Dekker, New York, pp 371–414

Smith DC, Simon M, Aldedge AL, Azam F (1992) Intense hydrolytic enzyme activity on marine aggregates and implications for rapid particle dissolution. Nature 359:139–142

Sorokin YI (1977) The heterotrophic phase of plankton succession in the Japan Sea. Mar Biol 41: 107–117

Spencer CM, Cai Y, Martin R, Gaffney SH, Goulding PN, Magnolato D, Lilley TH, Haslam E (1988) Polyphenol complexation-some thoughts and observations. Phytochemistry 27: 2397–2409

Stabel HH, Steinberg C (1976) Cleavage of macromolecular allochthounus soluble organic matter. Naturwissenschaften 63:533

Stabel HH, Moaledj K, Overbeck J (1979) On the degradation of dissolved organic molecules from Plußsee by oligocarbophilic bacteria. Arch Hydrobiol Ergeb Limnol 12:95–104

Steinberg C, Münster U (1985) Geochemistry and ecological role of humic substances in lakewater. In: Aiken GR, McKnight DN, Wershaw RL, MacCarthy P (eds) Humic substances in soil, sediment and water. Geochemistry, isolation, and characterization. Wiley, New York, pp 105–145

Steiner M (1938) Zur Kenntnis des phosphatekreislaufs in Seen. Naturwissenschaften 26: 723–724

Stevens SE Jr, Patteron CO, Myers J (1973) The production of hydrogen peroxide by blue-green alage: a survey. J Phycol 9:427–430

Stevenson FJ (ed) (1982) Humus Chemistry; genesis, composition, reactions. Wiley, New York

Stewart AJ, Wetzel RG (1982) Influence of dissolved humic materials on carbon assimilation and alkaline phosphatase activity in natural algal-bacterial assembedges. Freshw Biol 12:369–380

Swift G (1992) Biodegradability of polymers in the environment: complexities and significance of definitions and measurements. FEMS Microbial Rev. 103:339–346

Thurman M (ed) (1985) Organic geochemistry of natural waters. Nijhoff/Junk, Boston

Tien M, Kirk TK (1984) Lignin-degrading enzyme from *Phanerochaete chrysosporium*: purification, characterization, and catalytic properties of a unique H_2O_2-requiring oxygenase. Proc Natl Acad Sci USA 81:2280–2284

Tien M, Kirk TK (1988) Lignin peroxidase of *Phanerochaete chrysosporium*. Methods Enzymol 161:238–249

Torriani-Gorini A, Yagil E, Silver S (eds) (1994) Phosphate in microorganisms. ASM Press, Washington DC

Tranvik L (1988) Availability of dissolved organic carbon for planktonic bacteria in oligotrophic lakes of differing humic content. Microb Ecol 16:311–322

Tranvik L (1989) Bacterioplankton growth, grazing mortality and quantitative relationship to primary production in a humic and clearwater lake. J Plankton Res 11:985–1000

Tranvik L (1990) Bacterial growth on fractions of dissolved organic carbon of different molecular weights from humic and clear waters. Appl Environ Microbiol 56:1672–1677

Tranvik L (1992) Allochthonous dissolved organic matter as energy source for pelagic bacteria and the concept of the microbial loop. Hydrobiologia 229:107–114

Tranvik L, Sherr EB, Sherr BF (1993) Uptake and utilization of 'colloidal DOM' by heterotrophic flagellates in sea water. Mar Ecol Prog Ser 92:301–309

Visser SA (1985) Physiological action of humic substances on microbial cells. Soil Biol Biochem 17:457–462

Vrba J, Nedoma J, Šimek K, Seda J (1992) Microbial decomposition of polymer organic matter related to plankton development in a reservoir: activity of α-, β-glucosidase, and β-N-acetylglucosaminidase and uptake of N-acetylglucosamine. Arch Hydrobiol 126:193–211

Westrich JT, Berner RA (1984) The role of sedimentary organic matter in bacterial sulfate reduction: the G model tested. Limnol Oceanogr 29:236–249

Wetzel RG (1981) Long term dissolved and particulate alkaline phosphatase activity in a hardwater lake in relation to lake stability and phosphorus enrichments. Verh Int Verein Limnol 21:369–381

Wetzel RG (ed) (1983) Limnology, 2nd edn. Saunders, Philadelphia

Wetzel RG (1984) Detrital dissolved and particulate organic carbon functions in aquatic ecosystems. Bull Mar Sci 35:503–509

Wetzel RG (1990) Land-water interfaces: metabolic and limnological regulators. Verh Int Verein Limnol 24:6–24

Wetzel RG (1991) Extracellular enzymatic interactions in aquatic ecosystems: storage, redistribution, and interspecific communication. In: Chróst RJ (ed) Microbial enzymes in aquatic environments. Brock/ Springer series in contemporary bioscience. Springer, Berlin Heidelberg New York, pp 6–28

Wetzel RG (1992) Gradient-dominated ecosystems: sources and regulatory functions of dissolved organic matter in freshwater ecosystems. Hydrobiologia 229:181–198

Wetzel RG, Likens GE (eds) (1991) Limnological analyses. Springer, Berlin Heidelberg New York

Wetzel RG, Hatcher PG & Bianchi TS (1995) Natural photolysis by ultraviolet irradiance of recalcitrant dissolved organic matter to simple substrates for rapid bacterial metabolism. Limnol Oceanogr 40:1369–1380

Wheeler PA, Gosselin M, Sherr EB, Thibault D, Kirchman DL, Benner R, Whiteledge TE (1996) Active cycling of organic carbon in the central Arctic Ocean. Nature 380:697–699

Williams PJ leB (1981) Incorporation of microheterotrophic processes into the classical paradigm of the planktonic food web. Kieler Meeresforsch Sonderh 5:1–28

Williams PJ leB (1990) The importance of losses during microbial growth: commentory on the physiology, measurement and ecology of the release of dissolved organic material. Mar Microb Food Webs 4:175–206

Winkelmann G (1992) Microbial degradation of natural products. VCH, Weinheim

Wissmar RC, Richey JE, Stallard RF, Edmond JM (1981) Plankton metabolism and carbon processes in the Amazon River, its tributaries, and floddplain waters, Peru-Brazil, May-June 1977. Ecology 62:1622–1633

Wood TM, Garcia-Canpayo V (1990) Enzymology of cellulose degradation. Biodegradation 1: 147–161

Wood TM, McCrae SI (1979) Synergism between enzymes involved in the solubilization of native cellulose. Adv Chem Ser 181:181–209

Zeikus JG (1981) Lignin metabolism and the carbon cycle. Adv Microb Ecol 8:211–243

Zubay G (1993) Biochemistry. Brown, Dubuque, Iowa

10 Degradation of Dissolved Organic Matter in Humic Waters by Bacteria

Lars J. Tranvik

10.1 Introduction

The standing stock of dissolved organic matter (DOM) in surface waters depends on import, washout, indigenous primary production and processes of internal loss, including abiotic mineralization (particularly photooxidation), microbial mineralization and flocculation followed by sedimentation. The DOM in such waters is a complex mixture of different compounds. Some of these, such as free and combined amino acids and carbohydrates, have in many cases been identified and quantified. Although the bulk of the DOM has not been described in detail, a major constituent of it is generally humic matter. The composition of the fraction of the DOM that is utilized and mineralized by bacteria, however, is poorly known. This chapter concerns both the importance of microbial utilization for the dynamics of DOM, and the importance of recalcitrant DOM as a substrate for microbial growth in humic waters. The impact of such factors as flocculation and photochemical processes upon the microbial degradation will also be discussed. The further consequences which the production of bacterial biomass can have on the structure and function of the ecosystem through the consumption of DOM will be considered as well, but is elucidated in greater detail in Chapter 11.

10.2 Strict Evidence for Bacterial Degradation of Humic Matter?

As shown in other chapters of this book (McKnight and Aiken , chap. 1, this vol.; Perdue, chap. 2, this vol.), humic compounds are very diverse in origin and structure. They are sometimes broadly defined as organic substances, yellow or brown in colour, possessing a high molecular weight (e.g. Ishiwatari 1992). Other characteristics of humic substances include their having polyelectrolytic acidic properties and a highly aromatic structure (Thurman 1985, Killops and Killops 1993). Humic matter has generally undergone considerable diagenetic change, is relatively recalcitrant towards further biogeo-

Ecological Studies, Vol. 133
Hessen, Tranvik (eds) Aquatic Humic Substances

chemical transformations, and in most natural waters constitutes the major fraction of the organic compounds present (Thurman 1985). The precise determination of what compounds are humic and how they can be delimited from other organic compounds relies on operational definitions, a widely adopted one being that they should adsorb to a hydrophobic macroporous resin (XAD-8) at low pH (Thurman and Malcolm 1981).

Answering the question of whether and to what extent humic substances are utilized by bacteria is affected by how humic matter is defined, and in an empirical sense requires the isolation of humic substances from other organic compounds that could promote bacterial growth. Using the XAD-8 separation technique to achieve such isolation, Moran and Hodson (1990) compared bacterial production on the humic and non-humic fractions of DOM from a lake and from a wetland. Both fractions supported bacterial growth, although the non-humic DOM carried bacterial numbers several times as high per unit of initial organic carbon (Moran and Hodson 1990). In a later study, Moran and Hodson (1994) demonstrated that humic substances isolated from several marine sites also stimulated bacterial growth. Leff and Meyer (1991) studied the growth of a mixture of 47 different bacterial strains isolated from a blackwater river. The mixed inoculum proliferated in deionized water amended with isolated humic and fulvic acids obtained from the International Humic Substances Society. The bacterial biomass yield was 4- to 14-fold lower than in cultures with a glucose-peptone-yeast extract medium of similar dissolved organic carbon (DOC) concentration. It is not clear to what extent this was due to differences in substrate quality and/or to differential limitations in essential nutrients (such as nitrogen and phosphorus). Reitner et al. (1997), using the XAD-8 separation technique, reported bacterial utilization of the humic and of the non-humic fractions of the DOC in a eutrophic, brownwater lake to be roughly similar.

The XAD-8 isolation procedure involves acid-base treatment of the humic compounds, which may affect their properties as microbial substrates. For example, the molecular aggregates in the isolated humic matter may be arranged differently from in the original DOM solution, resulting in the degree of steric hindrance of the potential sites for enzymatic attack of the substrate differing. However, Moran and Hodson (1990) showed the bacterial biomass produced from unmanipulated DOM to be similar to the sum of the biomass produced from the separate humic and non-humic fractions. Thus, it is likely that humic compounds are also utilized by bacteria in the form in which they appear before the isolation process.

The lysis of isolated fulvic acids by manganese oxides to form such labile substances as pyruvate (Sunda and Kieber 1994) is a further indication of the bacterial utilization of humic compounds. The bacteria deposit Mn oxides on their surfaces, which was suggested as a strategy for the conversion of recalcitrant humic matter to highly available carbon sources.

Growth experiments (e.g. Moran and Hodson 1990) have shown both humic and non-humic fractions to be largely recalcitrant towards microbial breakdown, only a minor part of both fractions being mineralized during the course of such experiments, although in some cases the part mineralized has been found to be greater in the non-humic fraction. Since both fractions are heterogeneous, their constituents varying in recalcitrance, it is difficult to extrapolate from results for the initial labile fraction so as to elucidate the turnover times of the entire humic and non-humic pools of DOC. At present, there is no conclusive evidence supporting the view that aquatic humic matter (e.g. as identified by the XAD-8 extraction procedure) is substantially more recalcitrant towards microbial degradation than aquatic non-humic matter. Although the dominant humic fraction of the total freshwater DOM suggests that humic compounds are degraded more slowly than non-humic compounds, this may partly be a consequence of the predominance of humic compounds in the organic matter imported from terrestrial environments.

Thus, microbes clearly utilize humic matter, *sensu* common operational definitions. The limited data present also indicate that much of the recalcitrant DOM resides in the non-humic pool. Accordingly, the classification of DOM into humic and non-humic fractions does not separate the microbially available compounds from those that are recalcitrant. The present chapter concerns the microbial utilization of DOM with slow turnover rates, which appears to have undergone considerable diagenetic change. The compounds of interest are older than other DOM compounds, and have been subject to considerable transport. Thus, in the case of open lake environments, they are predominantly of littoral or terrestrial origin.

10.3 Bacterial Utilization of Water-Column DOM

10.3.1 General Aspects – Systems with a Negligible Import of DOM

Planktonic algae and cyanobacteria are the major indigenous primary producers found in pelagic environments. Although they form the basis of the phagotrophic food web, they also give rise to detrital organic carbon. Much of this detritus is in the form of DOM, commonly defined as the organic matter not retained by filters with 0.2-μm pores, thus including colloidal organic matter. It is produced partly by direct leakage from phytoplankton cells (e.g. Larsson and Hagström 1979) and partly by organisms at various trophic levels, e.g. due to sloppy feeding or excretion, during transfer through the phagotrophic food web (Jumars et al. 1989). In markedly pelagic environments, such as in open ocean areas, most of the DOM originates from algal production (Nissenbaum and Kaplan 1972, Stuermer and Harvey 1974, Meyers-Schulte and Hedges 1986). The seminal papers by Pomeroy (1974), Williams (1981) and Azam et al. (1983) promoted the present view of pelagic ecosystems, according to which a large fraction of the primary production is lost from the phagotrophic food chain as DOM, but is also reincorporated into the food web via bacterial uptake and subsequent bacterivory.

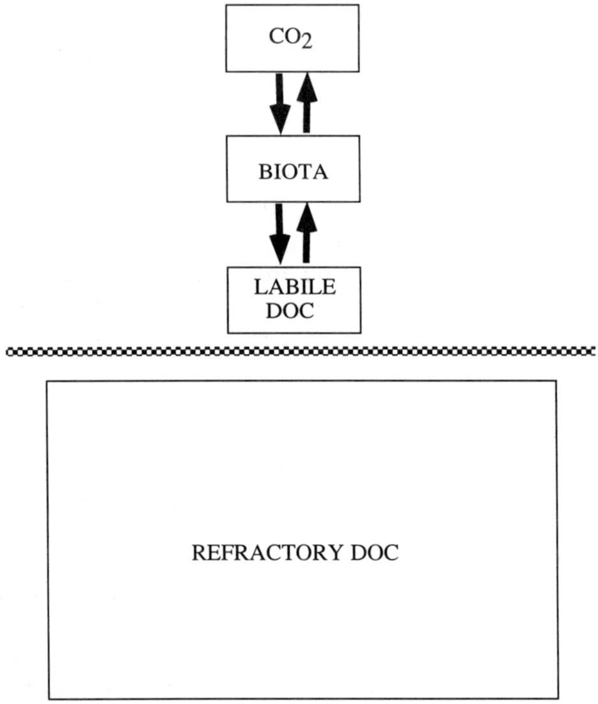

Fig. 10.1. An approximation of the water column carbon cycle, including a small pool of labile DOC and a large pool of DOC that is essentially inert

The labile pool of DOM, which remains at low concentrations due to continous high-affinity uptake by bacteria and turns over rapidly, constitutes a minor fraction of the total DOM. Typical components of this labile DOM are readily identifiable low-molecular-weight compounds, such as simple carbohydrates and amino acids. For an extensive review of specific compounds in DOM see Münster (1993).

Generally, only a small fraction of the total DOM (in lakes, average 14% of carbon; Søndergaard and Middelboe 1995) is readily available for bacterial uptake, the remainder sometimes being termed "refractory", implying that it is not significantly affected by microbial utilization. Accordingly, the oceanic DOM has been reported to have an average ^{14}C age of several thousand years (Williams and Druffel 1987). As a first approximation of the utilization of DOM by aquatic bacteria, this suggests there to be both a small labile pool of DOM that turns over rapidly and a large pool that does not take part in the microbial metabolism at any significant rate (Fig. 10.1). In most waters the total pool of organic matter is dominated by recalcitrant DOM, the particulate organic matter (POM, including both organisms and particulate detritus) constituting in most cases roughly 10% of the total organic matter (Wetzel 1984).

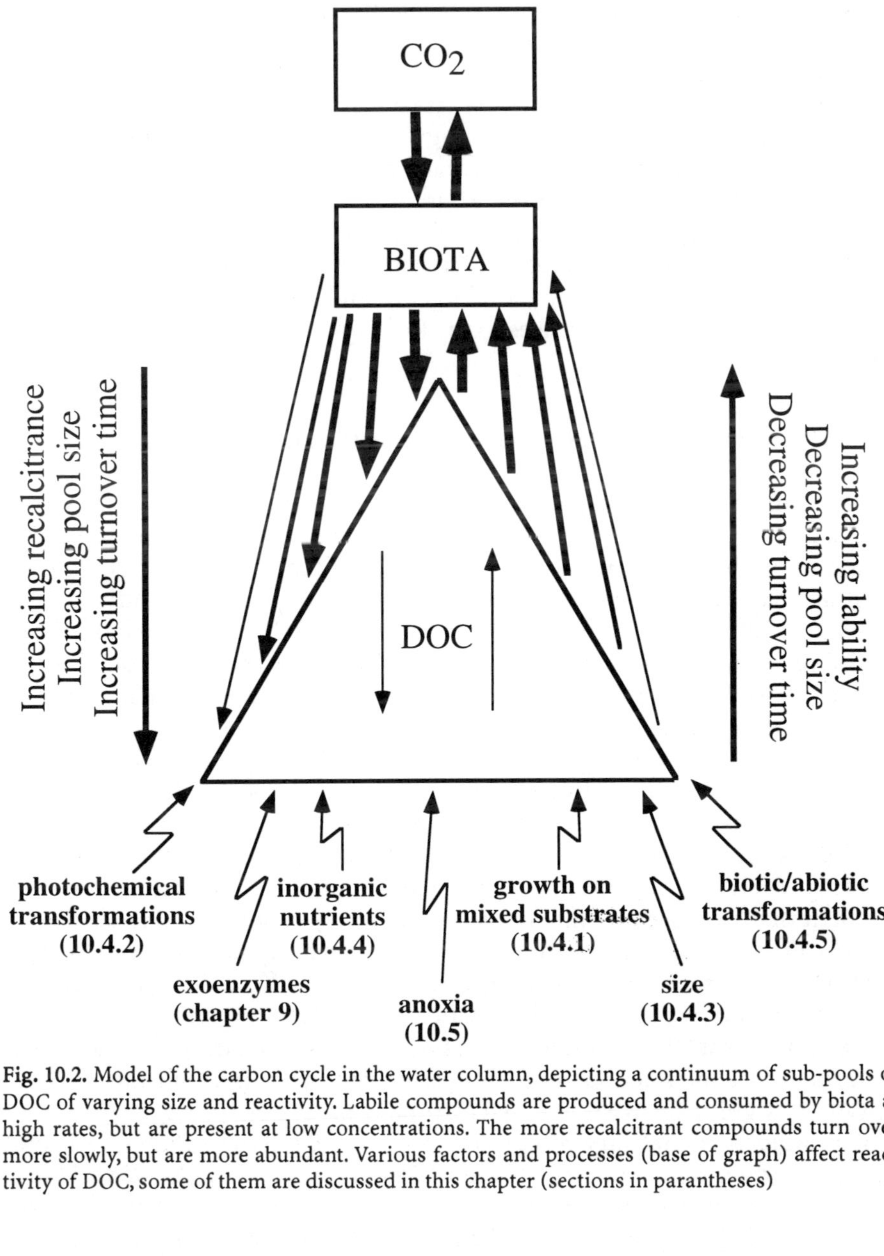

Fig. 10.2. Model of the carbon cycle in the water column, depicting a continuum of sub-pools of DOC of varying size and reactivity. Labile compounds are produced and consumed by biota at high rates, but are present at low concentrations. The more recalcitrant compounds turn over more slowly, but are more abundant. Various factors and processes (base of graph) affect reactivity of DOC, some of them are discussed in this chapter (sections in parantheses)

The simplified scheme shown in Fig. 10.1 ignores both that there are different degrees of recalcitrance and the transformations taking place which affect the recalcitrance or lability of the DOM. The recalcitrant DOM is produced by diagenetic changes in phytoplankton-derived DOM, a fraction of the labile DOM being deflected by various mechanisms into more recalcitrant forms (Fig. 10.2). Other simultaneous processes render the recalcitrant

DOM more labile and facilitate its entering the pool of DOM that is readily mineralized by bacteria. As shown in Fig. 10.2, a gradient of sub-pools of DOM that vary in their recalcitrance can be conceived. Such a model suggests there to be a small pool of very labile substrates (the top of the "DOM pyramid") that turns over rapidly due to high-affinity microbial uptake systems. The sub-pools of increasing recalcitrance are successively larger and increasingly slow in their turnover. Thus, they have a higher average age and tend to be transported further from their source. The predominance of recalcitrant substances and the presence of a minor labile fraction appear to be common features to all waters (Søndergaard and Middelboe 1995). However, there are considerable differences in the total amount, origin and structure of DOM.

10.3.2 Specific Properties of Aquatic Systems Rich in Humic Matter

Recalcitrant components generally dominate the DOM, both in environments heavily influenced by the import of humic matter and in habitats in which the influence of humic substances is minimal. However, despite their turning over slowly, recalcitrant substrates can contribute significantly to ecosystem dynamics due to their large pool size. Indeed, Wetzel has argued in several papers (e.g. 1984, 1992, 1995) that in most lakes DOM constitutes a major fraction of the total organic matter present, that it originates to a large extent in littoral and allochthonous sources and that it degrades slowly albeit continually. Accordingly, recalcitrant DOM provides "a stabile baseline" for lake metabolism and moderates the oscillations caused by the intermittent autochthonous production of labile organic matter.

The massive import of allochthonous DOM, typical of many lakes, affects the trophic structure. In humic lakes, there is much more detritus in the water than that produced by the indigenous primary producers. The utilization of this material by bacteria augments the heterotrophic processes further (Fig. 10.3), whereas the primary production is unaffected or is even suppressed, due to the rapid light attenuation in the water column caused by dissolved humic substances. Many studies of lakes that are rich in humic matter indicate that allochthonous organic matter (largely humic substances) constitutes a significant basis for heterotrophic processes, in addition to that which indigenous primary production provides. Bacterial abundance is correlated to the humic content of lakes (Hessen 1985; Tranvik 1988). Humic lakes have high respiration to production ratios (Salonen et al. 1983; del Georgio and Peters 1993). In a southern Swedish humic lake, the ratio of depth-integrated heterotrophic bacterial production to primary production was much higher than in an adjacent clearwater lake (Tranvik 1989). Similarly, Hessen et al. (1990) showed most of the zooplankton biomass in a humic lake to be derived from bacteria and detritus, only a minor fraction being attributable to feeding on phytoplankton. In a comparison of 18 lakes constituting a wide range of algal productivity, the ratio of heterotrophic to

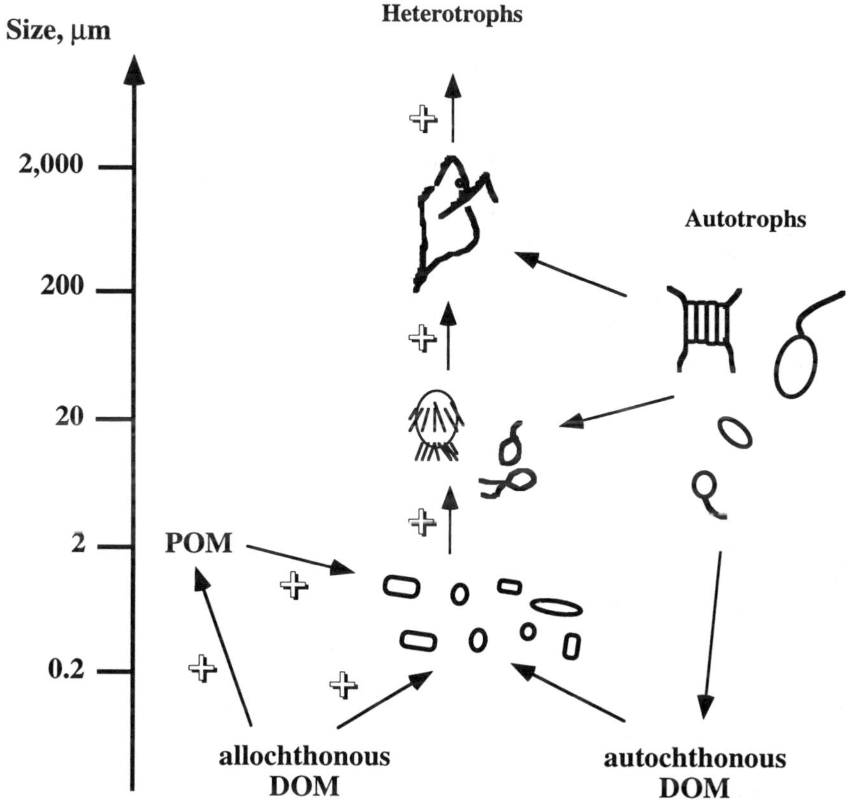

Fig. 10.3. The food web of a humic lake, including bacterial utilization of the DOM both from indigenous primary production (autochthonous) and from the watershed (allochthonous, largely humic matter). Fluxes that are enhanced by imported humic matter are indicted by plus signs (modified from Tranvik 1992)

autotrophic biomass was found to increase with decreasing primary productivity, probably due to the increasing importance of external energy sources entering the lakes as allochthonous DOM increased (Del Giorgio and Gasol 1995).

10.3.3 Estimating the Labile Fraction of DOC: The Regrowth Approach

Although DOC (dissolved organic carbon) is probably composed of a continuum of sub-pools that are degraded at different rates (Fig. 10.2), results of short-term degradation experiments (days or weeks) are often consistent with there being a small labile fraction and a larger refractory pool (Fig. 10.1). The bacterial utilization of DOC is commonly assessed in regrowth experiments in which natural water is first filtered, frequently

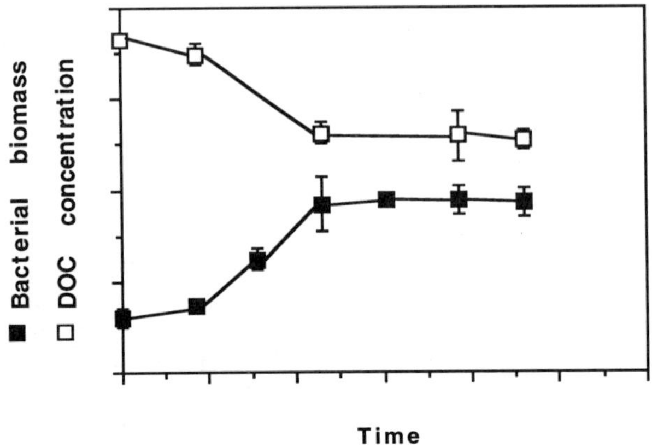

Fig. 10.4. A bacterial regrowth experiment in which lake-water bacteria are reinoculated into filtered water. Bacterial growth ceases when the labile fraction of DOC is exhausted, provided no other substances limit or inhibit the growth

through a 0.2-μm filter, so as to remove bacteria and other organisms, the water then being reinoculated with the indigenous bacteria by a small portion of water from the same source (usually ≤ 10% of the total volume) being added. The inoculum is filtered (pore size around 1 μm) to remove bacterivorous protozoa and other large organisms, while retaining the bacteria. The bacteria generally develop in a typical batch culture pattern, exhibiting a lag phase, an exponential growth phase and a stationary phase which suggests all the available substrate to be exhausted (Fig. 10.4). The fraction of DOC consumed until the stationary phase is reached is defined (in some cases implicitly) as the labile fraction, while the remainder is designated as recalcitrant. To specifically estimate the availability of DOC, excess inorganic nitrogen and phosphorus are added. Sometimes the bacterial abundance declines again, but this is usually coincident with the development of populations of bacterivorous flagellates in the culture.

Søndergaard and Middelboe (1995) compiled studies of this sort of DOC availability in lakes, rivers and seawater. The labile fraction of DOC in the 27 lakes studied was on average 14%. In a gradient of ten oligotrophic lakes, ranging from highly transparent clearwater to polyhumic water, the average labile fraction of DOC ranged from 6 to 14% (average 10%), there being no correlation between humic content and the labile fraction of DOC (Tranvik 1988). The relative similarity of the results for these lakes and the wide range of lakes involved in Søndergaard and Middelboe's (1995) overview (highly eutrophic lakes being included), together with the failure of the labile fraction to decrease with increasing humic content, do not support the idea that the DOC of humic lakes is less available than that found in other environments. However, in their comparison of data from 16 riverine sites, Søndergaard and Middelboe noted that sites along a blackwater river (from the study of Leff and Meyer 1991), i.e. habitats very rich in humic matter, deviated from sites along other rivers in their markedly lower fraction of labile DOC.

Sun et al. (1997) investigated the relationship between bioavailability and the share of aliphatic compounds estimated from basic compositional parameters such as elemental composition (H:C, N:C or O:C atomic ratios), carboxyl content and molecular weight. Bacterial growth was negatively correlated with O:C, as indicative of "old" humus with a higher content of COOH. Overall, humic matter with a high aliphatic content (as indicacted by the H:C ratio) appeared to be most available as a bacterial growth substrate. These data demonstrate a direct relationship between the structure of DOC and the ability of bacteria to utilize it for biomass production.

Many regrowth studies include calculations of growth efficiency, i.e. of the fraction of the organic carbon utilized that is converted to biomass. There is a wide range of reported growth efficiences, possibly to a large extent caused by methodological differences. In a study of lakes along a humic gradient (Tranvik 1988), growth efficiencies were derived from the consumption of DOC (indicated by the decrease in DOC) and the production of bacterial carbon (derived from bacterial biovolumes). The average growth efficiency for the different lakes was found to be 26%, no significant trends along the humic gradient being obtained. Thus, the available substrate in humic and in clearwater oligotrophic lakes appeared to be of similar substrate quality.

10.4 Some Factors Affecting Microbial Availability of DOC

Only a fraction of the DOC occurs as monomers that can be transported into bacterial cells. Utilization of the major part of the potentially available substrates requires initial hydrolysis by extracellular enzymes. The importance of extracellular enzymes is further elucidated in Chapter 9. The regrowth-culture studies of the microbial utilization of DOC that were presented above (Sect. 10.3.3) assume there to be a static system with a small labile pool of DOC and a large refractory pool of it. However, the microbial availability of DOC can be altered by several factors not operational in batch cultures, and not detectable at the time scales typical of regrowth experiments. Thus, there may be continual or recurrent transformations of recalcitrant DOM into more available forms. The next sections consider factors that affect the susceptibility of DOM to microbial degradation.

10.4.1 Growth on Mixed Substrates

The potential bacterial substrates in the water are complex mixtures of many substances, most of them present at very low concentrations. A successful bacterial growth strategy is likely to include the simultaneous utilization of several different substrates. Various experiments with pure cultures have shown that the threshold for growth on a specific organic substrate tends to be lower when another carbon source is present than when the substrate in question is the only carbon source available (reviewed by Egli 1996). For example, Law and Button (1977) reported that a particular marine bacterial strain did not grow on glucose as the sole carbon source when the glucose concentration was below some 0.5 mg/l, but that in the presence of alternative carbon sources it utilized glucose at concentrations as low as only a few μg/l. Thus, a combination of several substrates, each at low concentration, can promote bacterial growth that any given of the separate substrates could only support if present at much higher concentration.

Similar effects have been noted for the utilization of humic matter in laboratory cultures. De Haan (1974) found that a certain bacterial strain isolated from a humic lake did not grow on a medium in which fulvic acid was the only carbon source, but that in the presence of a second carbon source (lactate) the fulvic acid was degraded. In a later study (De Haan 1977), another strain from the same lake, able to grow on either fulvic acids or benzoate, grew much faster when both benzoate and fulvic acid were present in the medium. This was interpreted as a priming effect of the benzoate on the degradation of fulvic acid. Geller (1985) demonstrated that bacterial strains from lake water degraded high-molecular-weight DOC isolated from the lake by gel filtration to a greater extent in the presence of an additional substrate (glutamic acid). The increase in growth caused by the combination of substrates may have been due to a lowering of the threshold for the utilization of potentially available components within the fulvic acid fraction. Multi-substrate utilization, leading to a decrease in the threshold concentrations for the utilization of organic compounds, can explain the very low concentration of many substrates in natural waters (Egli 1996). It also implies that the presence of specific labile substrates facilitates the degradation of other, more recalcitrant compounds. The degradation of humic matter in natural waters may be a case of the simultaneous degradation of multiple substrates, such as that studied in the laboratory by De Haan (1974, 1977) and by Geller (1985).

10.4.2 Increased Degradability of Humic Matter Due to Photochemical Conditioning

As discussed in Chapter 5, DOM, and humic matter in particular, absorbs sunlight. This can result either in complete photochemical mineralization or in transformations yielding organic compounds of lower average molecular weight. Strome and Miller (1978) suggested that sunlight stimulates the bacterial degradation of natural DOC. Geller (1986) confirmed this, using pure bacterial strains and the high-molecular-weight fraction of lake-water DOC.

The photochemical production of a number of low-molecular-weight compounds from natural DOC has been demonstrated. Examples of photochemically produced substrates that are labile towards microbial decomposition include acetate and other carboxylic acids, aldehydes and pyruvate (Mopper and Stahovec 1986, Kieber and Mopper 1987, Kieber et al. 1990, Mopper et al. 1991, Backlund 1992, Allard et al. 1994, Wetzel et al. 1995).

Following the papers by Strome and Miller (1978) and Geller (1986), several studies corroborated the hypothesis of the sequential photolytic-microbial degradation of DOC. Kieber et al. (1989) reported there to be a close correlation between photochemical production and the microbial utilization of pyruvate in oceanic water. Bertilsson and Allard (1996) found several carboxylic acids (formic, acetic, malonic and oxalic acid) to be accumulated during intermittent radiation of sterile humic water, but the concentrations of carboxylic acids to remain low in parallel systems that were amended with bacteria. This suggested that the acids were decomposed by the bacteria. Lindell et al. (1995) found there to be an almost six-fold enhancement of bacterial biomass in regrowth cultures in which humic water was pre-exposed to artificial sunlight as compared with cultures in which humic water was preincubated in darkness. Similarly, Bushaw et al. (1996) reported a several-fold increase in bacterial growth on isolated humic substances that had been pre-exposed to sunlight as compared with dark controls. Similar results have been obtained in studies of the photolysis and microbial utilization of leachates from senescent littoral plants. Using both whole leachates and the isolated humic fraction of the leachate, Wetzel et al. (1995) showed a wide range of low-molecular-weight compounds stimulating bacterial production to be produced photochemically.

In addition to stimulating the bacterial degradation of humic matter by increasing the susceptibility of the substrate to microbial attack, sunlight can inhibit bacterial activity. The inhibitory mechanism involves damage to the DNA and other cell components (Karentz et al. 1994), but the inhibition may also be indirect, e.g. through the photochemical formation of toxic concentrations of free radicals and of hydrogen peroxide (Gjessing and Källqvist 1991). Direct inhibition is only effective during exposure to sunlight, and the radicals and the hydrogen peroxide are degraded relatively quickly in the water. Thus, if the bacterial utilization of photochemically produced DOC components is retarded due to the inhibition of bacteria by the sunlight, this probably only results in the degradation being temporally or spatially displaced. Utilization can occur subsequent to exposure to sunlight, either after mixing into deeper waters or during the night. Kaiser and Herndl (1997) suggested the net effect of sunlight on seawater bacteria to be stimulatory, since the bacteria recover from UV-stress rapidly after moving to deeper layers. Using incubations at fixed depths within the upper 2 m of the water column, Lindell et al. (1996) found depth-integrated bacterial inhibition to be 1–4% in humic lakes (up to 23% in clear lakes), the depth-integrated stimulatory effect due to transformations of DOC being 23–34%.

10.4.3 Effects of Size on the Recalcitrance of DOC

DOC is not dissolved in a strict sense, its definition being based on passage through certain filters. It can be visualized as a size continuum of monomers, polymers, macromolecular aggregates and colloids. The colloids are ubiquitous, both in marine (Koike et al. 1990; Wells and Goldberg 1991) and in freshwater environments (Burnison and Leppard 1983). In humic lakes, much of the DOC appears to be colloidal or of high molecular weight. The fraction of the total DOC of nominal molecular weight >10 kDa in lakes increases with increasing humic content, from 10 to 20% in clearwater lakes to some 50% in humic lakes (Tranvik 1990). Similarly, Middelboe and Søndergaard (1995) found the fraction of larger colloids (passing a 0.7-μm filter but retained on a 0.2-μm filter) to increase in the DOC of lakes as the total DOC concentration increased.

The coagulation of dissolved compounds into colloids and larger aggregates that can sink through the water column not only is a critical step in the loss of organic matter into the sediment, but also may affect the microbial degradability of the material. A reasonable hypothesis is that there is an inverse relationship between the size of the DOC components (colloids, molecules) and their degradability. Monomers are readily transported into bacterial cells, but macromolecules require an initial degradation by extracellular enzymes, which slows down the degradation process. Colloids are still slower to degrade, due to steric hindrance. With increasing size, colloids have a decreasing surface to volume ratio. Consequently, a large fraction of the potential bacterial substrates is not immediately accessible, being "hidden" by protective layers that must first be degraded (Lee and Wakeham 1992). In addition, the coagulation of DOC into larger but fewer colloids results in the rate of encounter of bacterial enzymes with substrate decreasing. This scenario is consistent with the model of Saunders (1976), which predicts that small molecules degrade faster than macromolecules.

At the same time, several studies demonstrate that the microbial availability of colloids and high-molecular-weight compounds is similar to or higher than the availability of smaller molecules. Meyer et al. (1987) showed that the >10-kDa fraction of the DOC of a blackwater river supports substantial bacterial growth. In a set of lakes that comprised a gradient in humic content, greater bacterial growth was found to occur per unit of >10 kDa DOC, compared with <10 kDa DOC (Tranvik 1990). On average, 52% of the bacterial growth on DOC from these lakes could be attributed to the utilization of >10-kDa compounds. Amon and Benner (1994, 1996) compared the utilization of high (>1 kDa) and low (<1 kDa) molecular weight DOM from several freshwater and marine sites. They found consistently higher bacterial growth per unit for the high-molecular-weight DOC. Middelboe and Søndergaard (1995) attributed 3–30% of the total DOC, as compared with 12–45% of the labile DOC, to 0.2 to 0.7-μm colloids. These studies support the size-reactivity continuum model of Amon and Benner (1996), which assumes the reactivity of DOM to increase with increasing nominal molecular size, being high for polymers that potentially are hydrolyzed and utilized by bacteria, and low for small molecules, that appear to have undergone extensive diagenetic change. The use of water from a humic lake for the experimental enhancement of the concentration of >100-kDa colloids, an enhancement representing a 92% increase in DOC concentration, resulted in a 63% increase in bacterial growth (Tranvik and Jørgensen 1995). This suggests these large colloids to be only slightly less available than average DOC. Similar manipulations of water from an oligotrophic clearwater lake resulted in an increase in DOC concentration but in no significant stimulation of bacterial growth (Tranvik 1994a, Tranvik and Jørgensen 1995). Hence, the colloidal DOC of the latter lake appeared to be highly recalcitrant.

The coagulation of colloids into aggregates larger than bacteria may further facilitate bacterial utilization. The coagulation of oceanic DOC on bubbles increases degradability (Kepkay and Johnson 1989, Kepkay 1994). A similar stimulation of the degradation of humic matter may take place in lakes. Tranvik and Sieburth (1989) compared the utilization of humic matter in dissolved form and after its partial flocculation into 10 to 1000-μm aggregates. The humic flocs were colonized by bacteria, more bacterial biomass being produced than in cultures in which all the humic matter was in a dissolved state. It is possible that in the humic aggregates the extracellular enzymes functioned more efficiently, because they diffused away from the cells more slowly.

There is considerable evidence, therefore, also from humic waters, that the degradability of DOM is relatively high for compounds of high nominal molecular weight. However, this is not necessarily due to larger molecules being more labile. Studies of the degradation of DOM fractions of different size rely on physical separation techniques, such as ultrafiltration. Aggregates of small molecules appear as high-molecular-weight compounds, and the apparent molecular weight of dissolved humic matter depends on pH and on ionic strength (e.g. De Haan et al. 1987). Small labile molecules may adsorb to larger, recalcitrant compounds (Carlson et al. 1985). Meyer et al. (1987) found that, although >10 kDa DOM was highly available initially, the availability decreased markedly after hydrolysis with hydrochloric acid. The authors suggested that this is due to the selective digestion of labile, low molecular weight matter attached to larger recalcitrant cores.

10.4.4 Dependence of DOC Degradation on Access to Inorganic Nutrients

Dissolved organic matter, in particular humic substances, exhibits high C:N and C:P ratios (Thurman 1985). Hence, heterotrophic bacteria encounter an organic substrate with excess C but a limited supply of other nutrients. This leads frequently to nutrient limitations on the degradability of DOC. A number of bioassay experiments involving the addition of inorganic nutrients demonstrate there being limitation by either N, P or both (e.g. Toolan et al.

1991, Coveney and Wetzel 1992, Morris and Lewis 1992, Zweifel et al. 1993). Hessen et al. (1994) found limitations of both N and P in two Norwegian humic lakes, especially during periods of high allochthonous DOC loading.

In bioassay experiments involving bacteria only, the remineralization of N and P is restricted, since the bacterivores that participate in the remineralization of nutrients bound within bacterial biomass are excluded. In a natural environment, N and P are remineralized and become available for the bacterial degradation of DOC repeatedly. Hence, although P (and in some cases N too, or a combination of both) limits the potential degradation of DOC, as based on regrowth experiments, this may not reflect the in situ situation. At times, the rates of remineralization of N and P may limit the rate of degradation of DOC. However, it is only when the rate of supply of potentially available DOC (e.g. through the photolysis of recalcitrant compounds or by external supply) exceeds the rate of supply of available N and P (dependent largely upon remineralization) that the ultimate extent of DOC mineralization is restricted by inorganic nutrients.

Empirical evidence for the enhanced degradation of DOC upon enrichment with inorganic nutrients derives from whole-lake experiments in the Experimental Lakes Area in Canada (Schindler et al. 1992). Organic carbon originating from autochthonous primary production was traced by the addition of dissolved inorganic ^{14}C. Measurements of the specific ^{14}C activity of DOC suggested there to be only a minor degree of autochthonous origin of the DOC. Upon fertilization with inorganic nutrients, however, the autochthonous fraction of DOC increased, without a corresponding increase in total DOC concentration. Thus, as Schindler et al. (1992) suggested, the addition of nutrients probably increased the mineralization of allochthonous DOC (humic matter). This may have been a direct effect of inorganic nutrients, but it may also have been due to the stimulated degradation of recalcitrant allochthonous DOC due to an increase in the access to labile autochthonous DOC (cf. Sect. 10.4.1).

Access to inorganic nutrients is partly regulated by DOM. Humic substances associate with phosphate, particularly in the presence of high concentrations of iron (Francko and Heath 1979; Jones et al. 1988; De Haan et al. 1990). This may restrict the bacterial and algal utilization of P (Stewart and Wetzel 1981). The return of the humic-bound phosphate to the free state is catalysed by sunlight (Francko and Heath 1982, Cotner and Heath 1990). A substantial pool of PO_4 that is bound to humic compounds and is slowly released, especially during sunlight hours, may serve as a "phosphorus buffering system". Recently, a similar process was described for nitrogen (Bushaw et al. 1996). Sunlight induced the transformation of recalcitrant dissolved organic nitrogen into ammonium, which could be utilized by heterotrophic bacteria.

10.4.5 Transformation of Labile DOC into Recalcitrant Forms

In oceanic environments in which the influence from terrestrial environments is negligible, the recalcitrant part of DOC originates mainly from indigenous primary production (Nissenbaum and Kaplan 1972). Initially, DOC from the algal primary producers is easily degraded, yet still gives rise to recalcitrant DOC components with ^{14}C ages of thousands of years (Williams and Druffel 1987). In lakes, DOC originates predominantly in the littoral and in the terrestrial and wetland vegetation of the watershed. This material, largely derived from lignocellulose, has been subject to extensive diagenetic change and is already recalcitrant before it reaches open water (Wetzel 1984, Hobbie 1988). Hence, the occurrence of recalcitrant humic matter in lake water is not dependent on internal processes within the water column, as is the case for oceans.

Nevertheless, the autochthonous fraction of DOC is subject to diagenetic modifications which can involve interactions with allochthonous humic matter. Schindler et al. (1992), adding ^{14}C-labelled inorganic carbon in a whole-lake experiment, found there to be a rapid increase in radioactive DOC initially, due to the excretion of ^{14}C that was fixed during primary production. Although most of this DOC was degraded during the following winter and spring, a substantial fraction remained in the water for at least an additional year without showing any tendency to decrease, except for the loss that could be attributed to hydrologic flushing of the lake. Thus, autochthonous DOC can enter the pool of recalcitrant DOC rather rapidly.

The labile compounds are transformed into recalcitrant forms either by microbial metabolism (Brophy and Carlson 1989; Tranvik 1993, 1994b, Lara and Thomas 1995) or abiotically through the adsorption of simple, labile compounds (such as carbohydrates or amino acids) to bulk DOM, followed by geochemical reactions (Keil and Kirchman 1994), possibly involving energy from solar UV light. Although the bacterial mineralization of DOC from algae decreased, due to the presence of humic matter combined with exposure to light, neither light nor humic matter separately had any effect on the recalcitrance of algal DOC (Tranvik and Kokalj 1998). This suggested that sunlight promotes reactions between humic matter and fresh DOC, rendering the DOC more recalcitrant. Accordingly, it has been suggested that photochemical reactions are involved in the formation of marine humic substances from labile precursors (Harvey et al. 1983).

The kinds of reactions suggested here do not explain the overall recalcitrance of freshwater DOM, which is largely due to processes occurring before DOM enters the water body. However, reactions with allochthonous humic matter may affect the fate of freshly produced autochthonous DOM. In this way, labile organic substrates from the primary production occurring in humic lakes are partly withdrawn from rapid microbial degradation.

10.5 Degradation in Anoxic Water Bodies and Sediments

Oligotrophic lakes heavily stained with brown humic matter comprise a significant fraction of the lakes in the northern boreal areas (Kortelainen 1993), and probably in the circumpolar northern temperate zone generally. Heating of the water is rapid, due to the absorption of solar radiation in the humic-stained water, and to the lakes frequently being small and sheltered from wind-induced turbulence. This results in shallow thermal stratification and in the volumes of hypolimnetic water being large (Bowling and Salonen 1990). Oxygen depletion occurs frequently, providing conditions for anaerobic microbial metabolism. Despite this, most studies of the microbial availability of aquatic humic matter focus on aerobic processes. The abundance of substrates for bacteria in the hypolimnion may be either similar to or higher than that in the overlying oxygenated waters. The hypolimnion is potentially enriched by substrates that diffuse out of the sediment, as well as by sedimentation from the epilimnion of algae and other organisms and by flocculated humic matter.

In situ bacterial production is studied less in anoxic than in oxic water. There are several reports, however, of high bacterial production in anoxic waters and in layers just below the oxic-anoxic interface (Lovell and Konopka 1985; McDonough et al. 1986; Pedros-Alio and Guerrero 1993; Ochs et al. 1995). Ochs et al. (1995) reported production to be several-fold higher in anoxic waters than in epilimnetic waters. Cole and Pace (1995) compared the bacterial production in the oxygenated epilimnetic and anoxic hypolimnetic waters of nine lakes. Although the bacterial specific growth rates were found to be lower in the hypolimnia, there was generally a greater amount of biomass there than in the epilimnia. The average rate of bacterial production per volume of water was 1.6 times higher in the hypolimnia, despite the temperature there being lower. The studies just cited suggest that bacterial production (and thus the bacterial mineralization of DOC) in anaerobic water is a significant process in the microbial processing of organic matter.

Anaerobic metabolism is less energy-efficient than oxygen respiration. Fermentative bacteria obtain only about one-quarter as much ATP per unit of substrate as aerobic bacteria (Fenchel and Finlay 1995). Hence, less bacterial biomass per unit of organic carbon that is mineralized is formed. The lesser growth efficiency during anaerobic metabolism results in a greater dissipation of energy at each trophic level, resulting in shorter food chains than in oxic environments (Fenchel and Finlay 1995). The lesser growth efficiency in anoxic environments, which results in less biomass per unit of DOC degraded being produced, is no a priori reason for the DOC degradation there being slower. Cole and Pace (1995), in estimating the growth efficiencies of hypolimnetic bacteria from the accumulation of inorganic carbon and bacterial production in the hypolimnion, found that although in anoxic environments the growth of bacteria is expected to be less efficient energetically, there were bacterial growth efficiencies there of 16–30%. This is roughly similar to the growth efficiencies reported in oxygenated waters. Although this method of calculating growth efficiency is not directly comparable with calculations obtained in confined experiments, the results do imply that the nutritional value of the substrates utilized by the anaerobic bacteria is high, inasmuch as such substrates would yield still higher growth efficiencies in aerobic metabolism.

Quantitatively, sulphate reduction and methanogenesis are the most important processes of terminal mineralization in anaerobic aquatic environments (Westermann 1993, Fenchel and Finlay 1995). The relative importance of each process depends on the availability of substrates and of electron acceptors. Sulphate reduction dominates in marine anoxic habitats, whereas in freshwater environments methanogenesis is generally a very important process. The substrate for these mineralizers is largely provided by fermentative bacteria (Westermann 1993). It has been suggested recently that humic substances themselves can serve as electron acceptors during the microbial degradation of organic matter (Lovley et al. 1996).

Investigators of wetlands (e.g. Boon and Mitchell 1995; Hamilton et al. 1995) and of sediments and hypolimnia of lakes (e.g. Rudd and Hamilton 1978; Fallon et al. 1980; Bédard and Knowles 1991) emphasize the role of methanogenesis in the degradation of organic matter. Several studies (reviewed by Wetzel 1983) suggest methanogenesis to have a substantial role in carbon cycling in lakes. Methanogenesis is also significant in habitats in which humic matter is dominant, e.g. blackwater wetlands (Happell and Chanton 1993; Pulliam 1993; Bianchi et al. 1996) and *Sphagnum* bogs (Krumholz et al. 1995). Thus, it is likely that humic compounds are mineralized via methanogenesis. Bianchi et al. (1996) tested specifically the availability to methanogens of laboratory grade tannic acid and of humic acids isolated from decomposing vascular plants. The results suggested that these substrates are utilized by methanogens. The pretreatment of humic acids by UV-B stimulated subsequent methanogenesis, analogous to the radiation-stimulated bacterial utilization of DOM in aerobic environments that was described above (Sect. 10.4.2). Acetic acid and other low molecular weight carboxylic acids are among the major products identified as being formed during the photolysis of humic matter (Wetzel et al. 1995; Bertilsson and Allard 1996). Acetic acid is also a major substrate for methanogenic bacteria, supplied normally by fermentative bacteria as an end product (Westermann 1993). Thus, sunlight may promote the formation of substrates for methanogenic bacteria. Stimulation may also be indirect, through the photochemically enhanced availability of the substrate to fermentative bacteria.

The formation of methane in sediments or in deeper anoxic layers of the water column provides a substrate for methane-oxidizing bacteria at the oxic-anoxic interface, in the upper oxygenated portion of the water column, or subsequent to mixing (Rudd and Hamilton 1978). Thus, methanogenesis followed by methanotrophy in the upper water column can serve as a benthic/hypolimnetic-epilimnetic link in the carbon cycle. This was borne out by ultrastructural studies of the bacterioplankton in the oxic waters of a humic lake in New York state, USA, which suggested that an appreciable fraction (20%) of the bacterial cells have internal membrane structures indicative of methylotrophic bacteria (Corpe and Jensen 1992). In a stratified humic lake in Norway in which there was vigorous methane production in the deeper anoxic layers, the carbon fixation by methane-oxidizing bacteria and heterotrophic bacteria in the oxygenated water column was found to be of the same order of magnitude (Hessen and Nygaard 1992).

Most of the metabolism in the anoxic zone probably takes place in the sediments, the contribution from the water column being less important (Rudd and Hamilton 1978). The pool size of sedimented organic matter is large compared with the pool size of dissolved organic matter in the overlying water column, on an areal basis. Tranvik et al. (1994) conducted a survey of water column and sediment microbial activity in humic lakes in southern Sweden. Their data indicate that, on an areal basis, sediment respiration (as measured by oxygen consumption) is about one-third as great as the respiration in the overlying water column (3 m water depth at sampling sites). One can reasonably assume that there was considerable additional anaerobic metabolism in the sediment, in addition to the oxygen respiration in the upper few millimetres that were oxygenated. In the upper centimetre of the sediment, the bacterial abundance per unit of lake area was about five-fold that in the whole overlying column of water. This illustrates that microbial metabolism in the sediment is a major component in whole-lake metabolism.

In oligotrophic humic lakes, the sediment is dominated by loose aggregates of humic matter, that flocculate in the water column and settle to the sediment. This can likewise be the fate of a considerable fraction of the water-column DOC (L. J. Tranvik, in prep.). In this way, some of the humic matter is retained in the lake and can slowly be degraded over long periods of time within the lake, the metabolism within the lake of the organic matter dissolved in the water being restricted by the hydrological retention time. The slow degradation in the sediment can be seen as an extension of the "pyramid model" of DOM dynamics presented in Fig. 10.2. The recalcitrant DOM is partly flocculated and becomes sedimented, turning over then very slowly. Nevertheless, the large pool size results in its making a significant contribution to the total mineralization of DOM, when the sedimentary particulate organic matter that is generated by the flocculation of DOM is taken into account.

10.6 Microbial Utilization of Humic-Bound Nitrogen and Phosphorus

Whereas most investigations of the degradation of DOM and humic matter focus on DOC, few studies have addressed similar aspects of the dissolved organic forms of other nutrients. Investigations of the microbial utilization of organic nitrogen address predominantly specific compounds, particularly nitrogen in free or combined amino acids (e.g. Hollibaugh and Azam 1983; Jørgensen et al. 1994; Kroer et al. 1994) or nitrogen and phosphorus in nucleic acids (Jørgensen et al. 1993; Turk et al. 1993). The availability of bulk organic nitrogen and phosphorus has been studied very little, in a way analogous to the investigations of DOC availability discussed above (Sect. 10.3–10.4).

In freshwater environments, primary production is often limited by the supply of phosphorus. Freshwater humic matter is largely exported from the terrestrial environment into coastal seawater, where primary production is frequently limited by a lack of nitrogen. Organically bound nutrients is a potentially important source of nitrogen for phytoplankton (Antia et al. 1991). Since riverine humic compounds stimulate the growth of algae (Granéli and Moreira 1990), it has been proposed that humic matter imported from terrestrial and freshwater environments constitutes a substantial supplementary nitrogen source for coastal phytoplankton (Carlsson et al. 1993). Direct algal utilization is one of the possible pathways of this nitrogen into the primary producers (Sherr 1988; Palenik and Morel 1990), although there is no conclusive evidence that it is quantitatively important. Algal utilization of humic-bound nitrogen may also be indirect, by means of bacterial (Carlsson et al. 1993, 1995) or photochemical mineralization (Bushaw et al. 1996) into inorganic or labile organic nitrogen sources.

10.7 Summary

This chapter has focused on the bacterial degradation of DOM, special emphasis being placed on humic waters. Although bacteria grow on isolated humic compounds, several studies indicate the bacterial yield per unit of DOC to be higher in the non-humic than in the humic fraction of lake-water DOC. However, most of the carbon in both fractions is recalcitrant. In this respect, humic matter does not appear to differ from other constituents of bulk DOC. The large size of the pool of humic matter implies it to be an important bacterial substrate, despite its turning over very slowly. Several factors, such as access to phosphorus and nitrogen, the size of molecules and colloids, flocculation into larger particles and sunlight-induced photochemical transformations of recalcitrant compounds into labile DOC affect the ability of the bacteria to utilize DOC.

Most investigations of the microbial degradation of DOM have concentrated on the DOC in oxic waters. However, in anoxic layers as well there is substantial degradation of DOC. Thus, anaerobic metabolism can contribute significantly to the degradation of freshwater humic matter. The microbial utilization of elements of DOM other than carbon, particularly organic-bound nitrogen and phosphorus, has received only limited attention. In coastal seawater where primary production is limited by nitrogen requirements, the import of freshwater humic matter by rivers may possibly be an important nitrogen source, readily mineralized by bacteria. A further aspect of the microbial utilization of humic matter, studied little and not reviewed in this chapter, is the carbon cycling that occurs within littoral habitats dominated by macrophytes, where fungal degradation plays a marked role.

It is characteristic of most of the studies summarized in this chapter that lumped parameters were employed. Hence, the degradation of organic matter is often described in terms of total DOC or rough characterizations (e.g. molecular weight fractions or the hydrophobic fraction of DOC). There is also little information available generally on the characteristics (e.g. the important taxa) of the microflora responsible for degradation. Current and future methodological developments will undoubtedly yield insight into intriguing ecological and biogeochemical issues here, such as those of adaptations and interactions within microbial communities found in environments rich in recalcitrant substrates, and of the specific structures in DOM that promote or limit microbial degradability.

References

Allard B, Borén H, Pettersson C, Zhang G (1994) Degradation of humic substances by UV radiation. Environ Int 20:97–101

Amon RMW, Benner R (1994) Rapid cycling of high-molecular-weight dissolved organic matter in the ocean. Nature 369:549–551

Amon RMW, Benner R (1996) Bacterial utilization of different size classes of dissolved organic matter. Limnol Oceanogr 41:41–51

Antia NJ, Harrison PJ, Oliveira L (1991) The role of dissolved organic nitrogen in phytoplankton nutrition, cell biology and ecology. Phycologia 30:1–89

Azam F, Fenchel T, Field JG, Gray JS, Meyer-Reil L-A, Thingstad F (1983) The ecological role of water-column microbes in the sea. Mar Ecol Prog Ser 10:257–263

Backlund P (1992) Degradation of aquatic humic material by ultraviolet light. Chemosphere 25: 1869–1878

Bédard C, Knowles R (1991) Hypolimnetic O_2 consumption, denitrification, and methanogenesis in a thermally stratified lake. Can J Fish Aquat Sci 48:1048–1054

Bertilsson S, Allard B (1996) Sequential photochemical and microbial degradation of refractory dissolved organic matter in a humic freshwater system. Arch Hydrobiol 48:133–141

Bianchi TS, Freer ME, Wetzel RG (1996) Temporal and spatial variability, and the role of dissolved organic carbon (DOC) in methane fluxes from Sabine River floodplain (southeast Texas, U.S.A.). Arch Hydrobiol 136:261–287

Boon PI, Mitchell A (1995) Methanogenesis in the sediments of an Australian freshwater wetland: comparison with aerobic decay, and factors controlling methanogenesis. FEMS Microbiol Ecol 18:175–190

Bowling LC, Salonen K (1990) Heat uptake and resistance to mixing in small humic forest lakes in southern Finland. Aust J Mar Freshw Res 41:747–760

Brophy JE, Carlson DJ (1989) Production of biologically refractory dissolved organic carbon by natural seawater microbial populations. Deep Sea Res 36:497–507

Burnison BK, Leppard GG (1983) Isolation of colloidal fibrils from lake water by physical separation techniques. Can J Fish Aquat Sci 40:373–381

Bushaw KL, Zepp RG, Tarr MA, Schulz-Jander D, Bourbonniere RA, Hodson RE, Miller WL, Bronk DA, Moran MA (1996) Photochemical release of biologically available nitrogen from dissolved organic matter. Nature 381:404–407

Carlson DJ, Mayer ML, Brann ML, Mague TH (1985) Binding of monomeric organic compounds to macromolecular dissolved organic matter in seawater. Mar Chem 16:141–153

Carlsson P, Segatto AZ, Granéli E (1993) Nitrogen bound to humic matter of terrestrial origin – a nitrogen pool for coastal phytoplankton. Mar Ecol Prog Ser 97:105–116

Carlsson P, Granéli E, Tester P, Boni L (1995) Influences of river transported humic substances and copepod grazing on a coastal plankton community. Mar Ecol Prog Ser 127:213–221

Cole JJ, Pace ML (1995) Bacterial secondary production in oxic and anoxic freshwaters. Limnol Oceanogr 40:1019–1027

Corpe WA, Jensen TE (1992) An electron microscopic study of picoplanktonic organisms from a small lake. Microb Ecol 24:181–197

Cotner JB, Heath RT (1990) Iron redox effects on photosensitive phosphorus release from dissolved humic materials. Limnol Oceanogr 35:1175–1181

Coveney MF, Wetzel RG (1992) Effects of nutrients on specific growth rate of bacterioplankton in oligotrophic lake water cultures. Appl Environ Microbiol 58:150–156

De Haan H (1974) Effect of a fulvic acid fraction on the growth of a *Pseudomonas* from Tjeukemeer (the Netherlands). Freshw Biol 4:301–310

De Haan H (1977) Effect of benzoate on microbial decomposition of fulvic acids in Tjeukemeer (The Netherlands). Limnol Oceanogr 22:38–44

De Haan H, Jones RI, Salonen K (1987) Does ionic strength affect the configuration of aquatic humic substances, as indicated by gel filtration? Freshw Biol 17:453–459

De Haan H, Jones RI, Salonen K (1990) Abiotic transformations of iron and phosphate in humic water revealed by double isotope labeling and gel filtration. Limnol Oceanogr 35:491–497

Del Giorgio PA, Gasol JM (1995) Biomass distribution in freshwater plankton communities. Am Nat 146:135–152

Del Giorgio PA, Peters RH (1993) Balance between phytoplankton production and plankton respiration in lakes. Can J Fish Aquat Sci 50:282–289

Egli T (1996) The ecological and physiological significance of the growth of heterotrophic microorganisms with mixtures of substrates. Adv Microb Ecol 14:305–386

Fallon RD, Harrits S, Hanson RS, Brock TD (1980) The role of methane in internal carbon cycling in Lake Mendota during summer stratification. Limnol Oceanogr 25:357–360

Fenchel T, Finlay BJ (1995) Ecology and evolution in anoxic worlds. Oxford University Press, Oxford

Francko DA, Heath RT (1979) Functionally distinct classes of complex phosphorus in lake water. Limnol Oceanogr 24:463–473

Francko DA, Heath RT (1982) UV-sensitive complex phosphorus: association with dissolved humic material and iron in a bog lake. Limnol Oceanogr 27:564–569

Geller A (1985) Degradation and formation of refractory DOM by bacteria during simultaneous growth on labile substrates and persistent lake water constituents. Schweiz Z Hydrol 47:27–44

Geller A (1986) Comparison of mechanisms enhancing biodegradability of refractory lake water constituents. Limnol Oceanogr 31:755–764

Gjessing ET, Källqvist T (1991) Algicidal and chemical effect of UV-radiation of water containing humic substances. Water Res 25:491–494

Granéli E, Moreira MO (1990) Effects of river water of different origin on the growth of marine dinoflagellates and diatoms in laboratory cultures. J Exp Mar Biol Ecol 136:89–106

Hamilton SK, Sippel SJ, Melack JM (1995) Oxygen depletion and carbon dioxide and methane production in waters of the Pantanal wetland of Brazil. Biogeochemistry 30:115–141

Happell JD, Chanton JP (1993) Carbon remineralization in a north Florida swamp forest: effects of water level on the pathways and rates of soil organic matter decomposition. Global Biogeochem Cycles 7:475–490

Harvey GR, Boran DA, Chesal LA, Tokar JM (1983) The structure of marine fulvic and humic acids. Mar Chem 12:119–132

Hessen DO (1985) The relation between bacterial carbon and dissolved humic compounds in oligotrophic lakes. FEMS Microbiol Ecol 31:215–223

Hessen DO, Nygaard K (1992) Bacterial transfer of methane and detritus: implications for the pelagic carbon budget and gaseous release. Arch Hydrobiol 37:139–148

Hessen DO, Andersen T, Lyche A (1990) Carbon metabolism in a humic lake; pool sizes and cycling through zooplankton. Limnol Oceanogr 35:84–99

Hessen DO, Nygaard K, Salonen K, Vähätalo A (1994) The effect of substrate stoichiometry on microbial activity and carbon degradation in humic lakes. Environ Int 20:67–76

Hobbie JE (1988) A comparison of the ecology of planktonic bacteria in fresh and salt water. Limnol Oceanogr 33:750–764

Hollibaugh JT, Azam F (1983) Microbial degradation of dissolved proteins in seawater. Limnol Oceanogr 28:1104–1116

Ishiwatari R (1992) Macromolecular material (humic substances) in the water column and sediments. Mar Chem 39:151–166

Jones RI, Salonen K, De Haan H (1988) Phosphorus transformations in the epilimnion of humic lakes: abiotic interactions between dissolved humic materials and phosphate. Freshw Biol 19: 357–369

Jørgensen NOG, Kroer N, Coffin RB, Yang XH, Lee C (1993) Dissolved free amino acids, combined amino acids, and DNA as sources of carbon and nitrogen to marine bacteria. Mar Ecol Prog Ser 98:135–148

Jørgensen NOG, Kroer N, Coffin RB (1994) Utilization of dissolved nitrogen by heterotrophic bacterioplankton: effects of substrate C/N ratio. Appl Environ Microbiol 60:4124–4133

Jumars PA, Pentry DL, Baross JA, Perry MJ, Frost BW (1989) Closing the microbial loop: dissolved carbon pathway to heterotrophic bacteria from incomplete ingestion and absorption in animals. Deep Sea Res 36:483–496

Kaiser E, Herndl GJ (1997) Rapid recovery of marine bacterioplankton activity after inhibition by UV radiation in coastal waters. Appl Environ Microbiol 63:4026–4031

Karentz D, Bothwell ML, Coffin RB, Hanson A, Herndl GJ, Kilham SS, Lesser MP, Lindell M, Moeller RE, Morris DP, Neale PJ, Sanders RW, Weiler CS, Wetzel RG (1994) Impact of UV-B radiation on pelagic freshwater ecosystems: report of working group on bacteria and phytoplankton. Arch Hydrobiol Beih 43:31–69

Keil RG, Kirchman DL (1994) Abiotic transformation of labile protein to refractory protein in sea water. Mar Chem 45:187–196

Kepkay PE (1994) Particle aggregation and the biological reactivity of colloids. Mar Ecol Prog Ser 109:293–304

Kepkay PE, Johnson BD (1989) Coagulation on bubbles allows the microbial respiration of oceanic dissolved organic carbon. Nature 385:63–65

Kieber DJ, Mopper K (1987) Photochemical formation of glyoxylic and pyruvic acids in seawater. Mar Chem 21:135–149

Kieber DJ, McDaniel J, Mopper K (1989) Photochemical source of biological substrates in sea water: implications for carbon cycling. Nature 341:637–639

Kieber DJ, Zhou X, Mopper K (1990) Formation of carbonyl compounds from UV-induced photodegradation of humic substances in natural waters: fate of riverine carbon in the sea. Limnol Oceanogr 35:1503–1515

Killops SD, Killops VJ (1993) An introduction to organic geochemistry. Longman, London

Koike I, Hara S, Terauchi K, Kogure K (1990) Role of sub-micrometre particles in the ocean. Nature:242–244

Kortelainen P (1993) Content of total organic carbon in Finnish lakes and its relationship to catchment characteristics. Can J Aquat Sci 50:1477–1483

Kroer N, Jørgensen NOG, Coffin RB (1994) Utilization of dissolved nitrogen by heterotrophic bacterioplankton: a comparison of three ecosystems. Appl Environ Microbiol 60:4116–4123

Krumholz LR, Hollenback JL, Roskes SJ, Ringelberg DB (1995) Methanogenesis and methanotrophy within a *Sphagnum* peatland. FEMS Microbiol Ecol 18:215–224

Lara R, Thomas DN (1995) Formation of recalcitrant organic matter: humification dynamics of algal derived dissolved organic carbon and its hydrophobic fractions. Mar Chem 51:193–199

Larsson U, Hagström Å (1979) Phytoplankton exudate release as an energy source for the growth of pelagic bacteria. Mar Biol 52:199–206

Law AT, Button DK (1977) Multiple-carbon-source-limited growth kinetics of a marine coryneform bacterium. J Bacteriol 129:115–123

Lee C, Wakeham SG (1992) Organic matter in the water column: future research challenges. Mar Chem 39:95–118

Leff LG, Meyer JL (1991) Biological availability of dissolved organic carbon along the Ogeechee River. Limnol Oceanogr 36:315–323

Lindell MJ, Granéli W, Tranvik LJ (1995) Enhanced bacterial growth in response to photochemical transformation of dissolved organic matter. Limnol Oceanogr 40:195–199

Lindell M, Granéli W, Tranvik L (1996) Impact of sunlight on bacterial growth in lakes of different humic content. Aquat Microb Ecol 11:135–141

Lovell CR, Konopka A (1985) Primary and bacterial production in two dimictic Indiana lakes. Appl Environ Microbiol 49:485–491

Lovley DR, Coates JD, Blunt-Harris EL, Phillips EJP, Woodward JC (1996) Humic substances as electron acceptors for microbial respiration. Nature 382:445–448

McDonough RJ, Sanders RW, Porter KG, Kirchman DL (1986) Depth distribution of bacterial production in a stratified lake with an anoxic hypolimnion. Appl Environ Microbiol 52: 992–1000

Meyer JL, Edwards RT, Risley R (1987) Bacterial growth on dissolved organic matter from a blackwater river. Microb Ecol 13:13–29

Meyers-Schulte KJ, Hedges JI (1986) Molecular evidence for a terrestrial component of organic matter dissolved in ocean water. Nature 321:61–63

Middelboe M, Søndergaard M (1995) Concentration and bacterial utilization of sub-micron particles and dissolved organic carbon in lakes and a coastal area. Arch Hydrobiol 133: 129–147

Mopper K, Stahovec WL (1986) Sources and sinks of low molecular weight organic carbonyl compounds in seawater. Mar Chem 19:305–321

Mopper K, Zhou X, Kieber RJ, Kieber DJ, Sikorski RJ, Jones RD (1991) Photochemical degradation of dissolved organic carbon and its impact on the oceanic carbon cycle. Nature 353: 60–62

Moran MA, Hodson RE (1990) Bacterial production on humic and non-humic components of dissolved organic carbon. Limnol Oceanogr 35:1744–1756

Moran MA, Hodson RE (1994) Support of bacterioplankton production by dissolved humic substances from three marine environments. Mar Ecol Prog Ser 110:241–247

Morris DP, Lewis WM (1992) Nutrient limitation of bacterioplankton growth in Lake Dillon, Colorado. Limnol Oceanogr 37:1179–1192

Münster U (1993) Concentrations and fluxes of organic carbon substrates in the aquatic environment. Antonie van Leeuwenhoek J Microbiol 63:243–274

Nissenbaum A, Kaplan IR. (1972) Chemical and isotopic evidence for the in situ origin of marine humic substances. Limnol Oceanogr 17:570–582

Ochs CA, Cole JJ, Likens GE (1995) Spatial and temporal patterns of bacterioplankton biomass and production in an oligotrophic lake. J Plankton Res 17:365–391

Palenik B, Morel FMM (1990) Comparison of cell-surface L-amino acid oxidases from several marine phytoplankton. Mar Ecol Prog Ser 59:195–201

Pedrós-Alió C, Guerrero R (1993) Microbial ecology in Lake Císo. Adv Microb Ecol 13:389–398

Pomeroy LR (1974) The ocean's food web: a changing paradigm. Bioscience 9:499–504

Pulliam WM (1993) Carbon dioxide and methane exports from a southeastern floodplain swamp. Ecol Monogr 63:29–53

Reitner B, Herzig A, Herndl GJ (1997) Role of ultraviolet-B radiation on photochemical and microbial oxygen consumption in a humic-rich shallow lake. Limnol Oceanogr 42:950–960

Rudd JWM, Hamilton RD (1978) Methane cycling in a eutrophic shield lake and its effects on whole lake metabolism. Limnol Oceanogr 23:337–348

Salonen K, Kolonen K, Arvola L (1983) Respiration of plankton in two small, polyhumic lakes. Hydrobiologia 101:65–70

Saunders G (1976) Decomposition in fresh water. In: Anderson J, Macfadyen A (eds) The role of terrestrial and aquatic organisms in decomposition processes. Blackwell, Oxford

Schindler DW, Bayley SE, Curtis PJ, Parker BR, Stainton MP, Kelly CA (1992) Natural and man-caused factors affecting the abundance and cycling of dissolved organic substances in precambrian shield lakes. Hydrobiologia 229:1–21

Sherr EB (1988) Direct use of high molecular weight polysaccharide by heterotrophic flagellates. Nature 335:348–351

Søndergaard M, Middelboe M (1995) A cross-system analysis of labile dissolved organic carbon. Mar Ecol Prog Ser 118:283–294

Stewart AJ, Wetzel RG (1981) Dissolved humic materials: photodegradation, sediment effects, and reactivity with phosphate and calcium carbonate precipitation. Arch Hydrobiol 92: 265–286

Strome DJ, Miller MC (1978) Photolytic changes in dissolved humic substances. Verh Int Verein Limnol 20:1248–1254

Stuermer DH, Harvey GR (1974) Humic substances from seawater. Nature 250:480–481

Sun L, Perdue EM, Meyer JL, Weis J (1997) Using elemental composition to predict bioavailability of dissolved organic matter in a Georgia river. Limnol Oceanogr 42:714–721

Sunda WG, Kieber DJ (1994) Oxidation of humic substances by manganese oxides yields low-molecular-weight organic substrates. Nature 367:62–64

Thurman EM (1985) Organic geochemistry of natural waters. Junk, Boston

Thurman EM, Malcolm RL (1981) Preparative isolation of aquatic humic substances. Environ Sci Technol 15:463–466

Toolan T, Wehr JD, Findlay S (1991) Inorganic phosphorus stimulation of bacterioplankton production in a mesoeutrophic lake. Appl Environ Microbiol 57:2074–2078

Tranvik LJ (1988) Availability of dissolved organic carbon for planktonic bacteria in oligotrophic lakes of differing humic content. Microb Ecol 16:311–322

Tranvik LJ (1989) Bacterioplankton growth, grazing mortality, and quantitative relationship to primary production in a humic and a clearwater lake. J Plankton Res 11:985–1000

Tranvik LJ (1990) Bacterioplankton growth on fractions of dissolved organic carbon of different molecular weights from humic and clear waters. Appl Environ Microbiol 56:1672–1677

Tranvik LJ (1992) Allochthonous organic matter as an energy source for pelagic bacteria and the concept of the microbial loop. Hydrobiologia 229:107–114

Tranvik LJ (1993) Microbial transformation of labile dissolved organic matter into humic-like matter in seawater. FEMS Microbiol Ecol 12:177–183

Tranvik LJ (1994a) Effects of colloidal organic matter on the growth of bacteria and protists in lake water. Limnol Oceanogr 39:1276–1285

Tranvik LJ (1994b) Colloidal and dissolved organic matter excreted by a mixotrophic flagellate during bacterivory and autotrophy. Appl Environ Microbiol 60:1884–1888

Tranvik LJ, Jørgensen NOG (1995) Colloidal and dissolved organic matter in lake water: carbohydrate and amino acid composition, and ability to support bacterial growth. Biogeochemistry 30:77–97

Tranvik LJ, Kokalj S (1998) Decreased biodegradability of algal DOC due to interactive effects of UV radiation and humic matter. Aquat Microb Ecol 32, in press

Tranvik LJ, Sieburth J McN (1989) Effects of flocculated humic matter on free and attached pelagic microorganisms. Limnol. Oceanogr.34:688–699

Tranvik LJ,. Granéli W, Gahnström G (1994) Microbial activity in acidified and limed humic lakes. Can J Fish Aquat Sci 51:2529–2536

Turk V, Rehnstam A-S, Lundberg E, Hagström Å (1993) Release of bacterial DNA by marine nanoflagellates, an intermediate step in phosphorus regeneration. Appl Environ Microbiol 58:3744–3750

Wells ML, Goldberg ED (1991) Occurrence of small colloids in sea water. Nature 353:342–344
Westermann P (1993) Wetland and swamp microbiology. In: Ford TE (ed) Aquatic microbiology – an ecological approach. Blackwell, Boston, pp 215 – 238
Wetzel RG (1983) Limnology. Saunders, Philadelphia
Wetzel RG (1984) Detrital dissolved and particulate organic carbon functions in aquatic ecosystems. Bull Mar Sci 35:503–509
Wetzel RG (1992) Gradient-dominated ecosystems: sources and regulatory functions of dissolved organic matter in freshwater ecosystems. Hydrobiologia 229:181–198
Wetzel RG (1995) Death, detritus, and energy flow in aquatic ecosystems. Freshw Biol 33:83–89
Wetzel RG, Hatcher PG, Bianchi TS (1995) Natural photolysis of recalcitrant dissolved organic matter to simple substrates for rapid bacterial metabolism. Limnol Oceanogr 40:1369–1380
Williams PJLeB (1981) Incorporation of microheterotrophic processes into the classical paradigm of the planktonic food web. Kieler Meeresforsch Sonderh 5:1–28
Williams PM, Druffel ERM (1987) Radiocarbon in dissolved organic carbon in the North Pacific Ocean. Nature 330:246–248
Zweifel UL, Norrman B, Hagström Å (1993) Consumption of dissolved organic carbon by marine bacteria and demand for inorganic nutrients. Mar Ecol Prog Ser 101:23–32

11 Food Webs and Carbon Cycling in Humic Lakes

Dag O. Hessen

11.1 Introduction

The concept of Eltonian pyramids has been a cornerstone in mainstream textbooks dealing with food webs. The central dogma is that there is an upward flow of energy based on primary production. The energy captured and stored in organic molecules by photosynthesis is sufficient to support a certain production of primary grazers, which again may support a lesser production of primary predators, and so on. Since there is a considerable "tax" paid in the form of non-digestible matter and respiration losses between and within each trophic level, the production (and frequently the biomass) of each level will have a pyramidal form, and the number of links in a chain will be finite. The food chain length and the food web complexity will be determined largely by the system productivity and the efficiency of energy transfer between levels of the food chain. Embedded in this theory is also the pedagogically attractive concept of distinct trophic levels. This simplified concept of energy flow has, over the past 30 years, been thoroughly criticized and reformulated (Pimm 1982, Polis and Winemiller 1994). Firstly, almost all organisms are omnivores in the sense that they feed on more than one trophic level. Secondly, in all aquatic ecosystems this grazer food chain based on autotrophic production will be supported by a detritus-based food chain utilizing dead organic matter either of autochthonous or allochthonous origin (Odum 1968; Wetzel et al. 1972; Rich 1984; Wetzel 1995). In fact, in most ecosystems only a modest share of primary production will be directly grazed, leaving a major share for the detritus food chain (Odum 1971). There is also a strong recycling component in most systems, and the "detrital loops" based on recycled organic carbon also contradict the conventional one-way flow diagrams (Patten 1985, Higashi et al. 1989, Gaedke et al. 1994).

While these aspects are of general validity to all aquatic systems, they are particularly well illustrated by the carbon (and nutrient) flow of allochthonously influenced systems. The extreme cases among these are the humic lakes, where the pyramidal structure may be entirely inverted, and secondary production may exceed primary production by a factor of 2 owing

Ecological Studies, Vol. 133
Hessen, Tranvik (eds) Aquatic Humic Substances

to the allochthonous supply of energy. To some extent the food webs of humic lakes resemble that of a saprobic system where heterotrophic processes are dominant. The low primary production to respiration ratio also supports a vigorous net export of carbon dioxide to the atmosphere. Benthic methanogenesis constitutes an important part of the detritus carbon pathway, and methane oxidation serves both as an important source of system CO_2-production and a major input of energy to the microbial loop.

This chapter will discuss how, or if, humic lakes deviate from classical food web assumptions and models derived from "clearwater" lakes, and how this affects various trophic levels, their interactions and the entire carbon and nutrient cycling. In doing so, it is important to remember that there is actually a gradual transition from "clearwater" to "brownwater" lakes, and that any lake will exhibit some "footprints" from humic dissolved organic carbon (DOC). However, using the ecosystems strongly influenced by humus as examples might help us understanding the general structuring role of humic DOC in pelagic ecosystems.

11.2 Humic Lake Communities

We may ask whether there exists such a thing as a "typical humic community". Apparently, there is little support for this idea among most taxa, but although there are no species that are clearly unique to humic lakes, there may be species that occur more commonly in humic lakes or rivers. For the phytoplankton, it is a common observation that flagellates in general and cryptomonads in particular tend to be abundant in humic lakes, yet they may also be dominant in non-humic lakes (Jones and Arvola 1984; Ilmavirta 1988; Brettum 1996; Jones, this Vol.). Small, soft-bodied and mixotrophic species may also be highly abundant in coloured lakes (Hessen et al. 1990). The success of these groups presumably depends on the abundance of bacteria as a source of both C and P, rather than the humic matter *per se*. More detailed discussion of these aspects is provided in Chapter 7.

As has been repeatedly demonstrated, heterotrophic bacteria clearly benefit from allochthonous sources of C (DeHaan 1974; Hessen 1985a; Moran and Hodson 1990; see Chap. 10). The taxonomy of free-living bacteria is a science still in its very beginning, yet by the use of RNA-probes, some rough assessments of diversity (although not "species" diversity) may be obtained.

Protozoans are key organisms in a number of humic lakes, and they may constitute a highly important link in the energy flow from bacteria to metazoans (see Chap. 10). As for the phytoflagellates, they may be equally important in non-humic lakes, so that their frequent abundance is not a unique property of humic lakes. For humic lakes, as for lakes in general, the success of microzooplankton depends heavily on the abundance of predators and competitors among the metazoans, and their abundance seems inversely related to the abundance of notably filter-feeding crustaceans like *Daphnia* and *Holopedium*. Their biomass and contribution to elemental cycling may range from negligible in lakes with high abundance of filter-feeders (Hessen et al. 1990; Salonen et al. 1994) to dominating in localities where crustaceans have low biomass (Kankaala et al. 1996). As with bacteria, the taxonomy and diversity of protozoans is still poorly understood, making strict statements on species affinities to humic lakes premature.

No particular metazoan zooplankton can be labelled as an indicator of humic content, either at the species level or at higher taxonomic levels. Berzins and Pejler (1989) reported a wide range of occurrence along a gradient in water colour for most rotifers, yet with most species "preferring" low-humus localities. Likewise, Sarvala and Halsinaho (1990) failed to demonstrate any clearcut pattern for crustacean zooplankton in a large number of Finnish lakes over a wide range of humus content. There may be pronounced differences between humic and non-humic acidified lakes, however. As recognized by Hobæk and Raddum (1980), a number of species sensitive to low pH may still persist at low pH in humic lakes, owing primarily to the complexation of labile aluminium by the organic acids (see Chaps. 1, 2). Yet the same species do occur at higher pH in non-acidified, clearwater lakes. The common notion that humic lakes have low plankton diversity does not hold true (Sarvala and Halsinaho 1990). In their multivariate analysis of 116 Finnish lakes, high diversity was found in humic lakes. Since high humus content in their study correlated well with high calcium and phosphorus content, the apparent positive effect of humus may be supported by Ca and P, since these parameters may favour zooplankton richness (cf. Hessen et al. 1995). Microfiltrators would supposedly benefit from the high biomass of bacteria and small-sized algae, and humic lakes may support a high biomass of large-sized metazooplankton owing to the frequent low biomass of vertebrate predators as well as the low light levels offering a refugium from visual predators.

Both production and diversity of planktonic communities may be as high as, or even higher than, that of oligo- and mesotrophic clearwater lakes, with the possible exception of the most colored lakes (cf. Salonen et al. 1992). This does not hold true for benthic invertebrates and fish communities, however. The strong light attenuation reduces the potential for benthic macrophytes, and in particular those lakes surrounded by *Sphagnum* mats and with well developed, fluffy dy-type (prefix dy from dystrophic) sediments and frequent anoxic hypolimnion provide few available niches and poor spawning opportunities. A few species of corixids, coleopterans, odonates and zygoptera do well in these habitats, while bottom-dwelling shredders and molluscs are poorly represented (Hargeby et al. 1992; Økland 1992). Due to the poorly developed benthic flora and fauna, the following account will mainly deal with the pelagic communities.

11.3 Carbon Pools and Carbon Fluxes

While most statements about humic lake food webs should be made with the proviso that there will always be gradients and exceptions, one property common to all humic lakes is the major flow of carbon entering the food web from allochthonous DOC to heterotrophic bacteria. Owing to this large DOC pool that serves as a major fuel for the heterotrophic base of the food web, the role of the *microbial food* web is clearly less disputable in humic lakes than clearwater lakes or marine waters (see Chaps 8 and 9). Including this pool of refractory, organic C as part of the food web, typical epilimnetic pools and fluxes of a humic lake are shown in Fig. 11.1. This inverted pyramid structure is largely founded on the formidable pool of terrestrially derived humus. Moderately to highly coloured lakes typically have concentrations of total organic carbon (TOC) ranging from 5 to 20 mg C l^{-1}, of which most is dissolved detritus of terrestrial origin. In general, the particulate pool is dominated by detritus, and constitute less than one third of the total carbon. The share of phytoplankton, bacteria, protozoans and metazoans may differ substantially between localities, however. Since bacteria in humic lakes are released from their intimate dependency on exudates from primary producers, they tend to have higher biomass and production relative to phytoplankton compared with clearwater lakes. The different partitioning of the other food-web compartments may be illustrated by the examples from enclosure experiments in highly coloured Lake Mekkojärvi, Finland (Salonen et al. 1992, 1994), moderately coloured Lake Kjelsåsputten, Norway (Hessen et al. 1990, and moderately coloured Lake Pääjärvi, Finland (Arvola et al. 1996; Kankaala et al. 1996).

In the Mekkojärvi experiment, Salonen and coworkers demonstrated the decoupling between bacterial production and phytoplankton production, and the effects of *Daphnia* removal. In both cases, flagellates were key components of the pelagic food web and the dominant grazer of bacteria. When *Daphnia* was present, they consumed ~35% of bacterial production (slightly more than the flagellates did), while flagellates probably constituted the most important food source for *Daphnia*. Thus, most of the bacterial production was channelled to *Daphnia*, either directly or via the flagellates. The role of metazoans may be more pronounced with regard to the P flux. Using P radiotracer (32P) in a later experiment in the same lake, the overwhelming role of daphnids for P flux was demonstrated (Salonen et al. 1994), where ~85% of all P in organisms was in *Daphnia*, and the major flux was from dissolved P, via bacteria to *Daphnia*.

➔

Fig. 11.1. Major pools and fluxes of carbon and phosphorus in summer in humic lake Kjelsåsputten. Relative sizes of pools (as mg C m^{-3}) are indicated by size of *boxes*. Direction of fluxes (as mg C m^{-3} day^{-1}) is shown by *arrows*. DIC Dissolved inorganic carbon (CO_2); DOC dissolved organic carbon; POC particulate organic carbon; A algae; B bacteria; Z zooplankton. Corresponding display for phosphorus pools and fluxes below. (Based on Hessen et al. 1990 and Hessen 1992a)

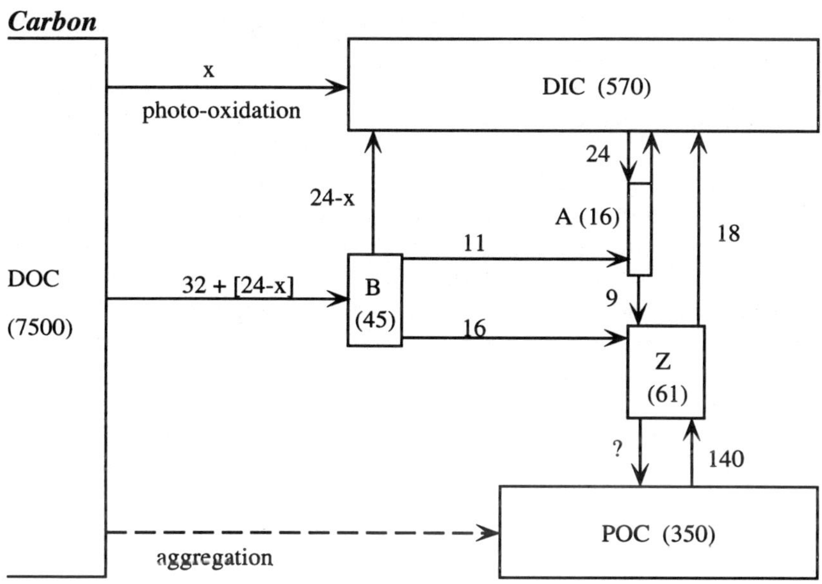
Carbon
x
DIC (570)
photo-oxidation
24
24-x
A (16)
18
11
DOC
32 + [24-x]
B
9
(7500)
(45)
16
Z
(61)
?
140
POC (350)
aggregation

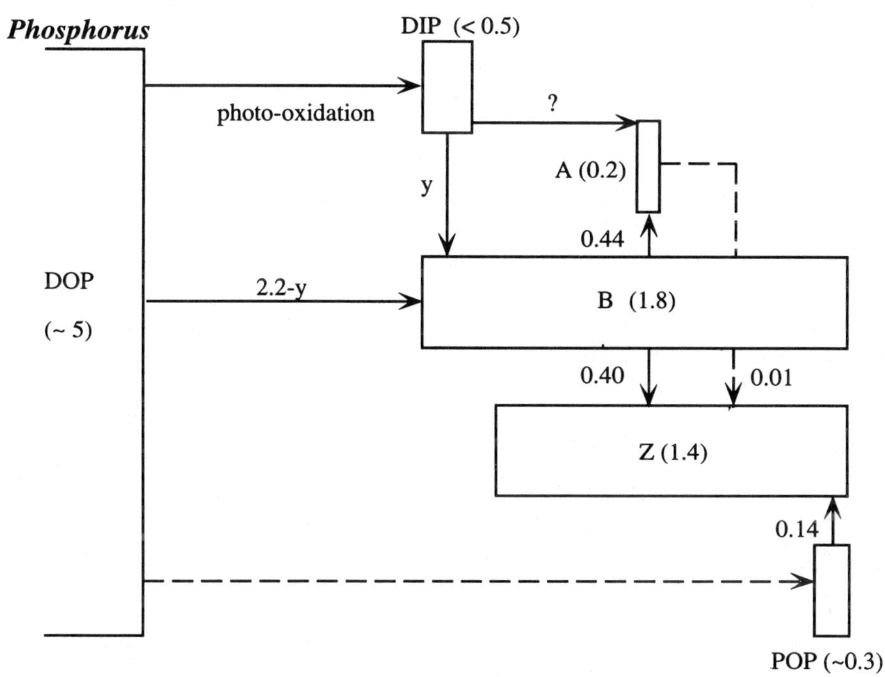
Phosphorus
DIP (< 0.5)
photo-oxidation
?
A (0.2)
y
0.44
DOP
2.2-y
B (1.8)
(~ 5)
0.40
0.01
Z (1.4)
0.14
POP (~0.3)

Tracer studies on carbon dynamics in the Norwegian humic lake Kjelsåsputten (~ 8 mg C l^{-1}, Hessen et al. 1990) clearly illustrated the heterotrophic nature of this allochthonously influenced system. Total phytoplankton biomass was 16 μg C l^{-1}, accounting for only 0.2% of TOC, or 3.4% of particulate organic carbon (POC), of which the particulate detritus was totally dominant. Total heterotrophic biomass (zooplankton and bacteria) was 106 μg C l^{-1}, i.e. a biomass ratio of 6.6:1 between consumers and producers. To this add that the phytoplankton pool comprised a large fraction of mixotrophs, and the tracer studies indicated that the phytoplankton obtained one-third of their carbon from heterotrophic activity (grazing on bacteria). Total gross heterotrophic production (bacteria, zooplankton and heterotrophic phytoplankton) was 51 μg C l^{-1} day^{-1}, while gross fixation of inorganic carbon was 24 μg C l^{-1} day^{-1}. Total system respiration was 44 μg C l^{-1} day^{-1} (net efflux to the atmosphere). This yields a production to respiration ratio of nearly 0.5. Even assuming an exudation of DOC from phytoplankton equalling 20% of gross primary production, and a 100% assimilation of these exudates, this could not sustain more than 6% of net bacterial production. This is in close agreement with the estimates of Jones and Salonen (1985).

A set of enclosure experiments in the Finnish humic lake Pääjärvi confirmed the flux from DOC to bacteria as the dominating carbon pathway (Arvola et al. 1996, Kankaala et al. 1996), and that community respiration far exceeded primary production. The fate of bacterial carbon differed from that of Lake Kjelsåsputten, however. In Lake Pääjärvi, daphnids were absent, and protozoans constituted the major link to metazoans. Owing to this dissipation of energy in the microbial food chain, additions of humus and phosphorus stimulated productivity of algae and bacteria, but only marginally influenced the metazoans. The addition of extra humus and phosphorus revealed a stimulation of primary production, and a subsequent stimulation of bacterial production that was not due directly to the addition of humus or phosphorus, but rather to increased exudation of organic carbon from algae. This contrasts with previous radiotracer studies from humic lakes (Hessen et al. 1990, Salonen et al. 1994) where a tight link from humus DOC via bacteria to metazoans was found. This again calls for caution when generalizing foodweb properties of humic lakes. In line with the findings of Hessen et al. (1990), Kankaala et al. (1996) noted a fairly high biomass of metazoans relative to protozoans, but very low production. The production to biomass ratio (P:B) for metazooplankton ranged from around 0.05 in autumn to 0.15 in summer, while the corresponding P:B ratio for protozoans was almost tenfold higher (~ 0.5–1.0). While this partly reflects the lower intrinsic growth rate of the metazoans relative to protozoans, it could also be indicative of nutrient deficiency in the metazoans.

Methane oxidation constitutes another major source of carbon for the pelagic food web of a number of lakes (Rudd and Hamilton 1978), and in particular in humic lakes that provide both a steady supply of organic matter and sufficiently low hypolimnetic oxygen levels for a vigorous methanogenesis. Estimates from Lake Kjelsåsputten indicated that methane oxidation amounted to 150 µg C l^{-1} day^{-1} over the o to 3 m depth in early summer (Hessen and Nygaard 1992). The proportion of oxidized CH_4-C converted to cell material is variable (Rudd and Hamilton 1978; Harrits and Hanson 1980). Assuming that 30% is converted to cell biomass, methane sustains a bacterial production of 45 µg C l^{-1} day^{-1}, i.e. exceeding the net production of heterotrophic bacteria (32 µg C l^{-1} day^{-1}). While the quantification is fairly preliminary, it nevertheless supports the view of a carbon metabolism in humic lakes that is largely driven by external supplies of carbon and thus differs fundamentally from the general concept of energy flow in clearwater lakes.

These aspects also have important bearings on the balance and exchange of inorganic carbon with the atmosphere. While the general picture of a lake dominated by autochthonous production is of a net flux of CO_2 from atmosphere to water, this is reversed in humic lakes (Salonen et al. 1983; Hessen et al. 1990; Kankaala et al. 1996). These lakes are net CO_2 conduits to the atmosphere owing to at least four reasons: the low primary production and carbon fixation, the vigorous respiration and dominance of heterotrophic processes, the frequently extensive methane oxidation and a pronouned photo-oxidation of organic molecules by short-wave light. The acid nature of these lakes implies low bicarbonate buffering and generally low levels of CO_2 in spite of supersaturation. The extent to which inorganic carbon in the form of CO_2 may be limiting for primary production in such lakes is not settled. These processes will also influence the entire food web, in particular the benthic fauna, by reducing ambient oxygen concentrations in the entire water column, and frequently inducing hypolimnetic anoxia.

Humic lakes frequently have higher concentrations of total phosphorus than corresponding clearwater lakes (see Chaps. 7, 8). Yet they are commonly assumed to have low phytoplankton yield per total P partly owing to low accessibility of P, chelating of Fe or other essential elements, low light levels or monopolizing of P by bacteria. These aspects depend primarily on the P concentration and humus concentrations, and reduced production of phytoplankton in coloured lakes will only occur if phytoplankton production actually is light-limited (see Chap. 7). A sample of 150 Norwegian lakes with dominance of low-coloured lakes (maximum 80 mg Platinum l^{-1}; Fig. 11.2) indicated that, although total P was the major determinant of phytoplankton biomass, colour (as milligrams of Platinum per litre) gave a significant negative contribution to the phytoplankton biomass yield per unit of P even for these low-coloured lakes (Hessen, unpubl. data). The contribution from colour strongly depends on the P-concentrations of the lake, however, as seen from a multiple analysis including both parameters. The phytoplankton biomass from the same sample of lakes may then be expressed as:

$$PB = (96.3 - 0.91 \times Pt) \times TP$$

where PB is phytoplankton biomass expressed as micrograms per litre, Pt is colour expressed as milligrams per litre and TP is total P expressed as micro-

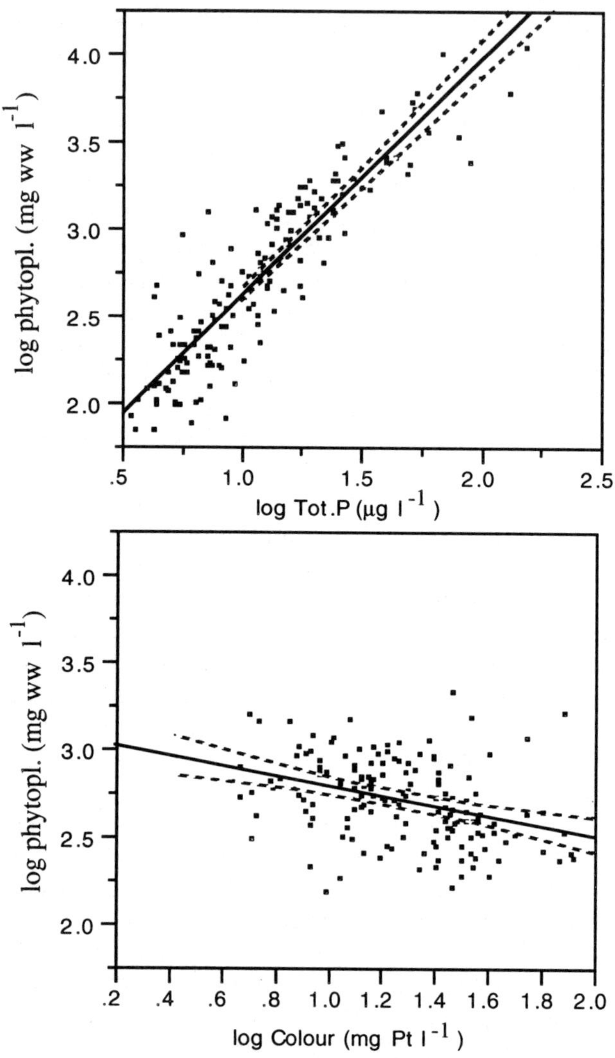

Fig. 11.2. Relationship between phytoplankton biomass and total P and colour from a survey of 150 Norwegian lakes. (unpubl. data)

grams per litre,. For more coloured lakes, this trend is even more clearcut (Heyman and Lundgren 1988, Chap. 8). On the other hand, Arvola et al. (1996) found no negative effect on phytoplankton biomass and production along a gradient of humus additions to enclosures. The negative effects of low penetration of photosynthetically active radiation (PAR-light) on photosynthesis, may partly be counteracted by even lower penetration of detrimental shortwaved UV-light (cf. Chaps. 6,7).

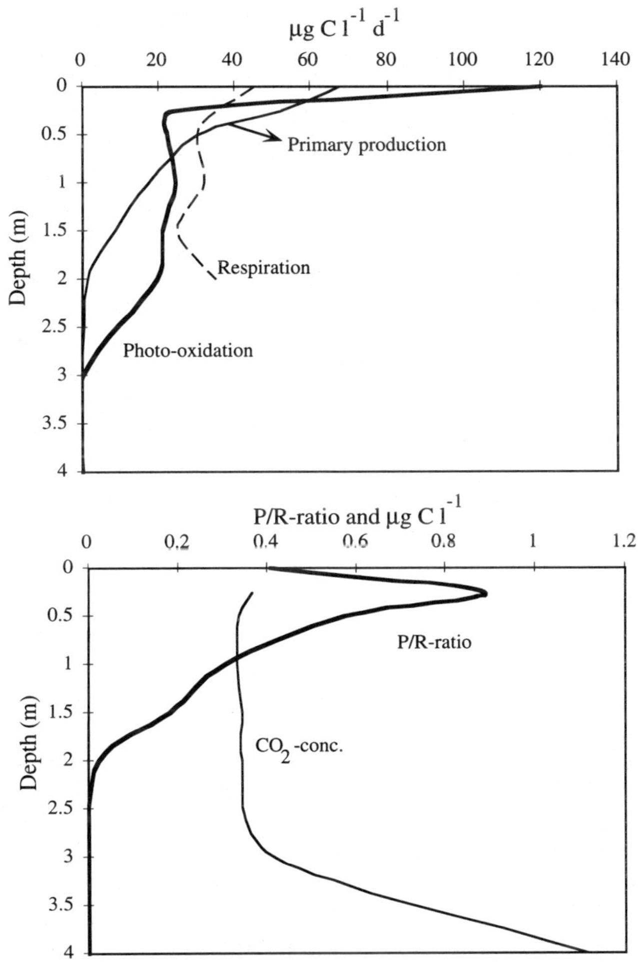

Fig.11.3. Vertical profiles of CO_2-production and consumption and the P:R ratio in a summer situation in humic lake Skjervatjern

The CO_2 fixation deficiency relative to actual community respiratory losses is not a unique property of strongly coloured lakes (Kling et al. 1992; del Giorgio and Peters 1993; del Giorgio et al. 1997). Based on a survey of published data, del Giorgio and Peters (1993) found that the epilimnetic P:R ratio (primary production to community respiration) rose above unity only in highly productive lakes (>17 µg chlorophyll a l^{-1}). If considering the mixed layer only, the P:R-balance was reached at about 10 µg chlorophyll a l^{-1}. At the oligotrophic end of the scale (0.5 µg chlorophyll a l^{-1}), the P:R ratio decreased to 0.5. They reached the conclusion that microbial respiration may strongly exceed current estimates based on bacterial production, and that heterotrophic processes are relatively more important in oligotrophic lakes.

Although the authors were not able to define the specific role of allochthonous DOC in their lakes, they suggested humic substances to be a major parameter explaining these patterns. The lower P:R ratio of 0.5 is similar to the integrated epilimnetic ratio of Lake Kjelsåsputten (see above). This ratio only included the epilimnetic processes, but when integrating over the entire water-column, the P:R ratio would decrease substantially (cf. Hessen and Nygaard 1992). This is seen clearly for the Norwegian humic lake Skjervatjern, where vertical profiles of primary production, photo-oxidation, dark respiration, as well as CO_2 concentrations were performed during summer 1992 (cf. Brettum 1994; Salonen and Vähätalo 1994; Hessen et al. 1997). While the P:R ratio was close to 0.5 at the surface, it increased to 0.8 just below the surface, and approached zero below 2 m (Fig. 11.3). This yields a mean P:R ratio of 0.2 over the upper 4 m, and far less if integrating over the entire water column. While this may underline some of the problems associated with obtaining reliable estimates of P:R ratios, it also supports the heterotrophic nature of these humic lakes. Though both factors contribute, the low P:R ratio in humic lakes thus depends more on high respiratory activity than depressed primary production.

11.4 Carbon Sources for Metazoans

There are at least two properties of humic lake food webs that differ essentially from clearwater lakes: their dependence on allochthonous energy and the overall importance of the microbial loop. The central role of bacteria relates to the fact that not only their biomass and production both often far exceed those of the phytoplankton, but also the humic environment often supports a highly efficient bacteria-consuming community. Mixo- and heterotrophic flagellates are reported to consume a significant share of bacterial production, eventually channelling it to higher trophic levels, yet there are conflicting reports on the relative importance of the unicellular bacterivores (Jones 1992; Chap. 9). Part of this apparent conflict may be owing to the abundance of crustacean zooplankton, particularly cladocera. Many of these species are highly efficient consumers of bacteria, and may constitute an important link in a three-step chain from DOC to fish. The efficiency of such a direct link is far higher than one with even one or two additional links (flagellates, ciliates and copepods) where a major portion of carbon entering the microbial food web is oxidized before entering the metazoan level. Moreover, the magnitude of this loop based chiefly on external sources of carbon will be far higher than one based solely on phytoplankton exudates.

The ability to utilize the bacterial pool is a cornerstone of energy transfer in humic lakes, and this property differs strongly within the crustacean zooplankton community. Of the common cladocerans in humic lakes, *Diaphanosoma brachyurum* and *Daphnia longispina* (and other *Daphnia* spp.)

are highly efficient bacterivores (Hessen 1985b, Kankaala 1988, Hessen et al. 1989). *Holopedium gibberum* and *Bosmina longispina* are far less efficient, yet they do utilize bacteria, while copepods and notably cyclopods barely utilize free-living bacteria. They may, however, graze on detrital particles that are colonized with bacteria. There is less information about the digestibility of heterotrophic bacteria. These are mostly Gram-negatives with heavy peptidoglycan cell-wall layers, offering some resistance to digestive enzymes. In general, bacteria are assimilated with lower efficiency than algae, yet at least for *Daphnia* and *Diaphanosoma*, assimilation efficiency was found to exceed 50% (Hessen et al. 1989). For the above mentioned species, there is a remarkable relationship between their specific P content and their ability to filter bacteria (Fig. 11.4), which could be an adaptive response to cope with high P demands.

While this and preceeding chapters support the view of a lesser importance of phytoplankton in strongly coloured lakes relative to clearwater lakes, there is no doubt that in particular qualitatively, but also quantitatively, the grazers still depend on a phytoplankton supply. While some mineral nutrients or organic N may be adsorbed or bound in humus aggregates, these are evidently poor food that is deprived of almost all essential macromolecules and easily degradable sources of C, yielding very low assimilation efficiencies of typically 0–20% (Sterner and Hessen 1994). While the bacteria may provide a major source of P for grazers (Hessen and Andersen 1991), they too are poorly assimilated relative to algae (Hessen et al. 1989) due the rigid peptidoglycan cell layer. Moreover, they lack most of the long-chained

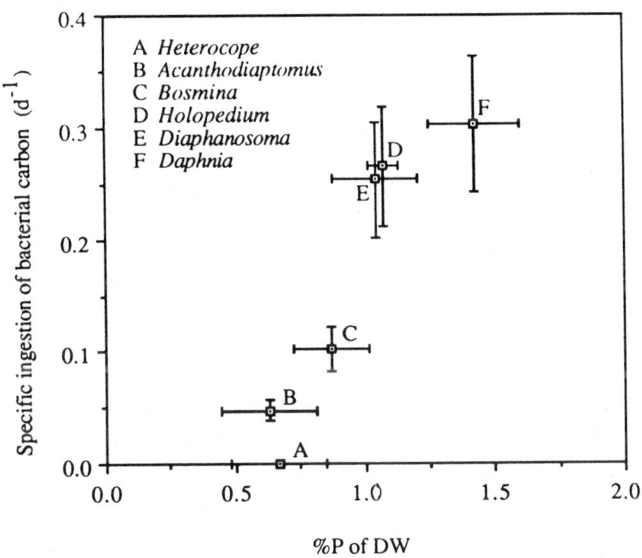

Fig. 11.4. Relationship between specific P content and specific ingestion of bacteria from humic lake Kjelsåsputten (Hessen and Andersen 1991)

unsaturated fatty acids that may be crucial for the zooplankton nutrition (Muller-Navarra 1995). A high proportion of the auto- or mixotrophic algae commonly recorded in humic lakes are grazing-resistant species or taxa, however (Hessen et al. 1990). A particular case was noted in a 7-year study of catchment and lake acidification of humic Lake Skjervatjern, western Norway (Hessen and Lydersen 1996; Chap. 3). While a minor depression in pH was observed in the acidified basin relative to the control (maximum 0.2 units from 4.6 to 4.4), changes in DOC and labile aluminium were not detected. The most striking change in the zooplankton community was the disappearance of the assumed acid-tolerant species *Holopedium gibberum* in the acidified basin. While this effect was not clearly attributable to changes in water chemistry, a remarkably close correlation was seen between the biomass of the dominant small chlorophycean *Oocystis* and that of *Holopedium*. The *Oocystis* group is fairly grazing-resistant, and may even benefit from grazing and gut passage (Porter 1975). Thus, in this case, the correlation could be an inverse to what would normally be assumed; *Oocystis* disappeared when the grazer disappeared.

Mixotrophic algae and heterotrophic flagellates are certainly a major link of carbon from DOC via bacteria in a number of humic lakes (Salonen et al. 1992, Kankaala et al. 1996), also providing an indirect source of bacterial carbon for species less capable of direct consumption of bacteria (*Bosmina*, *Holopedium*, copepods).

While dissolved organic carbon constitutes a major fuel for heterotrophic bacteria, the POC may constitute an almost equally important source of energy for the grazers. The potential role of detritus as a food source for zooplankton in humic lakes was noted already by Naumann (1918), who stated that "... the *Entomostraca* in these localities gain most of their nutrients from purely allochthonous detritus, while phytoplankton or detritus derived from plankton hardly contributes at all ...". The observation of predominating "amorphous" brown, humic compounds in the guts of cladocera in humic localities was taken as evidence for the above citation. Haney (1973) concluded from his extensive studies in humic Drowned Bog Lake, USA, that "..the high grazing and low algal productivity in Drowned Bog Lake indicates that allochthonous matter, for example from the *Sphagnum* mat, may provide an important input of nutrition into the bog's open-water system". A number of works have demonstrated the ability of various zooplankton to survive for periods on detrital particles as the sole diet. By use of two-compartment labelling (inorganic ^{14}C for algae and an organic ^{14}C amino-acid mixture for bacteria) of seston in enclosures, Hessen et al. (1990) estimated the relative share of bacteria, phytoplankton and detritus (unlabelled carbon) incorporated as somatic matter in the dominating crustacean zooplankters. Estimates based on both initial labelling and steady state conditions revealed a surprising pattern, with 50% or more of body C of all species originating from detrital C, and ~ 30% originating from bacteria, leaving a share of only ~20% of phytoplankton C in the diet (Fig. 11.5). The highest proportion of

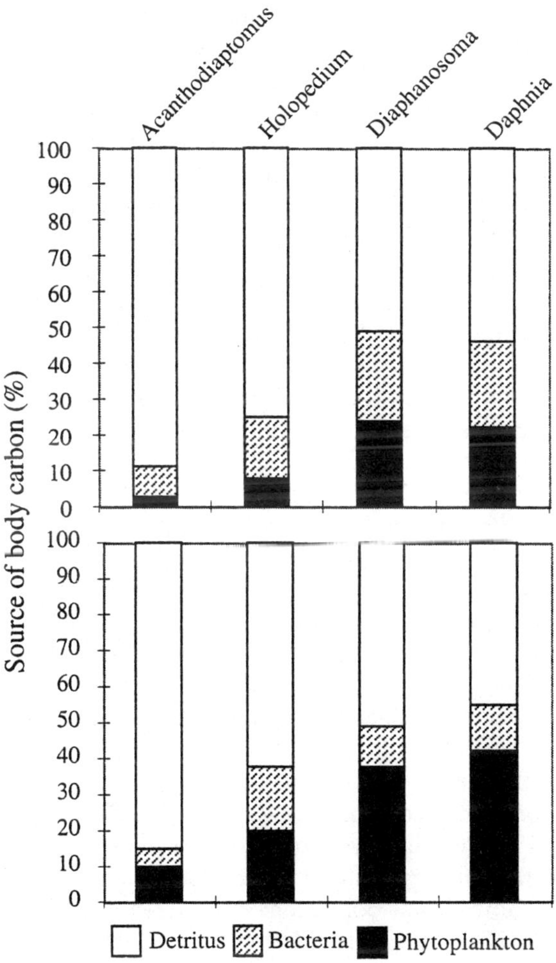

Fig. 11.5. Share of detritus, bacteria and phytoplankton in the diet of the dominating crustacean zooplankton in Lake Kjelsåsputten based on separate labelling of food compartments. Above based on initial labelling with net clearance rates. Below based on specific activity in sestonic fractions and animals at isotope equilibrium. (Hessen et al. 1990)

bacterial C was found in the two "microfiltrators" *Daphnia* and *Diaphanosoma*. Perhaps most suprising was the importance of detrital C for the copepod *Acanthodiaptomus denticornis*, which apparently gained more than 80% of its body C from detrital matter. For selectively feeding copepods, one would anticipate a higher percentage of high-quality food than for the non-selective cladocerans. These data are in support of Salonen and Hammar (1986) who, by tracer labelling of seston from different lakes, found that in excess of 80% of metazooplankton carbon in humic lakes was derived from non-living (detrital) C. There are obvious sources of error to this estimates, such as a large proportion of unlabelled protozoans in the diet, or leakage of label between compartments, yet the data indicate that metazoans may derive a substantial part of their body C directly from allochthonous detritus.

Further support for these radiotracer studies is given by stable isotope studies. The distinctive difference in $\delta^{13}C$ and $\delta^{15}N$ values between allochthonous C and phytoplankton in Loch Ness, Scotland, allowed Jones et al. (1998) to conclude that nearly 50% of the body C of zooplankton in this lake was derived from allochthonous C, either directly or via bacteria. This is particularly interesting because it demonstrates the potential importance of humic C also in large lakes like Loch Ness, and also because a strong direct link from allochthonous C to fish could be verified from this study.

A large share of such nutritionally poor food poses questions not only about the digestibility in terms of C, but also about how zooplankton cope with deficiencies in mineral nutrients or essential macromolecules. Assimilation of detrital matter will differ strongly with the detrital origin; recently dead algae may be assimilated with significantly higher efficiency than humus aggregates. In humic lakes, where the latter source may be dominant, a maximum 10% assimilation efficiency of ingested detritus is reasonable (Hessen 1992b). Owing to the mass of this reservoir relative to live seston, it will still constitute a potentially large pool of dietary C. On the other hand, even though bacteria and phytoplankton in extreme cases may constitute minor sources of dietary C, their qualitative importance as sources of P, N, fatty acids and other nutrients may be crucial to zooplankton productivity (cf. Sterner and Hessen 1994).

Though protozoans exploit both the phytoplankton and the bacterial food source, they too may benefit from a direct utilization of "dissolved" or particulate humus carbon. It is well known that protozoans may utilize a range or organic macromolecules and organic aggregates (Sherr 1988, Kristoffersen et al. 1996). The study of Sherr showed that polysaccharide molecules (dextran) of relatively large sizes (>500 000 Da) provided enhanced growth in heterotrophic flagellates, while smaller molecular sizes did not stimulate growth, suggesting phagocytosis as the major uptake mechanism. Yet, the role of recalcitrant humus aggregates as a dietary source for proto- and metazoans is far from conclusive.

The size of the DOC pool and the importance of the microbial loop can be labelled as specific properties of the humic lake relative to clearwater lakes. Yet the *efficiency* of C flow through the detritus food chain versus the grazer food chain will be determined more by the food web structure than the presence of humus DOC. A flagellate-dominated C pathway will increase dissipation of organic C to CO_2 (Kankaala et al. 1996). Gaedke et al. (1994) proposed to track a pulse of ^{14}C-label through the two food chains to determine the relative trophic transfer efficiency. A short residence time of the label (short half-lives) would indicate a substantial respiratory loss, and thus low efficiency. In line with these assumptions, they modelled far lower residence

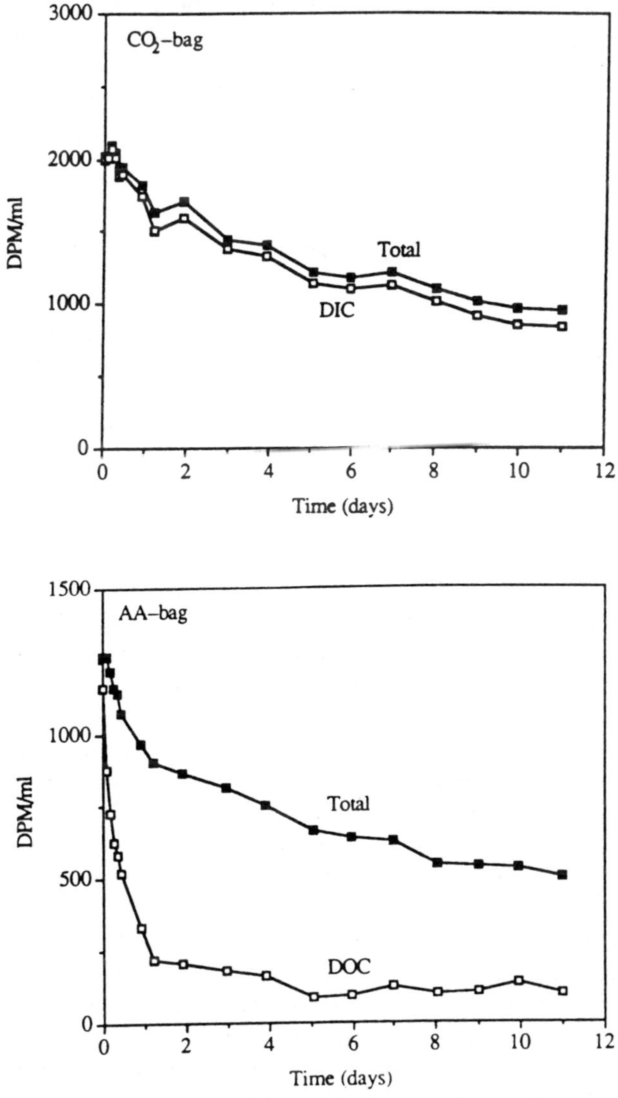

Fig. 11.6. Isotopic distribution and loss in a bag labelled with inorganic C (^{14}C bicarbonate); *CO_2-bag* and organic C (^{14}C amino acid mixture); *AA-bag.* DPM Desintegrations per minute. Label was lost at a higher rate from the detritus-chain (initial labelling of bacteria) relative to the grazer food chain (initial labelling of phytoplankton) (Hessen et al. 1990)

times for one unit of C in the grazer food chain relative to the detritus food chain in mesotrophic Lake Constance, Germany. Thus, while a large detritus pool (mainly autochthonous detritus) fuelled a high bacterial biomass, the detritus food chain (microbial loop) was still a very inefficient link of C due to the high respiratory losses. This was partly confirmed by a real tracer experiment of the same kind in humic Lake Kjelsåsputten (Hessen et al. 1990), where the bacteria was labelled by an organic ^{14}C-tracer and the phytoplankton by an inorganic ^{14}C-tracer in separate enclosures, showing that label was lost at a higher rate from the detritus food chain (initial labelling of bacteria) relative to the grazer food chain (initial labelling of phytoplankton) (Fig. 11.6).

11.5 Stoichiometry in the Humic Lake Food Web

While focusing on C in this context, C is but one of various elements that is needed for support of both autotrophs and heterotrophs. The skewed C:N and particularily C:P ratios of both dissolved and particulate pools in humic lakes strongly suggest that N or P may be limiting also for heterotrophs in these lakes. Though even pristine humic lakes commonly have fairly high concentrations of P (typical range 5–20 µg P l^{-1}) relative to their clearwater counterparts, the huge pools of DOC and POC still cause fairly high C:P ratios. Based on a large number of Swedish lakes, Meili (1992) depicted a range of C:P weight ratios ranging from 50 in biota to 100 in particulate detritus (of autochthonous origin), 350 in oligohumic epilimnion, 1500 in polyhumic epilimnion and up to 2000 in coloured forest streams. This is in line with data from a Norwegian humic lake, in which also an average epilimnetic C:P ratio of 1500 was reported (Hessen 1992a). In the epilimnion of an extremely coloured Finnish lake, Salonen et al. (1994) reported a C:P ratio of more than 1800:1, falling to 450:1 in the hypolimnion. Judging simply from stoichiometric ratios, bacteria with a C:P ratio of typically 25 would thus be expected to face a strict P limitation in these kinds of lakes (Vadstein and Olsen 1989; Hessen 1992b, Hessen et al. 1994). Such aspects have been more explored in multicellular heterotrophs such as crustacean zooplankton, which have a rather rigid stoichiometry (Sterner and Hessen 1994). Yet, even though bacteria have a more flexible stoichiometry, a corresponding approach may be fruitful also for this group. Tezuka (1990) demonstrated how bacteria change from mineralizers to consumers along a gradient of increasing C:P ratios in the medium, and clearly the role of bacteria in P cycling depends upon their stoichiometry relative to that of the medium (Elser et al. 1995). The extent to which the bacteria are C- or P-limited will depend not only on the elemental ratios, but also on their ability to assimilate the elements. P may be bound to Fe or adsorbed to humus aggregates, while C is largely bound up in complex, refractory macromolecules that are not easily assimilated. By assuming that bacteria maintain a rather rigid stoichiometry, a simple scenario for P release may be given: Q_d represents C:P in the medium (DOM), here set to 1500, Q_b represents C:P in bacteria, r_c the respiratory losses of C, and γ_c and γ_p represent assimilation of C and P, respectively, the corresponding biomasses being B_c and B_p. Then the bacteria will release P when $Q_b > Q_d \times B_c[\gamma_c - (\gamma_c\, r_{c)}]/B_p\gamma_p$ (cf. Hessen 1992b).

The fact that heterotrophs monopolize a major share of P in humic lakes has obvious implications for ecosystem structure. The importance of the heterotrophs relates both to their high proportions of total biomass and their high specific P content and thus P requirements. In culture, an extreme variability in C:P ratio may be induced in both bacteria and phytoplankton. A less flexible stoichiometry is normally encountered in nature. Even P-starved bacteria normally have a C:P ratio of less than 25 (Vadstein and Olsen 1989), while freshwater phytoplankton frequently have a C:P ratio far above 100 (Elser and Hassett 1994; Sterner and Hessen 1994). By contrast, crustacean zooplankton species have a more rigid stoichiometry, but with pronounced differences between species and taxa (Andersen and Hessen 1991). *Daphnia* spp. are characterized by a high specific P content, fairly stable around 1.5% of DW, or a C:P weight ratio of 30. Other cladocera that typically inhabit humic lakes, such as *Bosmina longispina* and *Holopedium gibberum*, have a lower specific P content and a typical C:P ratio around 45–50, while copepods normally have C:P ratios above 65.

The relative importance of bacteria in the P pool will tend to be high as a natural consequence of the high proportions of bacterial biomass to other biotic compartments. Hessen and Andersen (1991) estimated that approximately 55% of particulate P in a humic lake was bound in bacteria, and more than 45% in zooplankton, making the phytoplankton pool of P rather marginal. Salonen et al. (1994) found that 85% of total P was bound in the *Daphnia* biomass of a highly humic lake. These figures may be extremes, yet it is important to recall that when heterotrophic biomass is dominant in terms of C, this will be even more dramatic in terms of P. This is illustrated in Fig. 11.1, where epilimnetic pools and fluxes of C and P of a humic lake are compared. The relative pool sizes are fundamentally altered from C to P, and the predominant role of the heterotrophs is even more remarkable. It is important to notice that although the bacteria do not fill the conventional role of "remineralizers", depicted in most text-book food web diagrams, there is an important link from bacteria to phytoplankton also for P. This is not a loop via dissolved inorganic phosphorus (DIP), however, but direct ingestion by mixotrophic algae. Since the phytoplankton are doomed to lose the battle of P in such systems with low P and excess organic C (see Chaps. 7,8), they have apparently adopted the strategy that "if you can't beat them, eat them". Although there is a secondary gain in terms of C, it is reasonable to assume that shortage of P is the major cue for mixotrophy (Nygaard and Tobiesen 1993).

The higher trophic levels are also controlled by the deficiency in mineral nutrients relative to C in humic lakes. There is both theoretical and empirical support for the view that zooplankton, notably P-demanding species like *Daphnia*, may be directly limited by food quality in terms of C:P ratio (Hessen 1992b; Sterner and Hessen 1994). Thresholds for limitation depend on the relative assimilation efficiency for C versus P, but are normally predicted at C:P >65–120. To cope with their bodily demands, assimilated C: P must equal somatic C:P in the consumer. That is, when C is in excess more C must be disposed off (probably as increased defecation of C), and growth efficiency in terms of C decreases. On the contrary, animal utilization of P will be more efficient, and as the grazer approaches P limitation, P assimilation is maximized and P-release approaches zero. This is analogous with the arguments for bacterial P metabolism.

Using *Daphnia* as an example, relative P limitation can be estimated from the C:P ratio of the grazer relative to that of the food. Ingestion rates and assimilation efficiency also need to be known for both P and C for all food items. In order to balance somatic demands, the daily net ingestion of C (including respiratory losses) relative to net ingestion of P must equal the C:P ratio of the grazer. Using again the Lake Kjelsåsputten example, zooplankton in this lake had a net C ingestion of 19 μg C l^{-1} day^{-1} from algae and bacteria, barely balancing respiratory losses (18 μg C l^{-1} day^{-1}). Assuming a 5% assimilation efficiency for detritus, this would add another 7 μg C l^{-1} day^{-1} to zooplankton production. Maximizing P assimilation to 100% from all fractions, corresponding intake of P would be 0.63 μg P l^{-1} day^{-1}, or a C:P ratio in incorporated matter of 41 (by weight). Reducing P assimilation to 80% would yield a C:P-ratio of 52, still below what would be considered P-limiting for *Daphnia*. As long as zooplankton is on the brink of starvation in terms of food quantity (C limitation), the threshold C:P ratio for P-limitation is infinite (cf. Sterner 1997). With increasing food quantity, the potential for P limitation is introduced. Of 34 freshwater sites in northern Canada with sestonic C:P ratios ranging 75–360 (by weight), a majority were found to be P limited with regard to growth of *Daphnia* (Elser and Hassett 1994; Sterner 1997). Humic lakes with their high sestonic C:P ratios would be likely candidates for P-limited zooplankton, yet this may be obscured by the huge pool of detritus that is a major contributor to C, but which is assimilated with very low efficiency. The efficient consumption of P-rich bacteria may be seen as a strategy to counteract P deficiency.

The low light levels of humic lakes could, paradoxically, lead to high food quality of phytoplankton. Following the light to nutrient balance theory of Urabe and Sterner (1996), high light levels and low nutrient concentrations give high photosynthetic activity with C in excess relative to P. Converesly, a low light to nutrient ratio (like that of humic lakes) yields high food quality. Urabe and Sterner (1996) clearly demonstrated that such shifts in the light to nutrient ratio indeed influenced phytoplankton quality and thus the success of the grazer. Use of such models and stoichiometric models in general may not apply entirely in humic lakes. They may also be obscured by the relatively more complex P metabolism of these systems, since the major pool of dissolved organic phosphorus (DOP) is associated or complexed with iron and humus molecules, and probably not very accessible to organisms (Jones 1990; Salonen et al. 1994).

11.6 The Trophic Level Concept

Returning to the pyramidal structure and the trophic level concepts, these are not easily tracked in humic lakes. Most functional groups do operate on various trophic levels. The conventional food web diagram is drawn by inspection of species intersection points, reflected in a matrix where absence or presence of species interactions is qualitatively indicated (Cohen et al. 1986). Among the criticisms that have been raised against such simplified and mechanistic food web diagrams, the overall presence of omnivory is a major issue (Riley 1966; Odum 1968; Martinez 1991; Polis 1991; Morin and Lawler 1995), that is, animals feed on different trophic levels, and this is even more pronounced when seasonality, different ages, size groups or stages are included. In general, the degree of omnivory tends to increase the number of trophic levels in lakes (Sprules and Bowerman 1988). Omnivory is particularly important in humic lakes. The predominance of allochthonous inputs yields algal, bacterial and zooplankton communities that all feed on a multitude of trophic levels. The primary producers are commonly mixotrophs feeding on bacteria, i.e. they may be both "producers" and "consumers". The bacteria themselves are not "mineralizers" in the common sense; they fill a multitude of trophic levels. First they are chemoautotrophs like the methanotrophs, they may be consumers to the extent that they consume extracellular release from phytoplankton, or they may be producers with a functional role corresponding to that of phytoplankton, except that they convert DOC rather than DIC to cell material. One single *Daphnia* may share its intake of carbon between autotrophic and mixotrophic phytoplankton, heterotrophic and methanotrophic bacteria and detritus of mixed origin, from particulate feces to humus aggregates. Defecated matter apparently returns to the detrital pool, being reingested a number of times. This kind of "coprophagy" is noted in other aquatic systems, as well as from terrestrial herbivores, and is a way to extract a maximum of limiting elements from the food. This again makes it a difficult task to determine the real assimilation efficiency for detritus. The relative importance of the different carbon sources may vary seasonally.

The dietary flexibility may be even greater when considering species with a more pronounced life-cycle omnivory, e.g. nauplii vs. adult copepods. Stable isotope studies ($\delta^{13}C$ and $\delta^{15}N$) of planktonic food web confirm that actual food-web structure may be poorly predicted from simple considerations of species list and expected trophic interactions between species (Kling et al. 1992).

In line with Ulanowicz (see Halfon et al. 1996), Gaedke et al. (1994) proposed to distinguish trophic levels (the discrete values 1, 2, 3 etc.) from trophic position (non-integer values) where an organism is assigned to position from the average number of trophic transfers its food items have passed before consumption by the given predator. That is, if a copepod is satisfying

33% of its demands by grazing and 66% of its energy from consumption of a *Bosmina* (a strict grazer), its trophic position would be $2 \times 0.33 + 3 \times 0.66 = 2.64$. The problem with this, of course, is the assignment of trophic levels. In this case, primary producers are assigned to level 1, primary grazers to level 2 etc. The real problems arise from the fact that most compartments of the humic lake food web are fuelled from detritus of mixed origin, and the assignment of trophic level for detritus thus becomes critical. The importance of this pool to the whole food web was recognized by Lindeman (1942), who placed the huge pool of "ooze" central in his food web diagram, but still failed to account for its role in carbon cycling. Detritus may be classified as a dead source of energy and regarded as an equivalent to solar energy, or it may be classified as belonging to the first trophic level (like the autotrophs) owing to the fact that it is directly consumed by heterotrophs in accordance with food-web schemes from soil ecosystems (cf. Moore and DeReuter 1991). Burns et al. (1991) classified dead organic matter to the trophic position of its sources, but this creates a multitude of trophic levels of detritus that are not easily handled. Burns (1989) criticized the conventional assignment of species or populations to distinct trophic levels, yet argued for a trophic level concept as an abstract composite of those portions of the energy content that have been assimilated an equal number of times. This theoretical food network "unfolding" re-establishes the trophic level concept (Higashi et al. 1989), but requires intimate knowledge of energy flow in the system.

The calculation of trophic position for omnivorous species hinges on assignment to discrete trophic levels for phytoplankton, bacteria and detritus. The approach of Gaedke et al. (1994) distinguished between the detritus food chain and the grazer food chain. The base of each chain (detritus and phytoplankton, respectively) was assigned to trophic level 1, heterotrophic bacteria to level 2, and then the fractional contribution of both chains to each species may be computed according to its diet composition. This kind of definition still invokes a type of "catch 22" problem, since detritus is of a very heterogeneous origin, a large proportion of phytoplankton may be mixotrophs and heterotrophic bacteria derive their energy from a multitude of sources.

This may be illustrated by the methane pathway, a branch of the detritus chain. Particularly in humic lakes, the potential importance of methane calls for a clarification of trophic levels of the methanogens and the methanotrophs. Since methanogens derive their energy partly from detritus but also utilize by-products of photo-oxidation (like acetic acid), they should equal the heterotrophic bacteria with regard to trophic level, but then the methanotrophs (methane oxidizers) should be at a higher trophic level. If detritus is assigned to trophic level 1, then the heterotrophic bacteria (primarily utilizing detritus as an energy source) should be assigned to trophic level 2, as would be the methanogens, while the methanotrophs would be trophic level 3, one level above the zooplankton by which they are grazed.

The same problem was recognized by Martinez (1991). Despite a higly laboreous resolution of the food web of Little Rock Lake, USA, he assigned fine

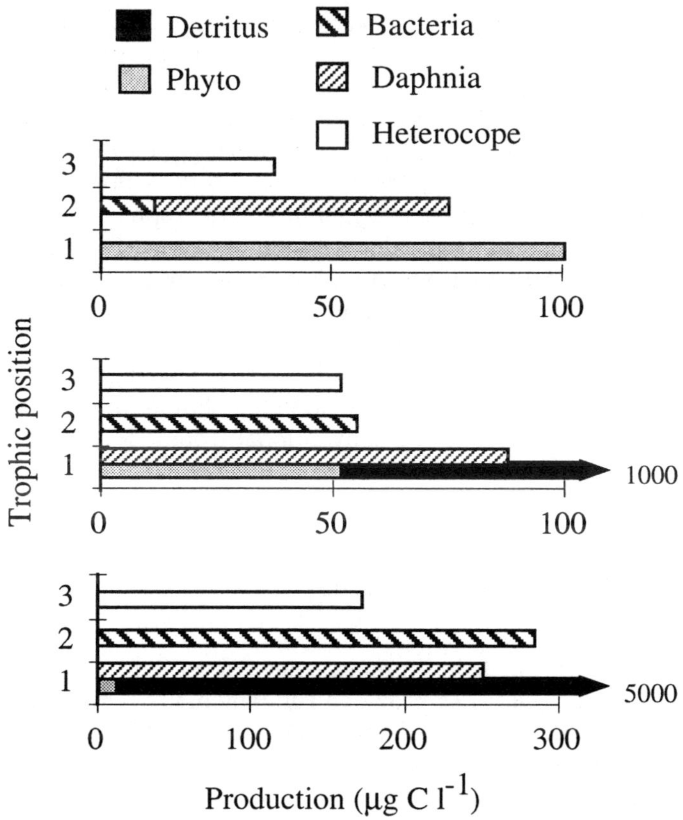

Fig. 11.7. Examples of trophic position assignments based on three scenarios. *Above* a pure grazing chain in which where bacteria derive their energy from phytoplankton exudates Primary production (100 µg C l^{-1} day^{-1}) may support a production of 10, 64 and 38 µg C l^{-1} day^{-1} of bacteria, Daphnia and Heterocope; respectively; *Middle* detritus is added to the diet (scenario 2), assuming a 7.5% growth efficiency for both bacteria and *Daphnia* feeding on detritus. *Daphnia* enters trophic position 1.32; *Below* increasing detritus (scenario 3) provides a fundamentally different carbon flow, but does not change scenario 2 trophic positions

organic matter into a single trophic level while acknowledging that this was "... a severely aggregated taxon assumed to be at trophic level 1". Some of these aspects are illustrated in by the three scenarios in Fig. 11.7. In the first scenario, a pure grazing chain, in which bacteria derive their energy from phytoplankton exudates, the primary production (100 µg C l^{-1} day^{-1}) may support a production of 10, 64 and 38 µg C l^{-1} day^{-1} of bacteria, *Daphnia* and the carnivorous copepod *Heterocope*, respectively. *Daphnia* is assumed to graze with equal efficiency on phytoplankton and bacteria, but having a growth efficiency (i.e. ingestion minus defecation and respiration losses) of 60% and 38% respectively (cf. Hessen et al. 1989). *Heterocope* is feeding solely on Daphnia with a growth efficiency of 60%. In this case, when daphnids obtain 6% of their carbon from bacteria and the rest from phytoplankton, their trophic position is 1.06 (1 × 0.94 + 2 × 0.06). Since *Daphnia* is assigned *a priori* to trophic level 2, *Heterocope* ends up at trophic position (and trophic level) 3. We may then in a second scenario add detritus to the diet (assuming a growth efficiency of 7.5% for both bacteria and *Daphnia* feeding on detritus). Since both phytoplankton and detritus are assigned to trophic level 1, their relative contribution to bacteria or *Daphnia* does not matter. Bacteria occupy trophic position 2, while *Daphnia* enters trophic position 1.32 (2 × 0.33 + 1 × 0.66). Increasing the detritus contribution (scenario 3) provides a fundamentally different carbon flow, but leaves the trophic positions unchanged from scenario 2. This is all a matter of definitions that will not have any practical implication for carbon flow studies based on tracers, but could have a great impact on food-web studies based on judgement of trophic positions. Allochthonously dominated systems should more than others call for precautions when using the conventional trophic level or the modified trophic position concepts.

11.7 Trophic Cascades and Energy Cycling

The exploration of cascading effects in limnetic, pelagic food webs (see Carpenter et al. 1985; Carpenter 1988) has provided fundamentally new insight into pelagic food-web interactions. Based on numerous lake manipulation and enclosure studies, top-down (piscivorous fish, planktivorous fish, herbivorous metazooplankton, phytoplankton biomass and composition, inorganic nutrients) or *vice versa* bottom-up mechanisms have been explored. Almost all these experiments have been performed in meso- and eutrophic clearwater lakes. The trophic cascade effects fade as one proceeds from the initial signal (i.e. manipulated fish biomass or nutrients; McQueen et al. 1986), and large *Daphnia* appears to be a key component for an efficient signal transfer between the top and the base of the food web and vice versa. Such trophic cascades have scarcely been explored in strictly humic lakes, yet there are several reasons to assume *a priori* that they would not work in line with general predictions from studies of clearwater lakes. First, there is a scarcity of planktivorous fish relative to eutrophic lakes and these are supposedly less efficient predators due to low light intensities. Second, the microbial food web plays a fundamentally more important role in humic relative to clearwater lakes and, from a bottom-up perspective, increased load of inorganic nutrients will not be directly converted to phytoplankton biomass.

This simplified scheme of energy flow and alternating regulation of food-web compartments via predation and competition deliberately omits recycling processes. The food-web diagram based on direct interactions is further obscured by the fact that recycling is such a pronounced property of aquatic ecosystems. Carbon is recycled at each and any level in the food web, meaning that not only the direct links between prey and consumer must be included, but also the feedbacks via recycling that may be equally important. This implies that the total amount of energy flowing through the system is far higher that that estimated from primary production or heterotrophic production alone. The recycling aspect of matter and energy in allochthonously influenced lakes is explored in a number of model papers (Patten 1985; Higashi et al. 1989; Patten et al. 1990), all conflicting the central dogma that mineral nutrients are recycled while energy is not (cf. Odum 1971, 1983). Inclusion of carbon (and thus energy) cycling such as in network trophic dynamics accounts for all the energy in a system until it dissipates as heat or CO_2.

In all food webs, the direct interactions are accompanied by a set of feed back processes due to mineral cycling, an *interactive* system in the terminology of Caughley (1976). While there is growing knowledge of these processes in lakes (Sterner et al. 1994, Hessen 1997), they are not well documented in humic lakes, where a large reservoir of humus-bound P and N also may participate in the recycling processes.

Some of these assumptions were partly contradicted by an experiment in which planktivorous whitefish (*Coregonus lavaretus*) were added to the strongly coloured Lake Mekkojärvi, resulting in an almost complete disappearance of the previously dominant *Daphnia longispina* (Järvinen and Salonen, 1997) . The vulnerability of daphnids in this lake was partly accredited to the extreme stratification of the lake, with anoxia below 0.6–0.9 m and thus no depth refugium. The shift from a *Daphnia*-dominated to a flagellate-dominated grazer chain did invoke changes in stoichiometry and phytoplankton nutrient limitation as well. In the absence of *Daphnia*, more P was recycled (less P bound in *Daphnia*) and the phytoplankton was more N-limited in line with stoichiometric theory (Sterner and Hessen 1994).

11.8 The Role of Detritus in Ecosystem Stability

Detritus was first defined by Odum and de la Cruz (1963) as "biodetritus" in the form of dead, particulate matter, including its associated microflora and fauna. Rich and Wetzel (1978) redefined this to non-predatory losses of organic carbon from all trophic levels, which would also include dissolved, allochthonous sources. The role of detritus in ecosystems was recognized in the pioneering works of Odum (1968, 1971), yet did not until recently fully merge with the conventional food web analysis based on distinct trophic levels. This is in spite of the fact that Lindeman (1942) developed the concept based on studies in Cedar Bog Lake (Minnesota), a highly detritus influenced ecosystem. The importance of detritus food webs and carbon recycling via detritus is now well recognized (Patten et al. 1990, DeAngelis 1992), but for the coming discussion it should be remembered that most of these detritus studies are based on autochthonous production. Allochthonous particulate matter (such as leaf litter) may be a major source of energy particularily for stream ecosystems, and also for some clearwater lake ecosystems (Larsson and Tangen 1975). Both these types of predominantly particulate detritus are essentially different from humus DOC.

A major question is whether the presence of a large (and relatively stable) pool of recalcitrant, organic carbon can help stabilize the entire food web. On simply intuitive grounds, there is reason to believe that such systems with a continuous supply of high quantities of low quality food and a pronounced degree of omnivory would be more stable than ecosystems based on a short spring burst of high-quantity and high-quality food and a preceding clear-water phase (Hessen 1990, Jones 1992). This allochthonous pool could thus violate general rules of aquatic ecosystem stability and "resilience" based on number of species and their interactions (cf. Sprules and Bowerman 1988, Havens 1993). This intuitive reasoning is partly contradicted by modelling prey, predator, detritus interactions along gradients of increasing detritus pools (DeAngelis 1992). The tendency for limit cycling (local instability) increases with increasing detritus fraction in simple four-compartment model systems (nutrients, autrotroph biomass, heterotroph biomass and detritus). Loss of bound nutrients from detritus into heterotroph biomass was suggested (DeAngelis 1992) as a potential reason for this apparent paradox, meaning that detritus in this case fuelled primary production, and that detritus enrichment would equate to a situation with nutrient enrichments. Such enrichments invariably yield decreased instability (cf. Rosenzweig 1971). It should be stressed, however, that systems dominated by a food source deprived of phosphorus (such as detritus) would not support recycling of P (Sterner and Hessen 1994). Hessen and Bjerkeng (1997) also found that P limitation could induce decreased stability by analysing limit cycles in a three-compartment system along gradients of the C:P ratio in the autotroph biomass. At sufficiently high C:P in the food, the most P-demanding grazers show pronounced oscillations, and finally die off. Note, however, that such simple models with dissolved P, algae and grazer will not reflect the dynamics of complex food webs, and particularly not the food webs of humic lakes where a major share of production is fuelled by heterotrophic processes. Moreover, systems dominated by allochthonous detritus may also behave quite differently from those dominated by autochthonous detritus.

What is clear is that detritus has a strong impact on ecosystem resilience. In systems where organic C is dominated by the detritus pool, changes in food-web compartments will yield less effect than changes in the detritus compartment (Dudzik et al. 1975; DeAngelis 1992), that is, such systems are resilient to disturbance of conventional food-web compartments, while less resilient to changes in the detritus pool. This is fairly intuitive, since detritus makes up by far the larger pool of C.

Though fairly obvious, it should also be stressed that allochthonous systems are "donor"-fed, i.e. they rely on import from neighbouring ecosystems, and that production by species in the recipient ecosystems is "subsidized" by

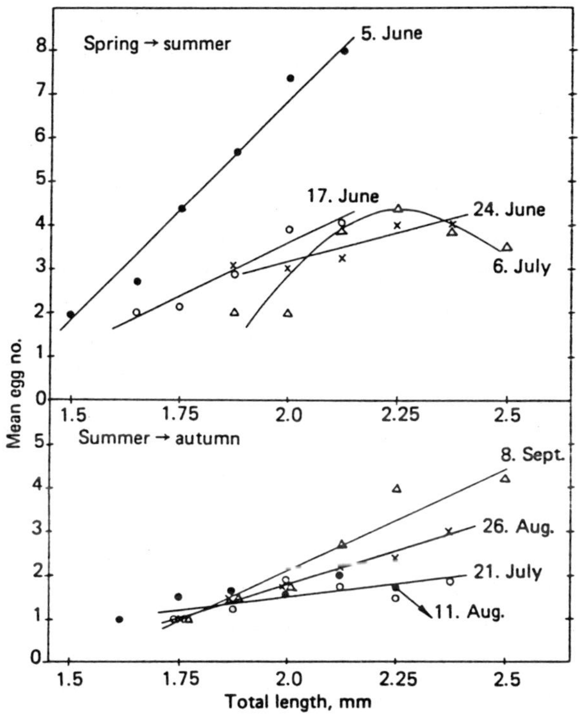

Fig. 11.8. Seasonality in size-specific egg production of Daphnia longispina in humic lake Kjelsåsputten (Hessen 1989)

allochthonous matter. Thus, they are bottom-up controlled in a manner that makes them rather insensitive to within-ecosystem changes and trophic cascades (Persson et al. 1994; Polis and Hurd 1994), but would be sensitive to drainage basin changes such as increased temperature and reduced precipitation, with subsequent effects on DOC supply to watersheds and increased transparency to short-waved radiation (Schindler and Curtis 1997).

While the excessive supply of recalcitrant C would be one major stabilizing factor, the P dynamics in humic systems would add to this. Planktonic freshwater communities are characterized by strong dynamics and rapid species replacements owing to high turnover rates, high growth capacities for all planktonic compartments and thus strong effects on competition, grazing and predation (Hairston and Hairston 1993). The strong oscillations hinge on a dynamic P pool, with a high pulse of accessible P in spring, and a subsequent depletion of this pool during summer. In contrast, humic lakes frequently have a substantial reservoir of P in the form of humus, Fe, PO_4 complexes (Chap. 7) that may support a steady production of phytoplankton. Also, the large store of P in a relatively stable bacterial biomass probably provides a sufficient source of P for the microfiltrators (Hessen and Andersen 1991) and promotes increased ecosystem stability. The seasonal succession of clearwater lakes, as described by the PEG-model (Plankton Ecology Group) (Sommer et al. 1986), does not apply well to humic lakes, in which the seasonal dynamics normally are less pronounced (Hessen 1990). In contrast to the rapid oscillations promoted by a dynamic pool of highly edible, high-quality food like (most) phytoplankton, the large pool of recalcitrant C in excess will tend to dampen these oscillations, and channel a steady flow of energy through the heterotrophs (Wetzel 1995). Preliminary ecosystem stability models support this view (Carpenter and Pace 1997), showing that pulses of humus to lakes tend to reverse very slowly, in contrast to the P-mediated pulses of chlorophyll a in clearwater lakes.

Though this chapter has focused on the stability of humic lakes, such lakes are of course also subject to seasonal variability. The input of allochthonous matter in spring and autumn provides the ecosystem with not only a fresh supply of inorganic nutrients, but also a fresh source of not yet degraded humus. In spite of minor changes in total P and chlorophyll over the ice-free season in humic Lake Kjelsåsputten, Hessen (1989) found a strong seasonal variability in egg production and nutritional status (length to dry-weight ratio, bodily C:N-ratio) of all cladoceran species, with peak productivity in spring, a mid-summer depression, and again increased productivity towards autumn (Fig.11.8). This seasonality was not correlated with phytoplankton biomass, but corresponded well with periods of snowmelt or rainfall, supplying fresh fuel of humus. Biomass of both bacteria and (to a lesser extent) phytoplankton increased also during these periods, underlining their possible qualitative importance. A similar pattern was seen in humic Lake 979 in the Exerimental Lakes Area (ELA), Canada, where population growth and biomass of *Daphnia rosea* were associated with flooding periods and thus assumed P mobilization from peat or otherwise increased food quality (Paterson et al. 1997). Seasonal changes in quantity and quality of humus may thus affect seasonal production of bacteria. In nutrient enrichment assays of water from humic lakes, Hessen et al. (1994) found most response to N and P additions in spring and autumn, while glucose additions only stimulated bacterial production when N and P were added simultaneously. Thus, they suggested that during spring and autumn, fresh inputs of humic matter gave C in excess and mineral nutrient deficiency in the bacteria, while during summer with low allochthonous inputs and predominance of "old" refractory humus, bacteria were co-limited by C, N and P. Owing to the importance of the microbial food chain, seasonal effects in bacterial production will affect the entire food web. Hessen (1990) suggested that zooplankton communities in humic lakes in general would be less susceptible to seasonal oscillations and competitive exclusion over the season, due to the large and stable supply of low-quality food. In such a situation, there will always be sufficient C from humus and bacteria for maintenance metabolism, allowing coexistence of species at the brink of starvation at very low growth rates, even during periods with low phytoplankton biomass.

11.9 Summary

In the search for unique properties of humic lakes, it should again be emphasized that there is a transition from "clearwater" to "humic lakes", and that humic matter is ever present in natural surface water. To pinpoint the particular features of humic localities, this chapter has dealt with real "brownwater" lakes, focusing on the pelagic carbon flow. The unifying feature of these localities is the huge pool of allochthonous recalcitrant humus DOC that provides the base of an entirely heterotrophic food chain. This detritus-based food chain and the grazer food chain (based on autotrophic production) merge at the protozoan or metazoan level. The relative importance of the carbon flux in these two food chains will differ between lakes, but the detritus food chain will in general be far more important in humic lakes relative to "clearwater" lakes. The predominance of heterotrophic processes violates the general concept of the pyramidal structure in the Lindeman tradition with a large phytoplankton pool at the base. It also creates systems with production to respiration ratios far below unity and hence systems that are net conduits of CO_2 to the atmosphere. The fact that the production of these systems is fuelled by detritus of mixed origin, including the methanogens, undermines the trophic level and trophic position concept of the foodweb. Ecosystem stability tends to increase with increasing influence of humus. In contrast to the rapid oscillations promoted by a dynamic pool of highly edible, high-quality food like (most) phytoplankton, the large pool of recalcitrant carbon in excess will tend to dampen these oscillations, and channel a steady flow of energy through the heterotrophs.

References

Andersen T, Hessen DO (1991) Carbon, nitrogen and phosphorus content of freshwater zooplankton. Limnol Oceanogr 36:807–814

Arvola L, Kankaala P, Tulonen T, Ojala A (1996) Effects of phosphorus and allochthonous humic matter enrichment on the metabolic processes and community structure of plankton in a boreal lake (Lake Pääjärvi). Can J Fish Aquat Sci 53:1646–1662

Berzins B, Pejler B (1989) Rotifer occurrence in relation to water colour. Hydrobiologia 184: 23–28

Brettum P (1994) Acidification of the humic lake Skjervatjern; effects on the volume and species composition of phytoplankton. Environ Int 20:313–320

Brettum P (1996) Changes in the volume and composition of phytoplankton after experimental acidification of a humic lake. Environ Int 22:619–628

Burns TP (1989) Lindeman's contradiction and the trophic structure of ecosystems. Ecology 70:1355–1362

Burns TP, Higashi M, Wainwright SC, Patten BC (1991) Trophic unfolding of a continental shelf energy-flow network. Ecol Model 55:1–26.

Carpenter SR (ed) (1988) Complex interactions in lake communities. Springer, Berlin, Heidelberg, New York

Carpenter SR, Pace ML (1997) Dystrophy and eutrophy in lake ecosystems: implications of fluctuating inputs. Oikos 78:3–14

Carpenter SR, Kitchell JF, Hodgson JR (1985) Cascading trophic interactions and lake productivity. BioScience 35:634–639

Caughley G (1976) Plant herbivore systems. In: May RM (ed) Theoretical ecology: principles and applications (Ed. R M May) Saunders, Philadelphia pp 94–113

Cohen JE, Briand F, Newman C (1986) A stochastic theory of community food webs. III. Predicted and observed length of food chains. Proc R Soc Lond [Biol] 228:317–353

DeAngelis DL (1992) Dynamics of nutrient cycling and food webs. Chapman and Hall, New York

DeHaan H (1974) Effects of fulvic acids, amino acids, and carbohydrates on the growth of a pseudomonas from Tjeukemeer (The Netherlands). Freshw Biol 4:301–310

del Giorgio PA, Peters RH (1993) Balance between phytoplankton production and plankton respiration in lakes. Can J Fish Aquat Sci 50:282–289

del Giorgio PA, Cole JJ, Cimleris A (1997) Respiration rates in bacteria exceed phytoplankton production in unproductive aquatic systems. Nature 385:148–150

Dudzik M, Harte J, Levy D, Sandusky J (1975) Stability indicators for nutrient cycles in ecosystems. LBL-3265 Berkeley, California. Lawrence Berkeley Laboratory, University of California, Berkeley

Elser JJ, Hassett RP (1994) A stoichiometric analysis of zooplankton phytoplankton interactions in marine and freshwater systems. Nature 370:211–213

Elser JJ, Stabler LB, Hassett RP (1995) Nutrient limitation of bacterial growth and rates of bacterivory in lakes and oceans: a comparative study. Aquat Microb Ecol 9:105–110

Elton C (1927) Animal Ecology. Macmillan, New York

Gaedke U, Straile D, Pahl-Wostl C (1994) Trophic structure and carbon flow dynamics in the pelagic community of large lakes. In: Polis G, Winemiller K (eds). Food webs. Integration of patterns and dynamics. Chapman and Hall, New York, pp 60–71

Hairston NG Jr, Hairston NG Sr (1993) Cause effect relationships in energy flow, trophic structure, and interspecific interactions. Am Nat 142:379–411

Halfon E, Schito N, Ulanowicz RE (1996) Energy flow through the Lake Ontario food web: Conceptual model and an attempt at mass balance Ecological Modelling 86:1–36

Haney JF (1973) An in situ examination of the grazing activities of natural zooplankton communities. Arch Hydrobiol 72:87–132

Hargeby A, Petersen RC Jr, Kullberg A, Svensson M (1992) Benthic macroinvertebrates along the soil/water interface of the HUMEX Lake 1989–1991. Environ Int 18:659–667

Harrits SM, Hanson RS (1980) Stratification of aerobic methane-oxidizing organisms in Lake Mendota, Madison, Wisconsin. Limnol Oceanogr 25:412–421

Havens K (1993) Effect of scale on food web structure responses. Science 260:243–244

Hessen DO (1985a) The relation between bacterial carbon and dissolved humic compounds in oligotrophic lakes. FEMS Microbiol Ecol 31:215–223

Hessen DO (1985b) Filtering structures and particle size selection in coexisting cladocera. Oecologia 66:368–372

Hessen DO (1989) Factors determining the nutritive status and production of zooplankton in humic lakes. J Plankton Res 11:649–664

Hessen DO (1990) Niche overlap between herbivorous cladocerans; the role of food quality and habitat homogeneity. Hydrobiologia 190:61–78

Hessen DO (1992a) Dissolved organic carbon in a humic lake; effects on bacterial production and respiration. Hydrobiologia 229:115–123

Hessen DO (1992b) Nutrient element limitation of zooplankton production. Am Nat 140: 799–814

Hessen DO (1997) Stoichiometry in food webs – Lotka revisited. Oikos 79:195–200

Hessen DO, Andersen T (1991) Bacteria as a source of phosphorus for zooplankton. Hydrobiologia 206:217–223

Hessen DO, Andersen T (1992) The algae grazer interface: feedback mechanisms linked to elemental ratios and nutrient cycling. Arch Hydrobiol Beih Ergeb Limnol 35:111–120

Hessen DO Bjerkeng B (1997) A model approach to planktonic stoichiometry and predator-prey stability. Freshw Biol 38:447–471

Hessen DO, Lydersen E (1996) The zooplankton story of humic lake Skjervatjern during whole catchment acidification. Environ Int 22:643–652

Hessen DO, Nygaard K (1992) Bacterial transfer of methane and detritus; implications for the pelagic carbon budget and gaseous outputs. Arch Hydrobiol Beih 37:139–148

Hessen DO , Nygaard K , Salonen K, Vähätalo A (1994) The effect of substrate stoichiometry on microbial activity and carbon degradation in humic lakes. Environ Int 20:67–76

Hessen DO, Andersen T, Lyche A (1989) Differential grazing and resource utilization of zooplankton in a humic lake. Arch. Hydrobiol 114:321–347

Hessen DO, Andersen T, Lyche A (1990) Carbon metabolism in a humic lake: pool sizes and cycling trough zooplankton. Limnol. Oceanogr. 35:84–99

Hessen DO, Faafeng BA, Andersen T (1995) Replacement of herbivore zooplankton species along gradients of ecosystem productivity and fish predation pressure. Can J Fish Aquat Sci 52:733–742

Hessen DO , Gjessing E T , Knulst J, Fjeld E (1997) DOC-fluctuations in a humic lake as related to catchment acidification, season and climate. Biogeochemistry 36:139–151

Heyman U, Lundgren A (1988) Phytoplankton biomass and production in relation to phosphorus: some conclusions from field studies. Hydrobiologia 170:211–228

Higashi M, Burns TP, Patten B C (1989) Food network unfolding: an extension of trophic dynamics for application to natural ecosystems. J Theor Biol 140:243–246

Hobaek A, Raddum GG (1980) Zooplankton communities in acidified regions of south Norway. SNSF-project, IR75/80, Ås, Norway, 132 pp

Ilmavirta V (1988) Phytoflagellates and their ecology in Finnish brown-water lakes. Hydrobiologia 161:255–270

Järvinen M, Salonen K (1998) Nutrient limitation of phytoplankton production under changing food web structure in a highly humic lake. Can J Fish Aquat Sci (in press)

Jones R I (1990) Phosphorus transformations in the epilimnion of humic lakes: biological uptake of phosphate. Freshwat. Biol. 23:323–337

Jones RI (1992) The influence of humic substances on lacustrine planktonic food chains. Hydrobiologia 229:73–91

Jones RI, Arvola L (1984) Light penetration and some related characteristics in small forest lakes in southern Finland. Verh Int Verein Theor Angew Limnol 22:811–816

Jones RI, Grey K, Sleep D, Quarmly C (1998) An assessment, using stable isotopes, of the importance of allochthonous organic carbon sources to the pelagic food web in Loch Ness. Proc R Soc Lond B 265:105–111

Jones RI, Salonen K (1985) The importance of bacterial utilization of released phytoplankton photosynthate in two humic forest lakes in southern Finland. Holarct Ecol 8:133–140

Kankaala P (1988) The relative importance of algae and bacteria as food for Daphnia longispina. Freshw Biol 19:285–294

Kankaala P, Arvola L, Tulonen T, Ojala A (1996) Carbon budget for the pelagic food web of the euphotic zone in a boreal lake (Lake Pääjärvi). Can J Fish Aquat Sci 53:1663–1674

Kling GW, Fry B, O´Brien JW (1992) Stable isotopes and planktonic trophic structure in Arctic lakes. Ecology 73:561–566

Kristoffersen K, Bernard C, Ekebom J (1996) A comparison of the ability of different nanoflagellates to incorporate dissolved macromolecules. Arch Hydrobiol 48:73–84

Larsson P, Tangen K (1975) The input and significance of particulate terrestrial organic carbon in a subalpine freshwater ecosystem. In: Wielgolaski FE (ed) Fennsocandian tundra ecosystems. I. Plants and microorganisms. Ecological studies vol 16. Springer , Berlin Heidelberg New York, pp 351–359

Lindeman RI (1942) The trophic dynamic aspect of ecology. Ecology 23:399–418

Lotka AJ (1925) Elements of physical biology. Williams and Wilkins, Baltimore

Martinez ND (1991) Artifacts or attributes? Effects of resolution on the Little Rock Lake food web. Ecol Monogr 61:367–392
McQueen DG, Post JR, Mills EL (1986) Trophic relationships in freshwater pelagic ecosystems. Can J Fish Aquat Sci 43:1571–1581
Meili M (1992) Sources, concentrations and characteristics of organic matter in softwater lakes and streams of the Swedish forest region. Hydrobiologia 229:23–41
Moore JC, DeReuter PC (1991) Temporal and spatial heterogeneity of trophic interactions within below-ground food-webs. Agric Ecosyst Environ 34:371–397
Moran MA, Hodson RE (1990) Bacterial production on humic and non-humic components of dissolved organic carbon. Limnol Oceanogr 35:1744–1756
Morin PJ, Lawler SP (1995) Food-web architecture and population-dynamics; theory and empirical evidence. Annu Rev Ecol Syst 26:505–529
Muller-Navarra DC (1995) Evidence that highly unsaturated fatty acids limit Daphnia growth in nature. Arch Hydrobiol 132:297–307
Naumann E (1918) Über die natürliche Nährung des limnischen Zooplanktons. Lunds Univ Årsskr, NF Avd 2, 14:1–47
Nygaard K, Tobiesen A (1993) Bacterivory in algae: a survival strategy during nutrient limitation. Limnol Oceanogr 38:273–279
Odum EP (1968) Energy flow in ecosystems; a historical review. Am Zool 8:11–18
Odum EP (1971) Fundamentals in ecology. Saunders, Philadelphia
Odum EP, de Lacruz AA (1963) Detritus as a major component of ecosystems. Bull Am Inst Biol Sci 13:39–40
Odum HT (1986) Energy in Ecosystems. In: N Polunin (Ed.): Ecosystem Theory and Application, John Wiley and Sons Ltd., Chichester. pp 337–369
Økland J (1990) Lakes and snails. Universal Book Services/Bachuys, Oegstgeest
Paterson MJ, Findlay D, Beaty K, Findlay W, Hendzel L, Schindler E, Stainton M, McCullough G (1997) Changes in the planktonic food web of a new experimental reservoir. Can J Fish Aquat Sci 54:1088–1102
Patten BC (1985) Energy cycling in the ecosystem. Ecol Model 28:1–71
Patten BC, Higashi M, Burns T P (1990) Trophic dynamics in ecosystem networks: Significance of cycles and storage. Ecol Model 51:1–28
Persson L, Anderson G, Hamrin SF, Johanson L (1988) Predator regulation and primary production along the productivity gradient of temperate lake ecosystems. In: Carpenter S R (ed) Complex interactions in lake communities. Springer, Berlin, Heidelberg, New York, pp 45–65
Persson L, Bengtsson J, Menge BA, Power ME (1995) Productivity and consumer regulation – concepts, patterns and mechanisms. In: Polis G, Winemiller K (eds) Food Webs. Integration of patterns and dynamics. Chapman and Hall, New York pp 396–434
Pimm SL (1982) Food webs. Chapman and Hall, London
Pimm SL, Lawton JH (1978) On feeding on more than one trophic level. Nature 275:542–544
Polis G (1991) Complex trophic interactions in deserts: an empirical critique of food web theory. Am Nat 138:123–155
Polis G, Hurd S (1994) Allochthonous inputs across habitats, subsidized consumers, and apparent trophic cascades: Examples from the ocean-land interface. In: Polis G and Winemiller K (eds) Food webs. Integration of patterns and dynamics. Chapman and Hall, New York, pp 275–285
Polis G, Winemiller K (eds) (1994) Food Webs. Integration of patterns and dynamics, Chapman and Hall, New York
Porter KG (1975) Viable gut passage of gelatinous green algae ingested by Daphnia. Nature 244:179–180
Raubenheimer D, Simpson SJ (1994) The analysis of nutrient budgets. Funct Ecol 8:783–791
Rich P H (1984) Trophic detrital interactions: vestiges of ecosystem evolution. Am Nat 123: 20–29
Rich PH, Wetzel RG (1978) Detritus in the lake ecosystem. Am Nat 112:57–71

Riley GA (1966) Theory of food-chain relations in the ocean. In: Hill M N (ed)The sea : Interscience, London, pp 438–463
Rosenzweig ML (1971) Paradox of enrichment: destabilization of exploitation ecosystems in ecological time. Science 171:385–387
Rudd JWM, Hamilton RD (1978) Methane cycling in a eutrophic shield lake and its effects on whole lake metabolism. Limnol Oceanogr 23:337–348
Salonen K, Hammar T (1986) On the importance of dissolved organic matter in the nutrition of zooplankton in some lake waters. Oecologia 8:246–253
Salonen K, Vähätalo A (1994) Photochemical mineralization of dissolved organic matter in Lake Skjervatjern. Envir Int 20:307–312
Salonen K, Kononen K, Arvola L (1983) Respiration of plankton in two small, polyhumic lakes. Hydrobiologia 101:65–70
Salonen K, Kankaala P, Tulonen T, Hammar T M, James, Metsälä T-R, Arvola L (1992) Planktonic food chains of a highly humic lake. Hydrobiologia 229:143–157
Salonen K, Jones RI , DeHaan H, James M (1994) Radiotracer study of phosphorus uptake by plankton and redistribution in the water column of a small humic lake. Limnol Oceanogr 39: 69–83
Sarvala J, Halsinaho S (1990) Crustacean zooplankton of Finnish forest lakes in relation to acidity and other environmental factors. In: Kauppi P, Anttilla P, Kenttämies K (eds) Acidification in Finland. Springer, Berlin Heidelberg New York, pp 1009–1027
Schindler DW, Curtis PJ (1997) The role of DOC in protecting freshwaters subjected to climate warming and acidification from UV-exposure. Biogeochemistry 36:1–8
Sherr EB (1988) Direct use of high molecular weight polysaccharide by heterotrophic flagellates. Nature 335:348–351
Sommer U, Gliwicz GM, Lampert W, Duncan A (1986) The PEG-model of seasonal succession of planktonic events in fresh waters. Arch Hydrobiol 106:433–471
Sprules WG, Bowerman JE (1988) Omnivory and food chain length in zooplankton food webs. Ecology 69:418–426
Sterner RW (1997) Modelling interactions of food quality and quantity in homeostatic consumers. Freshw Biol 38:473–481
Sterner RW, Hessen DO (1994) Algal nutrient limitation and the nutrition of aquatic herbivores. Ann Rev Syst Ecol 25:1–29
Sterner RW, Elser JJ, Hessen DO (1992) Stoichiometric relationships among producers, consumers and nutrient cycling in pelagic ecosystems. Biogeochemistry 17:49–67
Sterner RW, Elser JJ, Chrzanowski TH, Schampel JH, George, NB (1994) Biogeochemistry and trophic ecology: a new food web diagram. In: Polis G, Winemiller K (eds) Food webs. integration of patterns and dynamics. Chapman and Hall, New York, pp 72–80
Tezuka Y (1990) Bacterial regeneration of ammonia and phosphate as affected by the carbon:nitrogen:phosphorus ratio of organic substrates. Microb Ecol 19:227–238
Urabe J, Sterner R W (1996) Regulation of herbivore growth by the balance of light and nutrients. Proc Natl Acad Sci USA 93:8465–8469
Vadstein O, Olsen Y (1989) Chemical composition and PO4 uptake kinetics of limnetic bacterial communities cultured in chemostat under P-limitation. Limnol Oceanogr 34:939–946
Wetzel RG (1995) Death, detritus, and energy flow in aquatic ecosystems. Freshw Biol 33:83–89
Wetzel G, Rich PH, Miller MC, Allen HL (1972) Metabolism of dissolved and particulate carbon in a temperate hard-water lake. Mem Ist Ital Idrobiol Suppl 29: 185–143

12 Cycling of Dissolved Organic Matter in the Ocean

Ronald H. Benner

12.1 Introduction

The ocean is a major global reservoir of organic matter. Photosynthetic microalgae and bacteria are the major sources of organic matter in the ocean, and it is estimated that 45% (0.5×10^{17} g C year^{-1}) of global primary production occurs in the euphotic zone of the ocean (Hedges 1992). The contribution of organic matter from the continents to the ocean is ~100-fold lower than the input from marine primary production. Rivers discharge 0.4×10^{15} g C annually to the ocean, and aeolian transport contributes another 0.1×10^{15} g C annually (Hedges 1992). About 8×10^{17} g C resides in ocean waters and surface sediments (Hedges 1992), and, given the above mentioned annual fluxes, a mean residence time of 16 years is calculated for organic carbon in the ocean. This relatively short mean residence time indicates that most organic matter cycles rapidly in the ocean. In contrast, however, the average ^{14}C "age" of dissolved organic carbon (DOC) in the deep ocean is 4000–6000 years B.P. (Williams and Druffel 1987; Bauer et al. 1992), indicating that some organic matter has a long residence time in the ocean. These apparently contradictory observations indicate that residence times of the diverse mixture of organic molecules in seawater are highly variable. This chapter will review various aspects of the marine carbon cycle and provide new insights into the composition and transformations of dissolved organic matter (DOM) in the ocean.

12.2 Distribution and Abundance of Marine Organic Matter

The bulk of the organic carbon in the ocean is distributed throughout the water column as particulate (POM) and dissolved organic matter (DOM). A much smaller fraction (<20%) of the organic carbon in the ocean resides in surface sediments, predominantly in deltaic-shelf environments (Hedges 1992). Particulate organic carbon (POC), which is operationally defined as

Ecological Studies, Vol. 133
Hessen, Tranvik (eds) Aquatic Humic Substances

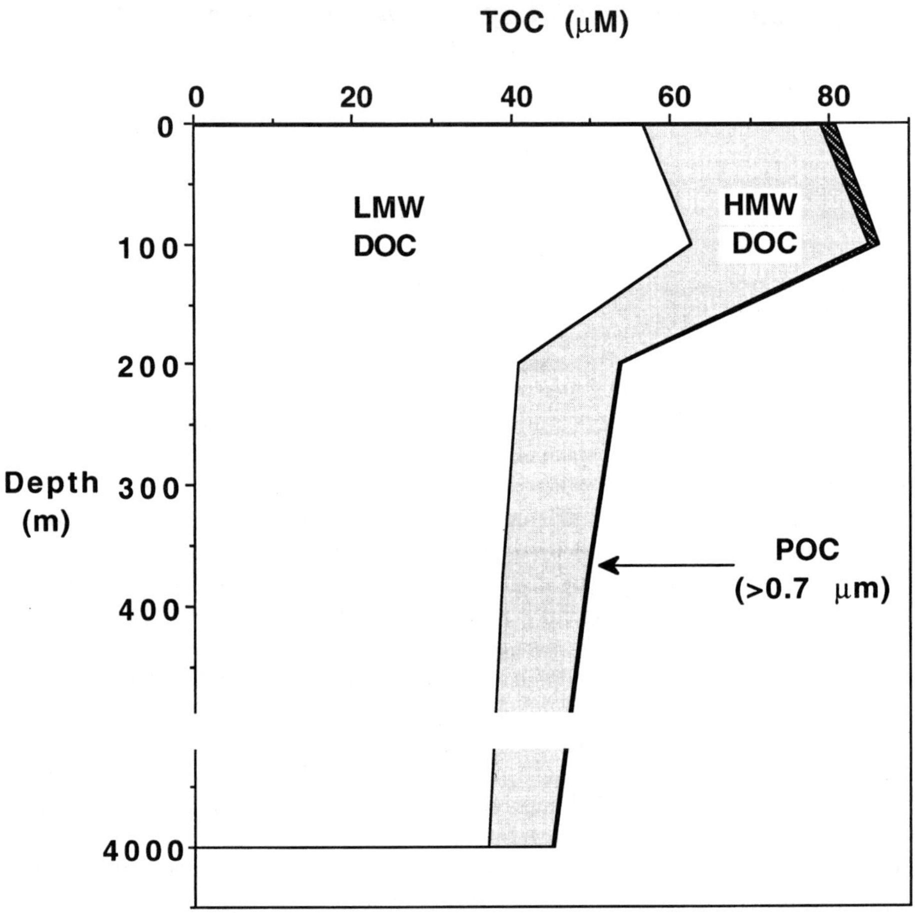

Fig. 12.1. Depth distribution of organic carbon in three size classes of organic matter at 12 S 135 W in the Pacific Ocean. LMW DOC <1 nm; HMW DOC > 1 nm <700 nm; POC >700 nm; TOC total organic carbon (Benner et al. 1997)

organic carbon retained on filters with a pore size of ~0.5 µm, accounts for <5% of the organic carbon in seawater. Most of the organic carbon in seawater is dissolved and is not associated with living organisms.

The concentrations of DOC in seawater are very dilute: ~80 µM in surface water and ~40 µM in deep water (Sharp et al. 1993, 1995). For comparison, distilled water produced in a typical research laboratory has a similar DOC concentration to deep ocean water. About 20–40% of the DOC in seawater is of high molecular weight (HMW; >1000 Da); most DOC resides in small, low-molecular-weight (LMW) molecules (Carlson et al. 1985; Benner et al. 1992, 1997; Ogawa and Ogura 1992; Guo et al. 1995). There is evidence that a small fraction of the operationally-defined DOC reservoir actually exists as submicron particles or colloids (Koike et al. 1990; Wells and Goldberg 1991). Submicron particles and colloids that pass filters used to retain particulate material would reside in what is referred to herein as the HMW fraction of DOC.

The size distribution and concentrations of organic carbon in a representative profile from the Pacific Ocean are presented in Fig. 12.1. The concentration of organic carbon in surface water (85 μM) is about 2-fold higher than that in deep water, indicating that most organic carbon originates in the surface ocean. The size distribution of organic carbon also changes with depth. The larger size classes of organic matter, POC and HMW DOC, comprise ~31% of the carbon in surface water and ~19% of the carbon in deep water. Most organic carbon in the ocean resides in deep water, and ~80% of deep water carbon occurs as LMW DOC.

As mentioned in Section 12.1, DOC in deep water is relatively old (Williams and Druffel 1987; Bauer et al. 1992) and resistant to microbial degradation (Barber 1968). Assuming that organic carbon in deep water is refractory and distributed throughout the water column, we can calculate the reactive fraction of surface water carbon by subtracting deep water carbon concentrations from surface water concentrations. If we do this calculation for each size class of organic carbon we find that 23 μM LMW DOC, 14 μM HMW DOC, and 1.6 μM POC are reactive. Only 38% of LMW DOC in surface water is reactive, whereas 64% of HMW DOC and 85% of POC are reactive. This analysis of the size and depth distribution of organic carbon in the ocean indicates that there is a general relationship between the size of organic matter and its reactivity. Naturally occurring reservoirs of POC and HMW DOC appear to be much more reactive than the more abundant LMW DOC.

12.3 Origin of Dissolved Organic Matter

Most DOM in the ocean appears to be of marine origin. Terrestrial contributions of organic matter to the ocean only account for about 1% of the total annual input. Thus, on the basis of annual fluxes, terrestrially-derived carbon should comprise a very small fraction of the organic carbon in the ocean if terrestrial and marine organic matter are of similar reactivity. Stable carbon isotope ratios ($\delta^{13}C$ notation) of DOC in the ocean are typically in the range of –20 to –22‰ (Williams and Gordon 1970; Druffel et al. 1992). These values are similar to those for marine plankton (Fry and Sherr 1984), but they are enriched in ^{13}C relative to riverine organic carbon (–27 to –29‰; Hedges et al. 1994). The stable carbon isotope values of marine DOC confirm that it is predominantly of marine origin.

Analyses of lignin-derived phenols, unique biomarkers of terrestrial plant carbon, further indicate that terrestrially-derived carbon accounts for <5% of marine DOC (Meyers-Shulte and Hedges 1986; Opsahl and Benner 1997).

Lignin-derived phenol concentrations are 2- to 3-fold higher in the Atlantic than in the Pacific Ocean (Opsahl and Benner 1997). Differences in concentrations of lignin-derived phenols between the Atlantic and Pacific Oceans appear to reflect the disproportionally large share of global riverine discharge into the Atlantic, which amounts to a 3.6-fold higher input to the Atlantic on a volume basis (Sverdrup et al. 1942; Baumgartner and Reichel 1975). Based on the concentration and distribution of lignin-derived phenols in rivers and the ocean, most terrestrially-derived DOC in the ocean is reactive and cycles more rapidly than bulk DOC in the deep ocean (Opsahl and Benner 1997).

Marine phytoplankton are the most important ultimate source of DOM in the ocean, but the relative importance of the various mechanisms of DOM formation are not well known. Potentially important mechanisms of DOM release include direct exudation by phytoplankton (Baines and Pace 1991), grazing activity (Strom et al. 1997), and viral lysis (Fuhrman and Suttle 1993). A review of the considerable literature on this topic is beyond the scope of this chapter, but it is important to recognize that the mechanism of DOM formation is likely to affect its chemical composition. Non-structural polysaccharides are important exudates of most phytoplankton (Decho 1990), whereas grazing and viral lysis may contribute more structural components of cells to the DOM reservoir. Tanoue et al. (1995) recently identified specific cell membrane proteins in seawater DOM, and it seems likely that grazing or viral lysis was responsible for their occurrence in DOM.

12.4 Chemical Composition of Dissolved Organic Matter

Much can be learned about the origin and cycling of DOM from the analysis of its chemical and isotopic composition. There has been considerable progress in the characterization of marine DOM during the past decade, but there is much yet to be learned. Fundamental questions about the sources, pathways of formation, and mechanisms of consumption remain to be answered. Recent progress in DOM characterization includes the development of new methods for the concentration and isolation of DOM from seawater and for the direct measurement of constituents in seawater. These new methods have increased knowledge of the bulk characteristics of DOM, such as elemental composition and the abundance of specific functional groups, as well as molecular composition.

A relatively wide range of C:N values have been reported for marine DOM (Jackson and Williams 1985; Hansell et al. 1993, 1997; Williams 1995), but most values are in the range of 10–20. These values indicate that DOM is relatively depleted in nitrogen compared with POM, which has values in the range of 6–10 (Bishop et al. 1977). It is not known whether DOM is depleted in nitrogen because it is naturally produced with carbon-rich components, such as carbohydrates, or because nitrogenous components are selectively removed during decomposition. The C:N values of different fractions of DOM isolated from seawater by various methods are quite variable. Humic substances isolated from seawater by adsorption onto XAD resins have C:N values ranging from 25–50 (Druffel et al. 1992; Hedges et al. 1992; Lara et al. 1993), making these the most carbon-rich components of DOM. HMW DOM isolated from seawater by tangential-flow ultrafiltration has C:N values (15–18) that are similar to those of the entire DOM reservoir (Benner et al. 1992, 1997).

A variety of spectroscopic techniques have been applied to DOM isolated from seawater using XAD resins or tangential-flow ultrafiltration. Nuclear magnetic resonance (NMR) spectroscopy has been particularly useful in the structural analysis of marine humic substances and HMW DOM. Solid-state ^{13}C NMR spectra of marine humic substances indicate they are dominated by unsubstituted alkyl carbons (Harvey and Boran 1985; Malcolm 1990; Hedges et al. 1992). Oxygen-substituted alkyl carbons, carboxyl carbons, and unsaturated carbons are present in similar but much lower abundances. In contrast, humic substances from freshwater environments have a much higher content of aromatic carbon (Malcolm 1990; Hedges et al. 1992), reflecting the input of DOM derived from vascular plants. The concentrations and structural characteristics of marine humics isolated from different depths and locations in the ocean are similar (Malcolm 1990; Druffel et al. 1992; Hedges et al. 1992), indicating that dissolved humic substances are relatively unreactive components of marine DOM. The average ^{14}C ages of marine humic substances are also "older" than those of the total DOC (Druffel et al. 1992), further indicating the relative unreactivity of marine humic substances.

The structural features of HMW DOM isolated by tangential-flow ultrafiltration are strikingly different from those of humic substances. Solid-state ^{13}C NMR spectra of HMW DOM isolated from surface water reveal a predominance of oxygen-substituted alkyl carbons characteristic of carbohydrates (Benner et al. 1992; McCarthy et al. 1993). Estimates from NMR spectra indicate that ~50% of surface water HMW DOM, or ~11 μM DOC, is associated with polysaccharides. In contrast, the abundance of oxygen-substituted alkyl carbon (polysaccharides) in deep water HMW DOM is greatly reduced (~2.2 μM DOC). The 5-fold decrease in abundance of polysaccharides between surface and deep water indicates that polysaccharides comprise a major fraction of reactive DOM that is cycled in the upper ocean (Benner et al. 1992).

The ultrafiltration and XAD resin isolation techniques concentrate components of seawater DOM that clearly differ in chemical composition. Organic matter isolated by ultrafiltration is much richer in nitrogen and carbohydrates than humic substances isolated using XAD resins. The principles governing the retention of materials during ultrafiltration are fundamentally different from those affecting retention on XAD resins. The retention of materials during ultrafiltration is based primarily on physical characteristics, such as molecular size and shape (Cheryan 1986), whereas retention on nonionic XAD resins is based on chemical characteristics, such as hydrophobicity (Thurman 1985).

The above described methods for the isolation of DOM from seawater are very useful for providing large quantities of organic matter in relatively pure form for detailed chemical characterization. It is important to note, however, that both isolation procedures recover <50% of the DOC, so most organic matter escapes isolation and chemical characterization using these approaches. The XAD resin isolation techniques typically recover 5–25% of the DOC from seawater (Gagosian and Stuermer 1977; Druffel et al. 1992). Tangential-flow ultrafiltration using filters designed to retain molecules with molecular weights >1000 Da typically recover 20–40% of the DOC from seawater (Benner et al. 1992, 1997; Guo et al. 1995). To the author's knowledge, a combination of both techniques has not been employed for the recovery of DOM from seawater, so we can only speculate that perhaps as much as 50% of seawater DOC could be recovered using a combination of techniques. As mentioned above, there are some large differences in the chemical characteristics of XAD-isolated and ultrafiltered DOM, so it seems reasonable to assume that a combination of isolation procedures would recover >30% of the DOC from surface seawater, a fraction that is routinely isolated using ultrafiltration alone.

Direct analyses of the chemical composition of seawater DOM are very challenging due to the low concentrations of organic constituents and the high concentrations of salts. Sensitive methods have been developed for the analysis of two major classes of biochemicals in seawater, amino acids and carbohydrates. Dissolved free and combined amino acid concentrations typically range from 300–500 nM in the surface ocean and 100–150 nM in the deep ocean (Lee and Bada 1977; Druffel et al. 1992). These concentrations of amino acids represent ~2% of the DOC in surface water and ~1% of the DOC in deep water.

Several colorimetric analyses of carbohydrates in seawater have been utilized, but the MBTH method (Johnson and Sieburth 1977) has become the most widely used method. This method method in combination with a sulfuric acid hydrolysis procedure (Pakulski and Benner 1992) was shown to produce high yields from a diverse group of carbohydrate polymers and from natural seawater samples. Using this method, combined and free carbohydrate concentrations typically range from 2–5 μM glucose equivalents in the surface ocean and 0.8–1.1 μM in the deep ocean (Pakulski and Benner 1994). These concentrations of carbohydrates represent ~25% of the DOC in the surface ocean and ~10% of the DOC in the deep ocean. By this analysis, carbohydrates are the single most abundant class of biochemicals in seawater.

Carbohydrates are a diverse group of biochemicals that cannot be easily quantified at the molecular level in a single analysis. However, a sensitive chromatographic method was recently introduced for the analysis of aldoses in seawater (Mopper et al. 1992, 1995). Aldose concentrations range from

300–800 nM in the surface ocean and 120–170 nM in the deep ocean (Borch and Kirchman 1997, Skoog and Benner 1997). Aldose concentrations represent ~3.5% of the DOC in surface water and ~1.5% of the DOC in the deep ocean. Aldoses represent 10–20% of the total carbohydrates measured by the MBTH method.

12.5 Microbial Utilization of Dissolved Organic Matter

Bacterioplankton are believed to be the dominant organisms responsible for the utilization of DOM in the ocean (Azam and Hodson 1977). During the past decade, studies of the growth rates of heterotrophic bacteria in the surface ocean indicate that bacteria consume a major fraction of primary production every day (Cole et al. 1988; Ducklow and Carlson 1992). Rates of bacterial production in oceanic surface waters are typically ~240 nM C day^{-1} (see Ducklow and Carlson 1992). Most estimates of bacterial growth efficiency in the surface ocean are <20% (Kirchman et al. 1991; Hansell et al. 1995; Pomeroy et al. 1995), indicating that the rate of remineralization of DOC by bacteria is at least 5-fold greater (~1.2 µM C day^{-1}) than the rate of bacterial production. Direct measurements of respiration in seawater also provide an estimate of the rate of DOC utilization. Most respiratory rates in the surface ocean fall within the range of 0.7–7 µM O_2 day^{-1} (Packard and Williams 1981; Williams 1984; Pomeroy et al. 1995; Biddanda and Benner 1997b). Bacteria appear to be responsible for at least 50% of oceanic plankton respiration (Williams 1984). Given a typical community respiratory rate of 2 µM O_2 day^{-1}, the author estimates a bacterial respiratory rate of ~1 µM O_2 day^{-1} in the surface ocean. This value is similar to that derived from measurements of bacterial production and growth efficiency.

Measurements of bacterial growth and respiration in the surface ocean indicate that bacteria remineralize ~1 µM C day^{-1}. Based on bacterial DOC remineralization rates, the mean residence time of DOC in the surface ocean is ~80 days. If other mechanisms of DOC removal are important in the surface ocean, then the mean residence time of DOC would be even shorter. This estimate of DOC mean residence time contrasts sharply with the much longer suggested residence time based on the average apparent ^{14}C age of DOC (~1000 year B.P.) in the surface ocean (Williams and Druffel 1987; Bauer et al. 1992). Estimates of residence time based on $\Delta^{14}C$ values are biased to longer-lived components that are more likely to be present in the DOC reservoir at any given time, whereas residence time estimates based on bacterial remineralization rates include compounds that cycle on time scales of hours to days and are therefore less likely to comprise a large fraction of the existing DOC reservoir. In combination, these independent estimates of cycling time indicate that marine DOM includes compounds with highly variable reactivities.

Studies of the microbial utilization of simple, LMW compounds, such as amino acids and sugars, indicate that these substrates are rapidly consumed in seawater (Suttle et al. 1991; Rich et al. 1996) and, to many, support the common belief that microbial metabolism in the ocean is mostly fueled by such simple substrates. While it may be true that free amino acids and sugars support much of the bacterial growth and metabolism in the ocean, these studies do not provide information about the immediate source(s) of the substrates. The investigator does not know whether the reactive pool of free amino acids and sugars was directly produced as monomers or whether the monomers are produced through the degradative action of exoenzymes acting on combined forms of amino acids and sugars.

The biological reactivity and availability of HMW and LMW DOM were directly compared in a series of experiments to address the concept that LMW DOM supports most of the microbial growth and metabolism in the ocean. The HMW and LMW fractions of natural DOM were separated by tangential-flow ultrafiltration and incubated with the natural bacterial community from the same water sample (Amon and Benner 1994, 1996). Rates of bacterial utilization of HMW DOM were up to 4-fold higher than rates of utilization of LMW DOM. Bacterial growth and respiration rates were higher in incubations with HMW DOM than with LMW DOM, but it appeared that LMW DOM was utilized more efficiently for bacterial growth (Amon and Benner 1994, 1996). Inorganic nitrogen was consumed during bacterial utilization of HMW DOM, whereas it was regenerated during the utilization of LMW DOM (Amon and Benner 1994; Gardner et al. 1996). Differences in bacterial growth efficiencies and nitrogen dynamics during these experiments indicate that the HMW and LMW substrates differed in chemical composition. The bioreactive components in HMW DOM are nitrogen poor, suggesting polysaccharides may be important substrates supporting bacterial carbon and energy demand in the ocean.

12.6 Relationship Between Size and Diagenetic State of Marine Organic Matter

We know that the DOM reservoir includes molecules with highly variable reactivities, but we know little about the factors that influence the reactivity and bioavailability of naturally occurring DOM. Chemical composition is certainly considered a major factor influencing the microbial utilization of DOM, but it is exceedingly difficult to study the changes in chemical composition of natural DOM under conditions and time scales that are biologically relevant. One simple approach is to study the composition of DOM collected from regions of the ocean that have sharp gradients in the concentrations of DOC and to relate changes in the composition and concentration of DOM to biological and physical processes. This approach was used earlier in Section

12.2. to describe the relative reactivities of different size classes of marine organic matter. This analysis revealed the following order of reactivity from most to least reactive: POM>HMW DOM>LMW DOM. This order of reactivity is inversely related to the naturally occurring concentrations of these reservoirs of marine organic matter.

Several recent studies have compared the chemical compositions of phytoplankton, POM, HMW DOM, and LMW DOM from various oceanic regions to investigate if diagenetic sequences are apparent within the size spectrum of marine organic matter. In the present example, aldoses are used to trace the diagenetic history of marine organic matter because we have a substantial amount of data on the aldose yields and compositions of different size fractions of marine organic matter. The fraction of organic carbon as aldoses has been shown to be a robust indicator of the diagenetic state of organic matter (Cowie and Hedges 1994). Freshly-produced organic matter has higher yields of organic carbon as aldoses than diagenetically-altered organic matter, because aldoses are preferentially removed during decomposition.

A large number of studies have analyzed the carbohydrate compositions of phytoplankton and POM (Romankevich 1984), but the methods of analysis and formats for reporting data are highly variable and therefore of less value for addressing diagenetic state and its relationship among size fractions of marine organic matter. The studies referred to herein used very similar sample collection techniques and analytical methods, and they provided sufficient information to present the data in a uniform format.

The aldose yields of POM and DOM from cultured phytoplankton and surface ocean water are presented in Fig. 12.2. A diverse group of phytoplankton, including *Synechococcus bacillaris*, *Phaeocystis* sp., *Emiliania huxleyi*, and *Skeletonema costatum*, were grown in synthetic media for 2 weeks, and the POM and DOM fractions were harvested using tangential-flow ultrafiltration (Biddanda and Benner 1997a). Seawater POM and DOM were collected from surface ocean water in the Pacific and Atlantic Oceans using the same ultrafiltration methods (Benner et al. 1992, 1997). The particulate and dissolved fractions isolated by ultrafiltration are referred to as UPOM and UDOM in Fig. 12.2. The UDOM fraction is the same as that described in Section 12.4 as HMW DOM. Aldose yields are presented as percentages of the total organic carbon in the sample.

Aldose yields are highest in the freshly-produced POM and DOM fractions from the phytoplankton cultures (Biersmith and Benner 1997; Fig. 12.2). It is interesting to note that HMW DOM had considerably higher aldose yields (~35%) than the phytoplankton cells (~20%), indicating that phytoplankton exudates are enriched with carbohydrates. The POM and HMW DOM fractions from surface ocean water had lower aldose yields than the respective fractions from the phytoplankton cultures. This is consistent with the fact that POM and HMW DOM collected from the surface ocean include materials that have undergone microbial decomposition. The differ-

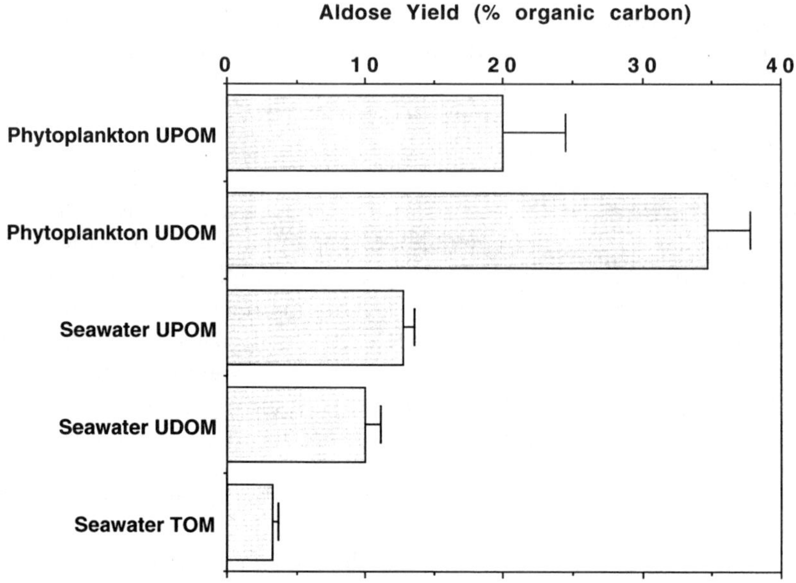

Fig. 12.2. Aldose yields in cultured phytoplankton and open ocean surface waters expressed as a percentage of organic carbon in the sample. Particulate and dissolved samples collected using tangential-flow ultrafiltration are designated UPOM and UDOM, respectively. Aldoses and organic carbon were directly measured in unfiltered seawater for the seawater TOM (total organic matter) samples. (Phytoplankton culture data from Biersmith and Benner 1997; seawater data from McCarthy et al. 1996 and Skoog and Benner 1997)

ence between aldose yields in phytoplankton and seawater POM was much less than the difference between phytoplankton and seawater HMW DOM. One explanation for this observation is that POM cycles more rapidly in seawater than HMW DOM and is therefore "fresher" than HMW DOM. This explanation is consistent with earlier arguments that POM is more reactive than HMW DOM, and it is consistent with the fact that seawater POM has a slightly higher aldose yield than seawater HMW DOM. The aldose yield in unfiltered surface ocean water is the lowest (~3.3%) of the measured yields (Fig. 12.2; seawater TOM). The major size fraction of organic matter in seawater is LMW DOM, and the low aldose yield in unfiltered seawater primarily reflects the aldose yield in LMW DOM. Thus, LMW DOM appears to be the most diagenetically altered size fraction in the ocean.

The aldose yields of different size fractions of organic matter collected from various locations and depths in the equatorial Pacific Ocean were recently reported (Skoog and Benner 1997) and are summarized in Fig. 12.3. Aldose yields were measured in unfiltered seawater, UPOM, and UDOM. The aldose concentration in LMW DOM was calculated as: [aldose in LMW DOM] = [aldose in unfiltered seawater] – [aldose in UPOM + UDOM]. Data for samples from three depths, surface water, the oxygen minimum, and deep

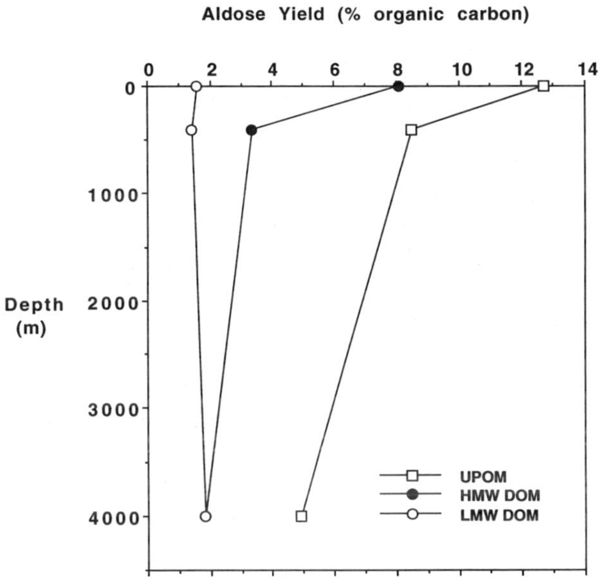

Fig. 12.3. Aldose yields in three size fractions of organic matter in seawater samples collected in the equatorial Pacific Ocean. UPOM and HMW DOM fractions were collected using tangential-flow ultrafiltration. Data for the LMW DOM fraction were calculated by mass balance. Each point represents the average of two samples collected at 2 S 140 W and 12 S 135 W. (Data from Skoog and Benner 1997)

water, are presented as average values of samples collected in two profiles at 2 S 140 W and 12 S 135 W.

The POM fraction had the highest aldose yields at all depths (Fig. 12.3). Aldose yields in POM decreased sharply from surface to oxygen minimum and deep water, indicating increasing diagenetic alteration with depth. The $\Delta^{14}C$ of POM also decreases with depth in the ocean, further indicating that surface water POM is "fresher" than deep water POM (Druffel et al. 1996). The aldose yields of HMW DOM fell between those of POM and LMW DOM (Fig. 12.3). As with POM, aldose yields in HMW DOM decreased sharply between surface and oxygen minimum waters. There was a slight decrease in the aldose yield of HMW DOM between the oxygen minimum and deep waters, suggesting that most of the diagenetic alteration of HMW DOM occurs in the upper ocean. Aldose yields in LMW DOM were low and uniform throughout the water column (Fig. 12.3), indicating that LMW DOM is relatively unreactive. These data indicate that the extent of diagenetic alteration of marine organic matter increases as the size of organic matter decreases.

12.7 Summary

This chapter has focused on a description of the abundance, composition, and distribution of organic matter in the ocean. Dissolved organic matter (DOM) is the most abundant form of marine organic matter and is a major reactive component of the global carbon cycle. Most marine DOM is derived from photosynthetic microorganisms, with only a small fraction (<5%) of terrigenous origin. Humic substances are in relatively low abundance in the ocean, accounting for about 15% of the carbon in marine DOM. Humic substances isolated from seawater are dominated by unsubstituted alkyl carbons and are low in aromatic carbon relative to humic substances from freshwater environments. Recent studies of the chemical composition of marine DOM used tangential-flow ultrafiltration for the isolation of DOM from seawater. About 25% of DOM in seawater can be isolated using an ultrafilter with a 1000 Da molecular weight cutoff. The high-molecular-weight (HMW) fraction of marine DOM isolated by ultrafiltration is rich in polysaccharides, which appear to be relatively reactive in the upper ocean. A variety of independent observations has led to the realization that low-molecular-weight (LMW) DOM is the least reactive and most diagenetically-altered component of organic matter in the ocean. This finding is somewhat surprising because of the common notions that small molecules cycle rapidly and macromolecules are the dominant form of organic matter that is selectively preserved in the environment. So, why does the bulk of LMW DOM in the ocean cycle relatively slowly? Chemical composition is undoubtedly important, but unfortunately we know little about the composition of LMW DOM in the deep ocean where most organic carbon in the ocean exists. We do know that low concentrations of chemically recognizable molecules, such as aldoses and amino acids, exist within the refractory reservoir of LMW DOM. The survival of these normally bioavailable molecules in LMW DOM suggests that they remain "hidden" from microbial enzymes. Whether they go unrecognized by microorganisms due to associations with other organic moieties or with inorganic components is not known, but the small overall size of this material (<1000 Da or ~1 nm) limits the possibilities. The observation that LMW DOM is more diagenetically altered than POM and HMW DOM suggests that LMW DOM consists of the degradation products of larger biomolecules.

References

Amon RMW, Benner R (1994) Rapid cycling of high-molecular-weight dissolved organic matter in the ocean. Nature 369:549–552

Amon RMW, Benner, R (1996) Bacterial utilization of different size classes of dissolved organic matter. Limnol Oceanogr 41:41–51

Azam F, Hodson RE (1977) Size distribution and activity of marine microheterotrophs. Limnol Oceanogr 22:492–501

Baines SB, Pace ML (1991) The production of dissolved organic matter by phytoplankton and its importance to bacteria: patterns across marine and freshwater systems. Limnol Oceanogr 36: 1078–1090

Barber RT (1968) Dissolved organic matter from deep waters resists microbial oxidation. Nature 220:274–275

Bauer JE, Williams PM, Druffel ERM (1992) ^{14}C activity of dissolved organic carbon fractions in the north-central Pacific and Sargasso Sea. Nature 357:667–670

Baumgartner A, Reichel E (1975) The world water balance. Oldenberg, Munich

Benner R, Pakulski JD, McCarthy M, Hedges JI, Hatcher PG (1992) Bulk chemical characteristics of dissolved organic matter in the ocean. Science 255:1561–1564

Benner R, Biddanda B, Black B, McCarthy M (1997) Abundance, size distribution, and stable carbon and nitrogen isotopic compositions of marine organic matter isolated by tangential-flow ultrafiltration. Mar Chem (in press)

Biddanda B, Benner R (1997a) Carbon, nitrogen, and carbohydrate fluxes during the production of particulate and dissolved organic matter by marine phytoplankton. Limnol Oceanogr 42: 506–518

Biddanda B, Benner R (1997b) Major contribution from mesopelagic plankton to heterotrophic metabolism in the upper ocean. Deep-Sea Res (in press)

Biersmith A, Benner R (1997) Carbohydrates in phytoplankton and freshly-produced dissolved organic matter. Mar Chem (in review)

Bishop JKB, Edmond JM, Ketten DR, Bacon MP, Silker WG (1977) The chemistry, geology, and vertical flux of particulate matter from the upper 400 m of the equatorial Atlantic Ocean. Deep-Sea Res 24:511–548

Borch NH, Kirchman DL (1997) Concentration and composition of dissolved combined neutral sugars (polysaccharides) in seawater determined by HPLC-PAD. Mar Chem 57:85–95

Carlson DJ, Brann ML, Mague TH, Mayer LM (1985) Molecular weight distribution of dissolved organic materials in seawater determined by ultrafiltration: a re-examination. Mar Chem 16: 155–171

Cheryan M (1986) Ultrafiltration handbook. Technomic Publishing, Pennsylvania

Cole JJ, Findlay S, Pace ML (1988) Bacterial production in fresh and saltwater ecosystems: a cross-system overview. Mar Ecol Prog Ser 43:1–10

Cowie GL, Hedges JI (1994) Biochemical indicators of diagenetic alteration in natural organic matter mixtures. Nature 369:304–307

Decho AW (1990) Microbial exopolymer secretions in ocean environments: their roles in food webs and marine processes. Oceanogr Mar Biol Annu Rev 28:73–153

Druffel ERM, Williams PM, Bauer JE, Ertel JR (1992) Cycling of dissolved and particulate organic matter in the open ocean. J Geophys Res 97:15639–15659

Druffel ERM, Bauer JE, Williams PM, Griffin, S, Wolgast D (1996) Seasonal variability of radiocarbon in particulate organic carbon in the northeast Pacific Ocean. J Geophys Res 101: 20543–20552

Ducklow HW, Carlson CA (1992) Oceanic bacterial production. In: Marshall KC (ed) Advances in microbial ecology, vol 12. Plenum Press, New York, pp 113–181

Fry B, Sherr EB (1984) $\delta^{13}C$ measurements as indicators of carbon flow in marine and freshwater ecosystems. Contrib Mar Sci 27:13–47

Fuhrman JA, Suttle CA (1993) Viruses in marine planktonic systems. Oceanography 6:51–63

Gagosian RB, Stuermer DH (1977) The cycling of biogenic compounds and their diagenetically transformed products in seawater. Mar Chem 5:605–632

Gardner WS, Benner R, Amon RMW, Cotner JB Jr, Cavaletto JF, Johnson JR (1996) Effects of high-molecular-weight dissolved organic matter on nitrogen dynamics in the Mississippi River plume. Mar Ecol Prog Ser 133:287–297

Guo L, Santschi PH, Warnken KW (1995) Dynamics of dissolved organic carbon (DOC) in oceanic environments. Limnol Oceanogr 40:1392–1403

Hansell DA, Bates NR, Gundersen K (1995) Mineralization of dissolved organic carbon in the Sargasso Sea. Mar Chem 51:201–212

Hansell DA, Williams PM, Ward BB (1993) Measurements of DOC and DON in the southern California bight using oxidation by high temperature combustion. Deep-Sea Res 40:219–234

Hansell DA, Waterhouse TY (1997) Controls on the distributions of organic carbon and nitrogen in the eastern Pacific Ocean. Deep-Sea Res 44:843–857

Harvey GR, Boran DA (1985) Geochemistry of humic substances in seawater. In:Aikin GR, McNight DM, Wershaw RL, MacCarthy P (eds) Humic substances in soil water and sediment: geochemistry, isolation, and characterization. Wiley-Interscience, New York, pp 233–248

Hedges JI (1992) Global biogeochemical cycles: progress and problems. Mar Chem 39:67–93

Hedges JI, Hatcher PG, Ertel JR, Meyers-Shulte KJ (1992) A comparison of dissolved humic substances from seawater with Amazon River counterparts by ^{13}C-NMR spectroscopy. Geochim Cosmochim Acta 56:1753–1757

Hedges JI, Cowie GL, Richey JE, Quay PD, Benner R, Strom M, Forsberg BR (1994) Origins and processing of organic matter in the Amazon River as indicated by carbohydrates and amino acids. Limnol Oceanogr 39:743–761

Jackson GA, Williams PM (1985) Importance of dissolved organic nitrogen and phosphorous to biological nutrient cycling. Deep-Sea Res 32:223–235

Johnson KM, Sieburth JMcN (1977) Dissolved carbohydrates in seawater. I. A precise spectrophotometric analysis for monosaccharides. Mar Chem 5:1–13

Kirchman DL, Suzuki Y, Garside C, Ducklow HW (1991) High turnover rates of dissolved organic carbon during a spring phytoplankton bloom. Nature 352:612–614

Koike I, Hara S, Terauchi K, Kogure K (1990) Role of sub-micrometre particles in the ocean. Nature 345:242–244

Lara RJ, Hubberton U, Kattner G (1993) Contribution of humic substances to the dissolved nitrogen pool in the Greenland Sea. Mar Chem 41:327–336

Lee C, Bada JL (1977) Dissolved amino acids in the equatorial Pacific, the Sargasso Sea, and Biscayne Bay. Limnol Oceanogr 22:502–510

Malcolm RL (1990) The uniqueness of humic substances in each of soil, stream and marine environments. Anal Chim Acta 232:19–30

McCarthy MD, Hedges JI, Benner R (1993) The chemical composition of dissolved organic matter in seawater. Chem Geol 107:503–507

McCarthy MD, Hedges JI, Benner R (1996) Major biochemical composition of dissolved high-molecular-weight organic matter in seawater. Mar Chem 55:281–298

Meyers-Schulte KJ, Hedges JI (1986) Molecular evidence for a terrestrial component of organic matter dissolved in ocean water. Nature 321:61–63

Mopper K, Schultz CA, Chevolot L, Germain C, Revuelta R, Dawson R (1992) Determination of sugars in unconcentrated seawater and other natural waters by liquid chromatography and pulsed amperometric detection. Environ Sci Technol 26:133–138

Mopper K, Zhou J, Ramana KS, Passow U, Dam HG, Drapeau DT (1995) The role of surface-active carbohydrates in the flocculation of a diatom bloom in a mesocosm. Deep-Sea Res 42: 47–73

Ogawa H, Ogura N (1992) Comparison of two methods for measuring dissolved organic carbon in sea water. Nature 356:696–698

Opsahl S, Benner R (1997) Distribution and cycling of dissolved organic matter in the ocean. Nature 386:480–482

Packard TT, Williams PJleB (1981) Rates of respiratory oxygen consumption and electron transport in surface seawater from the northwest Atlantic. Oceanol Acta 4:351–358

Pakulski JD, Benner R (1992) An improved method for the hydrolysis and MBTH analysis of dissolved and particulate carbohydrates in seawater. Mar Chem 10:55–66

Pakulski JD, Benner R (1994) Abundance and distribution of carbohydrates in the ocean. Limnol Oceanogr 39:930–940

Pomeroy LR, Sheldon JE, Sheldon WM Jr, Peters F (1995) Limits to growth and respiration of bacterioplankton in the Gulf of Mexico. Mar Ecol Prog Ser 117:259–268

Rich JH, Ducklow HW, Kirchman DL (1996) Concentrations and uptake of neutral monosaccharides along 140 W in the equatorial Pacific: contribution of glucose to heterotrophic bacterial activity and the DOM flux. Limnol Oceanogr 41:595–604

Romankevich EA (1984) Geochemistry of organic matter in the ocean. Springer, Berlin Heidelberg New York

Sharp JH, Benner R, Bennett L, Carlson CA, Dow R, Fitzwater SE (1993) A re-evaluation of high temperature combustion and chemical oxidation measurements of dissolved organic carbon in seawater. Limnol Oceanogr 39:1774–1782

Sharp JH, Benner R, Bennett L, Carlson CA, Fitzwater SE, Peltzer ET, Tupas LM (1995) Analysis of organic carbon in seawater: the JGOFS EqPac methods comparison. Mar Chem 48:91–108

Skoog A, Benner R (1997) Aldoses in various size fractions of marine organic matter: implications for carbon cycling. Limnol Oceanogr (in press)

Strom SL, Benner R, Ziegler S, Dagg MJ (1997) Planktonic grazers are a potentially important source of marine dissolved organic carbon. Limnol Oceanogr 42:1364–1374

Suttle CA, Chan AM, Fuhrman JA (1991) Dissolved free amino acids in the Sargasso Sea: uptake and respiration rates, turnover times, and concentrations. Mar Ecol Prog Ser 70:189–199

Sverdrup HO, Johnson MW, Fleming RH (1942) The oceans. Prentice Hall, Englewood Cliffs NJ

Tanoue E, Nishiyama S, Kamo M, Tsugita A (1995) Bacterial membranes: possible source of a major dissolved protein in seawater. Geochim Cosmochim Acta 59:2643–2648

Thurman EM (1985) Organic geochemistry of natural waters. Nijhoff/Junk, Dordrecht

Wells ML, Goldberg ED (1991) Occurrence of small colloids in sea water. Nature 353:342–344

Williams PJleB (1984) A review of measurements of respiration rates of marine plankton populations. In: Hobbie JE, Williams PJleB (eds) Heterotrophic activity in the sea, Plenum Press, New York, pp 357–389

Williams PJleB (1995) Evidence for the seasonal accumulation of carbon-rich dissolved organic material, its scale in comparison with changes in particulate material and the consequential effect on net C/N asimilation ratios. Mar Chem 51:17–29

Williams PM, Druffel ERM (1987) Radiocarbon in dissolved organic matter in the central North Pacific Ocean. Nature 330:246–248

Williams PM, Gordon LI (1970) Carbon-13:carbon-12 ratios in dissolved and particulate organic matter in the sea. Deep-Sea Res 17:19–27

Aquatic Humic Matter: from Molecular Structure to Ecosystem Stability

Dag O. Hessen and Lars J. Tranvik

IDENTIFICATION, DELIMITATION AND CHEMICAL PROPERTIES OF HUMIC SUBSTANCES. The most straightforward way to solve a problem is to first define the agents of interest. However, this is not a trivial matter when it comes to humic substances (HS). First of all, the definition of HS per se is not trivial, since the distinction between aquatic humus and other types of dissolved, colloidal or particulate matter is not unifying. Being composed of a multitude of complex molecules of different origin, structure, molecular size and age, HS also possess a number of different properties. Hence, at best, one can only hope to infer average structure and functionality from average properties (Perdue, this Vol.).Yet some basic features can be identified; the most obvious would be the presence of chromophores, typically giving from yellow to reddish or brownish colour owing to a pronounced light absorption in the UV, blue and green parts of the spectrum.

The composition of humic matter varies as a function of source and prehistory (McKnight and Aiken, this Vol.). Some of the HS derive from autochthonous sources (including wetlands and littoral zones), and others are of allochthonous (terrestrial) origin. In addition, the actual formation of aquatic humus includes a blend of different mechanisms. The old notion that humic matter simply is a degradation product of lignin has gradually been replaced by the recognition that HS are also formed by microbial activity in soils, where simple monomers from degraded plant material may polymerize into more recalcitrant macromolecules. Different polymerization mechanisms may operate in soil and water, and to which extent a given HS sample is chiefly a residual product of degradation or a subsequent polymerization of small, degraded molecules is unknown. To the heterogeneity with regard to origin and formation is added the heterogeneity with regard to age. The "age", as determined by the time between C fixation by photosynthesis and some arbitrary time for sampling, may differ from a few months to hundreds or even thousands of years. During spring melt, there may be a substantial release of fulvic acid from litterfall (McKnight et al. 1993; McKnight and Aiken, this Vol.), resulting in a dominance of humic matter less than 1 year old. Also HS from aquatic macrophytes may in some lakes contribute a significant source of fresh, and more degradable, humic matter (Wetzel 1992).

Ecological Studies, Vol. 133
Hessen, Tranvik (eds) Aquatic Humic Substances

HS are often operationally defined according to the methods used for their isolation. A number of different isolation procedures successfully remove the chromophoric portion of dissolved organic matter (DOM) from the water, but the "cutoff" between HS and other DOM may vary among methods. Hence, in addition to the variability that can be accredited to different types of biomass precursors and prehistory, different methods of isolation may contribute to the various properties that can be inferred for aquatic humus. For example, in the perhaps most commonly used isolation procedure/ operational definition, the hydrophobic characteristics of organic acids are utilized, as HS are adsorbed to a resin (XAD-8) at low pH (e.g. McKnight and Aiken, this Vol.).

One way of identifying the structural properties of humic, composite samples, is by randomly generating chemical parameters (e.g. elemental composition, carboxyl content and molecular weight) within the ranges found in natural samples, to obtain a probability distribution of structural features of humic matter (Perdue, this Vol). While this kind of analysis does not provide information on the actual composition of the chemical "species" of the composite sample, it provides a realistic estimate of variance and, more important, the most likely range of essential parameters such as level of unsaturation per atom of C, stoichiometry and structural features.

From an ecosystem productivity point of view, a "positive" attribute of aquatic humus is the ability to neutralize or buffer effects of acid precipitation, as well as chelating a number of toxic components. While the properties of metal complexation are widely recognized, the actual mechanisms are not settled. A closer examination of selected models for assessing these properties is given by Perdue (this Vol.). The overall ability of metal complexation may strongly reduce metal toxicity. Enhancement of iron uptake by phytoplankton has been demonstrated in the presence of humic matter, due to improved metal chelation. The other side of the coin is that also essential trace metals and even iron may be overchelated at high HS concentrations (reviewed in Jones, this Vol.), underlining the trade-off effect of aquatic humus over a concentration gradient, as also found with respect to survival of aquatic invertebrates (cf. Petersen and Persson 1987). Presence of humic matter affects the biota in numerous ways by chelating both essential and toxic compounds.

Similarly, aquatic humus plays a dual role with regard to the acidity of surface waters. Firstly, the humus molecules themselves are protonated, and contribute to acidity. A large survey of lakes (Lydersen, this Vol.) shows that organic acids decrease the pH of surface water by 0.5–2.5 units in the 0–50 $\mu eq\ l^{-1}$ range of ANC (acid neutralizing capacity). Thus, there is no doubt that organic acids, partly in contradiction to previous assumptions, may significantly contribute to low pH even in poorly buffered lakes impacted by anthropogenic acidification. On the other hand, organic acids play an important role in preventing further pH depressions of acidified systems through their pronounced buffering capacity.

The most striking property of HS in the context of acidification is their complex-binding of toxic cations, rendering them harmless to biota. This has been particularly demonstrated for Al (Lydersen, this Vol.), and is a likely cause of better survival among fish and invertebrates at low pH in humic waters than in clearwaters.

HS not only are ubiquitous, but also constitute a substantial fraction of the biospheres organic carbon. In boreal regions, soils are rich in C, and these C stores probably greatly outnumber the C biomass in plant matter. The dominance of detrital and/or humic C is even more pronounced in aquatic systems, where biomass C is a minor fraction of total organic C. The fate of this humic C is diverse. Some of it enters the food web, some is photo-oxidized into CO or CO_2, and some is buried in sediments. A major portion of the HS that enter the food web is subsequently mineralized, and a significant portion of HS in sediments is also converted to CH_4 or CO_2. Thus, ecosystems rich in HS will be a net source of greenhouse gases to the atmosphere. There is also a pronounced export of HS to marine areas, but most of this is oxidized or buried in sediments nearshore, so that HS from terrestrial sources constitute only a minor fraction of dissolved C in open oceans (Benner, this Vol.). Nevertheless, HS comprise a substantial fraction of marine dissolved organic carbon (DOC), suggesting that humic compounds are produced *in situ* in the oceans.

Weather conditions as well as more systematic climatic changes will strongly affect the fate of terrestrial primary production and soil stores of organic C. Warmer and drier climate results in less drainage of HS to aquatic ecosystems, and this in turn will affect not only the carbon flux to water bodies, but also the entire ecosystem in a multitude of ways. Such systematic changes over the past decades have been recorded over North America, causing increased transparency and particularly a higher exposure to detrimental shortwaved light for biota of many lakes (Schindler et al. 1996, Curtis, this Vol.). Changes in spectral properties and increased UV transparency may entail feedback effect by enhanced photodegradation of HS (Miller, this Vol.; Lean et al., this Vol.). To date, photochemical production of CO and CO_2 has been largely ignored in global carbon budgets, yet these could represent significant fluxes (Lean et al., this Vol.).

Absorption of Radiation and Photochemical Reactions in Humic Water. The UV-absorbing properties of HS have for long been recognized and spectrophotometric readings of UV absorbance is, in fact, the most widely used measure of aquatic HS. The absorptivity of humic compounds decreases with increasing wavelength, being highest in the UV-B, lower in the UV-A and least in the PAR (photosynthetic active radiation) region of the solar spectrum. *In situ* UV-A and UV-B attenuation coefficients in lakes are well described by power functions of DOC (HS), and attenuation coefficients may be well predicted from spectrophotometric readings (Scully and Lean 1994; Lean et al., this Vol.). The interactions of solar radiation and humic matter are of fundamental importance to both large-scale biogeochemical processes and food web dynamics. Thus, both the mechanisms and the effects deserve considerable attention. The dual role (positive/negative) of humic matter is also evident when it comes to radiation effects. The strong absorption of PAR implies the widely recognized lowered photosynthesis in humic-stained lakes. On the other hand, aquatic humus effectively blocks short wavelength radiation (UV-A, UV-B) that may be harmful to organisms. Simultaneously, this radiation causes breakdown of macromolecules or organic nutrients into compounds with enhanced bioavailability thereby stimulating bacterial growth. These stimulatory effects are counteracted by the photochemical production of CO and a set of highly reactive, strong oxidants (Miller, this Vol.; Lean et al., this Vol.) that interact with the chemistry of both detritus and live cells. Accordingly, the absorption of photons by dissolved humic substances results in a number of different phenomena having both positive and negative implications for primary and secondary production in the water. Recent aspects of the photochemical reactions with DOC like production of carbonyl suphide (a precursor for atmospheric sulphate aerosols) and cycling of mercury in lakes are also discussed in this book.

The suite of low molecular weight compounds that originate from sunlight degradation of aquatic humus has attracted considerable interest, owing to the potential effects on aquatic secondary production (Moran and Zepp 1997; Tranvik, this Vol.). While a number of studies report enhanced bacterial growth due to UV exposure of water, this is not always the case (Amon and Benner 1996a). The net response of bacteria reflects the tradeoff between the damage to cells by direct solar radiation and photochemically produced reactive species, and the stimulatory indirect effect due to enhanced substrate availability.

Direct cell damage due to sunlight and inhibition caused by reactions with highly reactive, short-lived radicals take place during the sunlit hours in the surface layers. Stimulation due to the utilization of photochemically produced substrates may occur also during night and after mixing to deeper layers protected from UV radiation. Due to this temporal and spatial displacement, it is probable that the positive effects of photochemical reactions on bacterial growth dominate. Furthermore, the attenuation of the active radiation with depth, in combination with the mixing depth of the upper irradiated water column, determines the significance of photochemical transformations at the ecosystem scale. Hence, strongly absorbing, deeply mixed layers of water experience only minor impact from photochemical reactions of DOM, since most of the radiation is absorbed (and thus photochemically active) in a small fraction of the water volume.

For photochemical breakdown, a key issue is the quantum yield, which expresses the ratio between the number of photoproduced events and the number of photons absorbed (Miller, this Vol.). The quantum yield is wavelength dependent. Photochemical reactions involving natural humic matter and solar radiation generally show high quantum yields in the UV-B, but the quantum yield may also be substantial in the UV-A, and even significant within the PAR (Miller, this Vol.), depending on the photochemical reaction studied. Hence, the photochemical formation of different products may depend in different ways on the radiation spectrum. Due to the higher absorbance in the humic water at shorter wavelengths (see above), the radiation climate will change downwards through the water column. Accordingly, the photochemical reactions will not only exhibit an overall decrease caused by decreasing light, but will also change quantitatively. This is due to altered spectral composition of the radiation, in combination with the differing wavelength dependence of quantum yields for different photoproducts. From the examples given in Miller (this Vol.) it can be deduced that HS give rise to formaldehyde only in the very surface water, while they would be oxidized to CO also at greater depths.

PRIMARY PRODUCTION IN HUMIC WATERS. The primary producers of humic lakes are challenged by a number of the physico-chemical processes described above, among which the availability of light, phosphorus and iron are key parameters (Jones, this Vol.). As has been extensively demonstrated, aquatic humus offer an efficient protection from short-wave radiation. Not only UV, but also blue and partly green light is strongly absorbed by HS. This could offer a strong selection pressure for the pigments that most efficiently harvest the long wavelengths (yellow to red). Yet, there is no strong support for this assumption. The strong absorption of photons could lead to a more general light limitation, restricting the primary production (Jones, this Vol.; Jansson, this Vol.). Accordingly, Carpenter and Pace (1997) presented a model of chlorophyll-DOC interactions in lakes, where one of the central parameters reflected the decrease in algal growth due to the presence of chromophoric DOC. However, at least in the case of phytoplankton, the widespread opinion of light limitation of primary production in humic waters does not seem to be generally valid. When supported by high P loads, even strongly stained lakes may be highly productive. Although light attenuation is strong in humic lakes, the phytoplankton development is sometimes constrained by the availability of nutrients rather than quanta (Jones, this Vol.).

Laboratory or mesocosm experiments testing additions of humic matter yield conflicting results with regard to the effect on primary production. Stimulation of primary production upon the addition of humic matter can be accredited to liberation of inorganic nutrients, or increased availability of iron. A most intriguing and largely unresolved matter is the various associations between humus, phosphorus and iron, which to a great extent are determined by short wavelength radiation. The equilibria and kinetics of this complex are key determinants of the complex P metabolism in humic lakes, and thus the phytoplankton community. In the absence of HS, phosphate can adsorb to ferric oxides and hydroxides and precipitate, while even small amounts of HS inhibit this loss of P from the euphotic zone. The phytoplankton access to P in such humus-Fe-PO_4 complexes is not settled, however. One mechanism that strongly promotes the access is photolysis, notably by UV-mediated photoreduction of Fe (Francko and Heath 1979, 1982).

Despite all the unique properties of humic lakes with regard to thermal regimes, light and nutrient availability, there seems to be no such thing as a unique "humus community" of phytoplankton species (Jones, this Vol.), although flagellate communties dominated by mixotrophic chrysophytes are common in boreal humic lakes (Jansson, this Vol.).

The key role of heterotrophic bacteria in humic waters (see below) is thoroughly addressed in this book. Their substantial biomass and activity have important bearings also on the phytoplankton (Jansson, this Vol.). Exudates or detritus originating from primary producers form the base of the conventional microbial food web. In most clearwater and marine localities, bacterial secondary production depends largely on organic C derived from phytoplankton, and a predominant C limitation can be expected for bacteria. This puts constraints on the bacterial uptake of P. Bacteria have a high relative P content and high P demands, but high affinities for P as compared with phytoplankton, and would thus be superior competitors for mineral nutrients when released from C limitation (cf. Vadstein and Olsen 1989). If an increasing share of the P pool is allocated into bacterial (heterotrophic) biomass, primary production will decline, less C will be released and the bacteria may finally become C-limited. Theoretically, bacteria should thus be balanced between P (or eventually N) and C limitation, and with regard to competition for P they would "learn the lesson" that "you can't bite (or beat) the hand that feeds you". When, however, the bacteria are supported by other sources of organic C, as they are in humic lakes, they are released from this intimate dependency on the algae. As a result, they become more efficient competitors for mineral nutrients, creating a competitive disadvantage for the primary producers. In such localities, a high ratio of heterotrophic:phototrophic biomass and production would be expected (Jansson, this Vol.), in accordance with empirical evidence (Salonen et al. 1983; Jones and Salonen 1985; Tranvik 1989; Hessen et al. 1990).

Microbial Utilization of Humic Matter. The availability of humic-bound energy and nutrients to heterotrophs is of major interest in the assessment of the biological role of humic substances. Large efforts have been devoted to the bioavailability of C, and a number of bioassays with heterotrophic bacteria inoculated into humic water or media with isolated humic substances as the single source of organic C unambiguously demonstrate the potential for aquatic bacteria to utilize these purportedly refractory molecules. Accordingly, separation of the humic and non-humic fractions of DOM demonstrates that most of the C in both fractions is recalcitrant, and in several cases the two fractions have been shown to be of roughly equal availability. Hence, the operational dichotomy between humic and non-humic matter does not overlap with the distinction between organic C that is readily mineralized and that which is recalcitrant (Tranvik, this Vol.).

A range of different factors determine the bacterial utilization of humic substances, including photochemical conditioning of the substrate (see above), access to essential nutrients (e.g. N and P) and average molecular size (Tranvik, this Vol.). Sun et al. (1997) investigated the relationship between bioavailability and the share of aliphatic compounds estimated from basic compositional parameters such as elemental composition (H:C, N:C or O:C atomic ratios), carboxyl content and molecular weight. Bacterial growth was negatively correlated with O:C, as indicative of "old" humus with a higher content of COOH. Overall, humic matter with a high aliphatic content (as indicated by the H:C ratio) appeared to be most available as a bacterial growth substrate. Such direct linkages between chemical measures of humus quality and bioavailability are very promising tools for the assessment of the biological properties of humic matter from various sources (Perdue, this Vol.).

Molecular size has received great interest as a factor of importance for the bioavailability of humic substances and dissolved organic matter in general. Most of the DOM occurs as polymers that can be incorporated into bacterial cells only after initial cleavage into smaller molecules, typically by extracellular enzymes. The action of these enzymes is affected by interactions with organic substances, a factor that is probably of particular importance in waters rich in humic substances (Münster and DeHaan, this Vol.). Hence enzymes may become inactivated by binding to humic matter. This inactivation may also be viewed upon as a preservation of the enzyme, allowing translocation and reactivation of the enzyme upon arrival in another habitat, possibly as a result of photolytic release of the enzyme from the humic substance (Wetzel 1992). The need for extracellular enzymatic cleavage suggests that the utilization of macromolecules requires considerable investments by the bacteria, rendering the substrate less "attractive". Thus, it is a generally thought that higher molecular weight moieties within the DOM are more recalcitrant towards microbial utilization than lower molecular weight ones. However, this opinion is poorly supported. On the contrary, investigations in humic waters suggest that the high molecular weight fraction is a readily available carbon source, as compared with smaller molecules (Meyer et al. 1987; Tranvik 1990). Amon and Benner (1996b) verified this observation also for seawater. They proposed a model of decreasing bioavailability and increasing diagenetic change with decreasing molecular size. Hence, the smaller molecules would be the most diagenetically altered and simultaneously the least bioavailable ones (Benner, this Vol.). Possibly, one can view the larger molecules as possessing substantial amounts of reduced energy that can be "recognized" and utilized by microbes, while the small organic molecules are largely deprived of energy that can be utilized and are to a great extent unrecognized by bacteria, for reasons that are poorly known (Benner, this Vol.).

The bioavailability of adsorbed or chelated N and P is largely an unresolved matter. Probably some of the P may be accessible to both bacteria and algae (Jones, this Vol.). While humus is poor in N, usually with C:N mass ratios of 18–30:1 for humic acids and 45–55:1 for fulvic acids (Thurman 1985, Perdue, this Vol.), it may still be a quantitatively important pool of N. Bioassays with aquatic humus from a number of freshwater sources (rivers) suggested that only 5–8% of humus-bound organic N was directly available for phytoplankton in 1-week tests (Hessen and Källqvist 1994), while obviously increased storage and/or presence of short-wavelength light strongly increase bioavailability not only of C (Lindell et al. 1995), but also of N and P (Francko and Heath 1979, 1982; Bushaw et al. 1996)

Proliferation of Humus-Derived Energy Through the Food Web: Effects of Humic Matter on Food Web Dynamics. The role of aquatic humus may be addressed from several points of departure, but the core issue is the carbon flux from the dissolved detritus pool to the heterotrophs, starting with the bacteria. Humic matter is by far the dominating pool of organic carbon in many lakes, and the fraction of DOC that is available for microbial utilization appears to be as high in humic lakes as in other aquatic habitats. Hence, the high concentrations of DOC in humic lakes constitute a carbon source to bacteria of paramount importance (Tranvik, this Vol.).

To the extent that HS support bacterial production, they will also support the bacterial consumers, providing a link from detritus to protozoan and metazoan bacterivores, and further up the food web. Grazing by protozoan bacterivores, predominantly flagellates, may constitute the major cause of mortality for bacteria in humic lakes, thereby channelling humic-derived energy to higher trophic levels (Tranvik 1989). Flagellates have the capacity to incorporate macromolecules directly into their food vacuoles (Sherr 1988, Tranvik et al. 1993), suggesting a shortcut in the food web, diminishing the respiratory losses at lower trophic levels. However, experiments with natural high molecular weight organic matter from humic lakes suggest that this pathway is at most of marginal importance (Tranvik 1994).

Metazoan zooplankton may, however, also gain a considerable portion of their body carbon directly from detritus (Salonen and Hammar 1986; Hessen et al. 1990), and the very fact that zooplankton gross production may exceed primary production in humic lakes clearly demonstrates the potential role of a direct and indirect detritus pathway in these systems (Hessen, this Vol.). Another indirect link from sediment detritus is bacterial oxidation of methane, which may provide a significant source of C to the pelagic food web of lakes (Rudd and Hamilton 1978), and in particular those with high concentrations of HS (Hessen and Nygaard 1992). A high share of nutritionally deficient detritus in the diet could pose dietary quality constraints on the pelagic grazers, and the ability to efficiently utilize the large bacterial food source would thus be supposedly beneficial. This holds in particular for the supply of P, since bacteria may provide by far the most important source of P for the zooplankton of these lakes (Hessen and Andersen 1991). While "microfiltrators" may be common among the metazoan zooplankton in the humic lakes, it is not possible to identify any typical "humic lake zooplankton community", however.

The predominance of heterotrophic processes not only implies that such systems are vigorous net producers of CO_2 and CH_4, but also contradicts the conventional notion of food web organization with productivity at succeeding trophic levels forming a pyramidal structure. Rather, pyramidal structures tend to be inverted, and omnivory/mixotrophy is a common property for all members of the pelagic food web, including those normally iconsidered as primary producers. Thus the application of the trophic level concept is a particular challenge in these systems (Hessen, this Vol.). It would also follow from this that top-down or in particular bottom-up generated trophic cascades would be strongly dampened and less effective in systems strongly influenced by HS.

A most important question is how the major impact of HS on biotic and abiotic properties affects the system stability. Intuitively, one would expect a dominant and fairly stable supply of low quality organic C to promote system stability (cf. Wetzel 1984, 1995). A relatively large but recalcitrant pool of humus-bound P and N would support this stability. In contrast, in clearwater systems the spring burst of phytoplankton leads to depletion of inorganic nutrients. In combination with increased grazing, this subsequently leads to a clearwater phase, before a second productivity peak may be reached in autumn (cf. Sommer et al 1986). These dynamics are buffered in humic lakes, and thus from an ecosystem point of view the stability of humic lakes clearly deviates from clearwater lakes and marine areas in the temperate region.

It is obvious that humic substances exhibit a multitude of effects on the biogeochemistry and ecology of lakes and other water bodies, ranging from their numerous impacts on the physico-chemical environment, to modifications of the structure and dynamics of biota at all trophic levels. Humic substances are major agents in aquatic ecology and biogeochemistry. Undoubtedly, they offer inspiring insights into the complexity of ecosystems.

References

Amon RMW, Benner R (1996a) Photochemical and microbial consumption of dissolved organic carbon and dissolved oxygen in the Amazon River system. Geochim Cosmochim Acta 60:1783–1792

Amon RMW, Benner R (1996b) Bacterial utilization of different size classes of dissolved organic matter. Limnol Oceanogr 41:41–51

Bushaw KL, Zepp RG, Tarr MA, Schulz-Jander D, Bourbonniere RA, Hodson RE, Miller WL, Bronk DA, Moran MA (1996) Photochemical release of biologically available nitrogen from dissolved organic matter. Nature 381:404–407

Carpenter SR, Pace ML (1997) Dystrophy and eutrophy in lake ecosystems: Implications of fluctuating inputs. Oikos 78:3–14

Francko DA, Heath RT (1979) Functionally distinct classes of complex phosphorus in lake water. Limnol Oceanogr 24:463–473

Francko DA, Heath RT (1982) UV-sensitive complex phosphorus: association with dissolved humic material and iron in a bog lake. Limnol Oceanogr 27:564–569

Hessen DO, Andersen T (1991) Bacteria as a source of phosphorus for zooplankton. Hydrobiologia 206:217–223
Hessen DO, Källqvist T (1994) Bioavailability of humic bound organic nitrogen for freshwater algae. Nitrogen from Mountains to Fjords. Newsletter 1/94. NIVA, Oslo.
Hessen DO, Nygaard K (1992) Bacterial transfer of methane and detritus: implications for the pelagic carbon budget and gaseous release. Arch Hydrobiol 37:139–148
Hessen DO, Andersen T, Lyche A (1990) Carbon metabolism in a humic lake; pool sizes and cycling through zooplankton. Limnol Oceanogr 35:84–99
Jones RI, Salonen K (1985) The importance of bacterial utilization of released phytoplankton photosynthate in two humic forest lakes in southern Finland. Holarct Ecol 8:133–140
Lindell MJ, Granéli W, Tranvik LJ (1995) Enhanced bacterial growth in response to photochemical transformation of dissolved organic matter. Limnol Oceanogr 40:195–199
McKnight DM, Smith RL, Harnish RA, Miller CL, Bencala KE (1993) Seasonal relationships between planktonic microorganisms and dissolved organic material in an alpine stream. Biogeochemistry 21:39–59
Meyer JL, Edwards RT, Risley R (1987) Bacterial growth on dissolved organic matter from a blackwater river. Microb Ecol 13:13–29
Moran MA, Zepp RG (1997) Role of photoreactions in the formation of biologically labile compounds from dissolved organic matter. Limnol Oceanogr (in press)
Petersen RC, Persson U (1987) Comparison of the biological effects of humic materials under acidified conditions. Sci Total Environ 62:387–398
Rudd JWM, Hamilton RD (1978) Methane cycling in a eutrophic shield lake and its effects on whole lake metabolism. Limnol Oceanogr 23:337–348
Salonen K, Hammar T (1986) On the importance of dissolved organic matter in the nutrition of zooplankton in some lake waters. Oecologia (Berl) 8:246–253
Salonen K, Kolonen K, Arvola L (1983) Respiration of plankton in two small, polyhumic lakes. Hydrobiologia 101:65–70
Schindler DW, Curtis JP, Parker BR, Stainton M (1996) Consequences of climate warming and lake acidification for UV-B penetration in North American boreal lakes. Nature 379:705–708
Scully NM, Lean DRS (1994) The attenuation of ultraviolet radiation in temperate lakes. Arch Hydrobiol Beih Ergeb Limnol 43:135–144
Sherr EB (1988) Direct use of high molecular weight polysaccharide by heterotrophic flagellates. Nature 335:348–351
Sommer U, Gliwicz GM, Lampert W, Duncan A (1986) The PEG-model of seasonal succession of planktonic events in fresh waters. Arch Hydrobiol 106:433–471
Sun L, Perdue EM, Meyer JL, Weis J (1997) Using elemental composition to predict bioavailability of dissolved organic matter in a Georgia river. Limnol Oceanogr (in press)
Thurman EM (1985) Organic geochemistry of natural waters. Dr W Junk, Boston
Tranvik LJ (1989) Bacterioplankton growth, grazing mortality, and quantitative
relationship to primary production in a humic and a clearwater lake. J Plankton Res 11: 985–1000
Tranvik LJ (1990) Bacterioplankton growth on fractions of dissolved organic carbon of different molecular weights from humic and clear waters. Appl Environ Microbiol 56:1672–1677
Tranvik LJ (1994) Effects of colloidal organic matter on the growth of bacteria and protists in lake water. Limnol Oceanogr 39:1276–1285
Tranvik LJ, Sherr EB, Sherr BF (1993) Uptake and utilization of "colloidal DOM" by heterotrophic flagellates in seawater. Mar Ecol Prog Ser 92:301–309
Vadstein O, Olsen Y (1989) Chemical composition and PO_4 uptake kinetics of limnetic bacterial communities cultured in chemostat under P-limitation. Limnol Oceanogr 34:939–946
Wetzel RG (1984) Detrital dissolved and particulate organic carbon functions in aquatic ecosystems. Bull Mar Sci 35:503–509
Wetzel RG (1992) Gradient-dominated ecosystems: sources and regulatory functions of dissolved organic matter in freshwater ecosystems. Hydrobiologia 229:181–198
Wetzel RG (1995) Death, detritus, and energy flow in aquatic ecosystems. Freshw Biol 33:83–89

Subject Index

Ecological Studies

Volumes published since 1992

Volume 94
The Ecology of Aquatic Hyphomycetes (1992)
F. Bärlocher (Ed.)

Volume 95
Palms in Forest Ecosystems of Amazonia (1992)
F. Kahn and J.-J. DeGranville

Volume 96
Ecology and Decline of Red Spruce in the Eastern United States (1992)
C. Eagar and M.B. Adams (Eds.)

Volume 97
The Response of Western Forests to Air Pollution (1992)
R.K. Olson, D. Binkley, and M. Böhm (Eds.)

Volume 98
Plankton Regulation Dynamics (1993)
N. Walz (Ed.)

Volume 99
Biodiversity and Ecosystem Function (1993)
E.-D. Schulze and H.A. Mooney (Eds.)

Volume 100
Ecophysiology of Photosynthesis (1994)
E.-D. Schulze and M.M. Caldwell (Eds.)

Volume 101
Effects of Land Use Change on Atmospheric CO_2 Concentrations: South and South East Asia as a Case Study (1993)
V.H. Dale (Ed.)

Volume 102
Coral Reef Ecology (1993)
Y.I. Sorokin

Volume 103
Rocky Shores: Exploitation in Chile and South Africa (1993)
W.R. Siegfried (Ed.)

Volume 104
Long-Term Experiments With Acid Rain in Norwegian Forest Ecosystems (1993)
G. Abrahamsen et al. (Eds.)

Volume 105
Microbial Ecology of Lake Plußsee (1993)
J. Overbeck and R.J. Chrost (Eds.)

Volume 106
Minimum Animal Populations (1994)
H. Remmert (Ed.)

Volume 107
The Role of Fire in Mediterranean-Type Ecosystems (1994)
J.M. Moreno and W.C. Oechel (Eds.)

Volume 108
Ecology and Biogeography of Mediterranean Ecosystems in Chile, California and Australia (1994)
M.T.K. Arroyo, P.H. Zedler, and M.D. Fox (Eds.)

Volume 109
Mediterranean-Type Ecosystems. The Function of Biodiversity (1995)
G.W. Davis and D.M. Richardson (Eds.)

Volume 110
Tropical Montane Cloud Forests (1995)
L.S. Hamilton, J.O. Juvik, and F.N. Scatena (Eds.)

Volume 111
Peatland Forestry. Ecology and Principles (1995)
E. Paavilainen and J. Päivänen

Volume 112
Tropical Forests: Management and Ecology (1995)
A.E. Lugo and C. Lowe (Eds.)

Volume 113
Arctic and Alpine Biodiversity. Patterns, Causes and Ecosystem Consequences (1995)
F.S. Chapin III and C. Körner (Eds.)

Volume 114
Crassulacean Acid Metabolism. Biochemistry, Ecophysiology and Evolution (1996)
K. Winter and J.A.C. Smith (Eds.)

Volume 115
Islands. Biological Diversity and Ecosystem Function (1995)
P.M. Vitousek, L.L. Loope, and H. Adsersen (Eds.)

Ecological Studies

Volumes published since 1992

Volume 116
High Latitude Rainforests and Associated Ecosystems of the West Coast of the Americas: Climate, Hydrology, Ecology and Conservation (1996)
R.G. Lawford, P. Alaback, and E.R. Fuentes (Eds.)

Volume 117
Anticipated Effects of a Changing Global Environment on Mediterranean-Type Ecosystems (1995)
J. Moreno and W.C. Oechel (Eds.)

Volume 118
Impact of Air Pollutants on Southern Pine Forests (1996)
S. Fox and R.A. Mickler (Eds.)

Volume 119
Freshwater Ecosystems of Alaska (1997)
A.M. Milner and M.W. Oswood (Eds.)

Volume 120
Landscape Function and Disturbance in Arctic Tundra (1996)
J.F. Reynolds and J.D. Tenhunen (Eds.)

Volume 121
Biodiversity and Savanna Ecosystem Processes. A Global Perspective (1996)
O.T. Solbrig, E. Medina, and J.F. Silva (Eds.)

Volume 122
Biodiversity and Ecosystem Processes in Tropical Forests (1996)
G.H. Orians, R. Dirzo, and J.H. Cushman (Eds.)

Volume 123
Marine Benthic Vegetation. Recent Changes and the Effects of Eutrophication (1996)
W. Schramm and P.H. Nienhuis (Eds.)

Volume 124
Global Change and Arctic Terrestrial Ecosystems (1997)
W.C. Oechel (Ed.)

Volume 125
Ecology and Conservation of Great Plains Vertebrates (1997)
F.L. Knopf and F.B. Samson (Eds.)

Volume 126
The Central Amazon Floodplain: Ecology of a Pulsing System (1997)
W.J. Junk (Ed.)

Volume 127
Forest Decline and Ozone: A Comparison of Controlled Chamber and Field Experiments (1997)
H. Sandermann, A.R. Wellburn, and R.L. Heath (Eds.)

Volume 128
Productivity and Sustainability of Southern Forest Ecosystems (1997)
R.A. Mickler and S. Fox (Eds.)

Volume 129
Pelagic Nutrient Cycles: Herbivores as Sources and Sinks (1997)
T. Andersen

Volume 130
Vertical Food Web Interactions: Evolutionary Patterns and Driving Forces (1997)
K. Dettner, G. Bauer, and W. Völkl (Eds.)

Volume 131
The Structuring Role of Submerged Macrophytes in Lakes (1998)
E. Jeppesen et al. (Eds.)

Volume 132
Vegetation of the Tropical Pacific Islands (1998)
D. Mueller-Dombois and F.R. Fosberg

Volume 133
Aquatic Humic Substances: Ecology and Biogeochemistry (1998)
D.O. Hessen and L.J. Tranvik (Eds.)

Volume 134
Oxidant Air Pollution Impacts in the Montane Forests of Southern California (1998)
P.R. Miller and J.R. McBride (Eds.)

Volume 135
Predation in Vertebrate Communities: The Białowieża Primeval Forest as a Case Study (1998)
B. Jędrzejewska and W. Jędrzejewski